SULFUR DIOXIDE REMOVAL FROM WASTE GASES

SULFUR DIOXIDE REMOVAL FROM WASTE GASES

SECOND EDITION

A.V. Slack
G.A. Hollinden

NOYES DATA CORPORATION
Park Ridge, New Jersey London, England
1975

Library of Congress Catalog Card Number: 79-167797
ISBN: 0-8155-0585-X
Printed in the United States

Published in the United States of America by
Noyes Data Corporation
Noyes Building, Park Ridge, New Jersey 07656

CONTENTS AND SUBJECT INDEX

PREFACE TO SECOND EDITION

Almost all of that which was said nearly four years ago in the preface to the first edition of this book applies equally well today. Sulfur oxide removal from waste gases is still a problem of major concern, and the technology is still changing very rapidly.

There have been several major changes and developments in the field. Some of the methods described in the first edition are omitted from the present one because they have been casualties of the competitive situation. Others that were mentioned only sketchily or not at all have come to the fore and now occupy significant positions.

Probably the main development is the rapid increase in the number of full-scale scrubber installations, of which there were very few in 1971. This is particularly true in Japan where regulations are fast becoming more stringent.

A better understanding of the basic chemical and physical factors in SO_2 absorption has also been developed over the past few years, but there is still some distance to go.

The authors appreciate the many indirect contributions by development, engineering, and operating people in the field. Since most of the work is not yet published, a book like the present one could not be written without such contributions.

A.V. Slack

SAS Corporation, Sheffield, Alabama
March 30, 1975

G.A. Hollinden

TVA Office of Power, Chattanooga, Tennessee
March 30, 1975

PREFACE TO FIRST EDITION

This book is not a reference work, but rather a status report on a technology that is being developed under forced draft. Sulfur oxide removal from waste gases is not a new problem. Smelter operators for more than a century have coped with the unwanted fumes produced by oxidation of the sulfur in the ore during roasting. Power plants, refineries, sulfuric acid plants, and Claus sulfur plants have perhaps a more recent history, but even for these industries the problem goes back for half a century or more. It is only in the last few years, however, that sulfur oxide removal has become a major research and development effort. Pressures to improve the environment have made it necessary to greatly expand activity in the field, in somewhat of crash program.

Presently there are hundreds of organizations around the world giving attention to the problem, in a general way or in developing removal methods and evaluating them for use. Processes are in all stages of development, from the idea stage to plant trials. Numerous and widely differing chemical paths are being explored, and process flowsheets ranging from very simple to very complicated are being developed.

In this context, a reference book does not seem appropriate because the technology is changing too fast. There is no basic body of information that could be presented, because the various processes under study not only cut across much of the entire field of sulfur chemistry but also involve many of the chemical engineering unit operations. Moreover, the nitrogen oxide and particulate matter problems cannot be dissociated from that of removing sulfur oxides.

Because of this, the book is intended to serve only as an overview of the current development situation, written from the viewpoint of one who is involved both in process development and in the selection of processes for use by a major utility. An effort has been made to list all the major processes (and as many of the minor ones as feasible), summarize the basic chemistry of each, analyze the engineering problems, identify advantages and disadvantages, evaluate the effect of major economic factors, and make a guess as to how each process compares with the others.

This is obviously an ambitious undertaking and one that cannot be wholly accurate and well-documented; it is, however, the type of analysis that must be carried out by those responsible for fitting plants around the world with sulfur oxide removal facilities. No process yet proposed is a completely satisfactory solution to the problem but many have promise; at some point it is necessary to set them all side by side and see which looks best, even though in a preliminary and incomplete way. This book represents such a beginning.

The information on which the analyses presented herein are based comes from many sources. Some of it is readily available as published material, but the bulk is from two sources: unpublished reports and published documents not widely distributed or publicized, and discussions with other workers in the field. There is a large amount of such information that is nonproprietary but not widely known. Of particular benefit have been visits to research organizations in all the countries of Europe and Asia (with the exception of the USSR) that have active programs in the field.

It should be emphasized that the analysis presented is for the current situation only. With the fast moving pace in the field, such an overview made a year from now may be quite different. New processes will be announced, older ones will change in status, and commercial units now being constructed will add the final proof to some processes. If this book has any value, it is in getting everything together to serve as a starting point in analyzing sulfur oxide removal technology, with adjustments to be expected as the technology matures. In spite of best efforts, errors no doubt have crept into the descriptions and evaluations of some of the processes. Hopefully, these will be few.

The list of acknowledgements due is too long to include. The main contributions have come from the large number of research and development people with whom I have discussed the problems in developing an effective and economical sulfur oxide removal technology (and whose indulgence I invite where my evaluations differ from theirs). Several of my coworkers have reviewed and corrected the text, and my secretary, Mrs. Juanita Hargett, carried out the mechanics of the project quite efficiently. Finally, without the patience and forbearance of my wife, this project would have failed in the early stages.

A.V. Slack

Muscle Shoals, Alabama
June 17, 1971

THE PROBLEM

Of the several air pollutants that plague the world, sulfur oxides (SO_2, sulfur dioxide, and SO_3, sulfur trioxide) have received special attention—more so than any other polluting material except perhaps particulate matter (dust). The severity of the problem—effects on people, animals, and vegetation—is a matter on which there are varying opinions and will not be debated here. A more important consideration for the present discussion is that control of sulfur oxide emission has come to be regarded as a matter of great importance both by governmental regulatory agencies and the general public.

A major concern is the very large amount of sulfur oxides emitted. Various estimates have been made; the general consensus is that the U.S. emission in 1970 was over 35,000,000 tons. Similar data for other parts of the world are not readily available but it is known that emission is heavy in all highly industrialized countries, particularly in areas such as the Ruhr Valley in Germany and the Tokyo area in Japan. Emission is also increasing at a rapid rate. The potential emission for the year 2000 has been estimated to be over three times that for 1970.

SOURCES

Practically all SO_2 (and SO_3) comes from combustion of naturally occurring sulfur compounds, either as elemental sulfur or as a constituent of an ore or a fossil carbonaceous material. Examples are organic sulfur compounds in coal and petroleum, pyrites (FeS_2) both as such and as a constituent of coal, hydrogen sulfide (H_2S) in sour natural gas, and metal sulfide ores. The reasons for oxidizing these materials are varied; elemental sulfur and pyrites are burned to make SO_2 in the process of producing sulfuric acid, metal ores are roasted to burn away the sulfur in recovering the elemental metal (e.g., zinc, copper, and lead), the H_2S is separated from natural gas and part of it burned to SO_2 as one step in conversion to elemental sulfur, and coal and oil are burned for heating and for power production.

In most cases the overall process includes a unit for recovering SO_2 from the gas stream emitted, either because sulfur or sulfuric acid is the primary product or because the SO_2 concentration in the gas is high enough to make recovery as a by-product economical. (However, there is still a pollution problem because recovery is never complete.) The burning of carbonaceous fuel is a major exception; the exit gases from heat recovery units are so dilute in SO_2 that no effort normally is made to remove the pollutant from the stack gas. The relative contributions of the various sources to the SO_2 emission problem are given in Table 1.1.

TABLE 1.1: WORLD-WIDE SO_2 EMISSION

Sources	Emissions, 10^6 tons per yr	
Coal	102	
Petroleum	28	
Nonferrous smelting	16	
Industrial	6	
Total man-generated SO_2:		152
Marine	60	
Land H_2S (SO_2 equivalent)	140	
Total natural sources oxidized to SO_2:		200
Sulfur in SO_4 from sea salt (SO_2 equivalent)		88
Total world-wide:		440

It is clear from these data that coal is the main culprit, contributing almost 70% of the total man-made emission. Moreover, other data show that production of power from coal and oil accounts for almost half the man-made total.

Power Production from Coal

Since the problem thus is centered in power plants and mainly in coal-burning power plants, the future of coal-based power becomes a major consideration. With the exception of oil, the other major competitors to coal as an energy source, hydroelectric and nuclear, do not produce SO_2 and therefore one way of reducing the problem is to adopt such alternate sources to a greater degree. The potential for hydroelectric is well nigh exhausted and further significant growth is not expected.

Nuclear, however, is increasing rapidly (Figure 1.1). It is estimated that by 2000 over half of the electrical power will be generated by the nuclear method, with coal (plus other fossil fuel) declining to about 30%. Nevertheless, overall power production is expected to increase so much by 2000 that fossil fuels, while losing percentagewise, will increase in total usage by almost twofold over current consumption. Hence the severity of the SO_2 problem will continue to grow.

FIGURE 1.1: PROJECTED SOURCES OF POWER IN THE U.S.

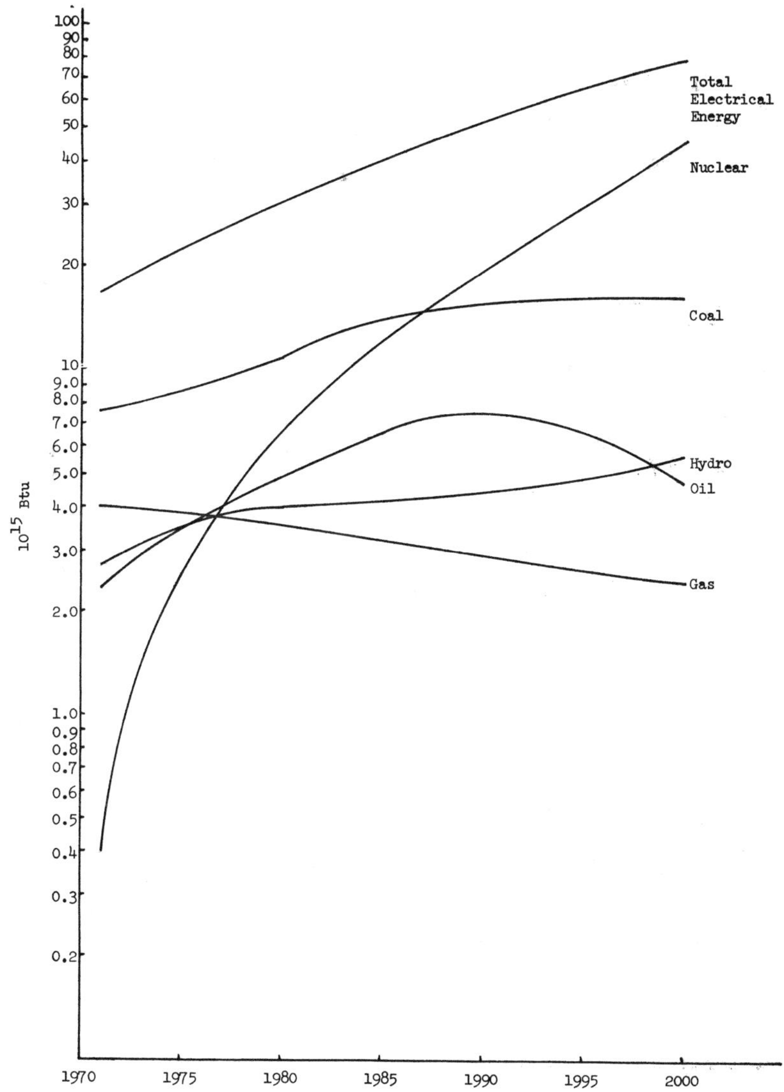

TABLE 1.2: ESTIMATED SULFUR DIOXIDE EMISSION IN U.S. WITHOUT ABATEMENT

	Annual Emission of Sulfur Dioxide (millions of tons)			
	1970	1980	1990	2000
Power plant operation (coal and oil)*	20.0	41.1	62.0	94.5
Other combustion of coal	4.8	4.0	3.1	1.6
Combustion of petroleum products (excluding power plant oil)	3.4	3.9	4.3	5.1
Smelting of metallic ores	4.0	5.3	7.1	9.6
Petroleum refinery operation	2.4	4.0	6.5	10.5
Miscellaneous sources**	2.0	2.6	3.4	4.5
Total	36.6	60.9	86.4	125.8

*With breeder reactor.

** Includes coke processing, sulfuric acid plants, coal refuse banks, refuse incineration, and pulp and paper manufacturing.

February 1970 estimates by National Air Pollution Control Administration, excluding transportation.

The changing distribution between sources is given in Table 1.2 above. It should be noted that even the major increase in this estimate may be conservative as it assumes development and wide adoption of the breeder reactor, which is generally considered essential for major conversion to nuclear power as an energy source. Potential power plant emission of SO_2 in the U.S. with and without the breeder reactor are compared in Figure 1.2.

FIGURE 1.2: PROJECTED SO_2 EMISSION AS AFFECTED BY DEVELOPMENT OF THE BREEDER-TYPE NUCLEAR REACTOR

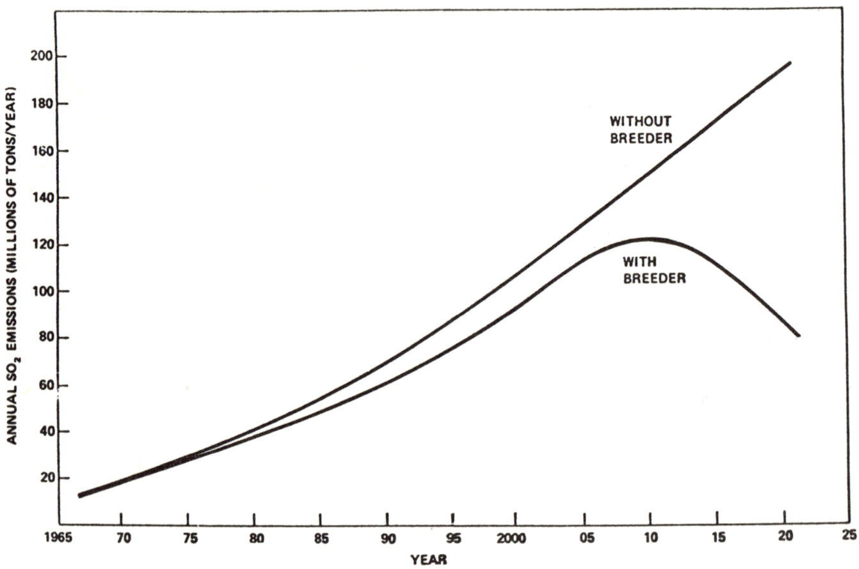

Source: NAPCA, February 1970

Another factor will also aggravate the situation. In the interest of economy, the evolutionary trend has been to larger and larger power production units and to greater production (by multiple units) at individual sites. Boiler size has reached the 1100 to 1300 Mw (megawatt; equals 1000 kilowatts or 1,000,000 watts) range and central stations producing over 2500 Mw are in operation; for a typical coal of the high-sulfur type, such stations emit over 1,500 tons of SO_2 per day. The significance of this is that damage from a pollutant such as SO_2 depends mainly on its concentration at ground level, and this depends much more on the amount emitted at a particular site than on the total amount over the entire country.

Increase in overall amount mainly produces more sites at which emission becomes a problem; larger central stations, however, make the individual problem more severe and intensify the need for remedial action. The amount of SO_2 emitted at a particular site depends, of course, on the sulfur content of the coal burned. Coals vary widely in this respect; the bituminous type generally used in power plants is highest in sulfur, although there is much bituminous coal of low sulfur content. Subbituminous, lignite, and anthracite generally contain less than 1% sulfur. The situation in regard to coal type and geographical location of reserves (in the U.S.) is summarized in Table 1.3. The totals given are for the total coal in the United States of the type listed. States with small amounts are not listed and for some of the states listed, the amount is insignificant as indicated by a dash.

TABLE 1.3: SULFUR CONTENT OF U.S. COAL RESERVES*

Coal Type and Location		Billion Tons of Coal with Indicated Sulfur Content		
		<1%	1 to 3%	>3%
Bituminous				
Alaska		21	--	--
Colorado		62	--	--
Illinois		--	21	119
Indiana		--	16	19
Kentucky (West)		--	5	32
Kentucky (East)		22	7	--
Missouri		--	--	79
New Mexico		11	--	--
Ohio		--	13	38
Pennsylvania		1	50	7
Utah		22	--	4
West Virginia		47	45	9
Wyoming		13	--	--
	Total	215	191	322
Subbituminous				
Alaska		71	--	--
Colorado		17	--	--
Montana		131	--	--
New Mexico		51	--	--
Wyoming		108	--	--
	Total	387	--	--

(continued)

TABLE 1.3: (continued)

Coal Type and Location		Billion Tons of Coal with Indicated Sulfur Content		
		<1%	1 to 3%	>3%
Lignite				
Montana		84	--	--
North Dakota		319	--	--
	Total	406	41	--
Anthracite				
Pennsylvania		12	--	--
	Total	15	--	--
	Grand Total	1,023	234	322
	Per cent	65	15	20

* Only major reserves are included.

From this it is evident that there is plenty of low sulfur coal (defined as coal of less than 1% sulfur content) in the ground. Unfortunately, most of it is of the type not usually burned in power plants. Moreover, by far the greatest tonnages are in the western states where population density is low; in the eastern areas, where power consumption is high and where most of the power is produced, only a relatively small area (West Virginia and eastern Kentucky) has significant low sulfur coal reserves.

The nonbituminous, low sulfur types can be burned in power plants, of course, but normally are not because of factors such as low heating value, difficulty in getting good combustion, and location. Recent advancements have reduced the boiler design problem, but the other drawbacks still remain.

The composition of the combustion gases evolved in the burning of coal is an important factor in pollution abatement, not only because of pollutant concentrations but also because nonpollutants in the gas stream can have major effects on processes or techniques for pollutant removal. Only nitrogen (from the combustion air) passes through the removal system without effect.

Coal is such a complex and variable raw material (Table 1.4) that the combustion gas contains many compounds, both gaseous and solid. Moreover, the variability in coal composition is so great that a generally applicable analysis of the stack gas cannot be given.

The composition given in Table 1.5 is fairly typical of stack gases evolved in burning western Kentucky coal. Variations from this occur for various reasons, including composition of the coal, type of boiler, and the way in which the boiler is operated; typical effects of such factors are shown in Table 1.6.

TABLE 1.4: EXAMPLE OF COAL COMPOSITION*

Constituent	Percent by Weight
Volatile matter	32.9
Fixed carbon	46.4
Sulfur	3.3
Moisture	5.4
Ash	12.0
	100.0
H	4.5
N	1.3
C	66.0
O	7.5
S	3.3
Ash	12.0
Moisture	5.4
	100.0

* Coal from western Kentucky. Coals vary so widely in composition that this example should not be regarded as typical.

TABLE 1.5: COMPOSITION OF STACK GAS EVOLVED IN BURNING COAL FROM WESTERN KENTUCKY*

Constituent	Percent by Volume**
Nitrogen (N_2)	74.56
Carbon dioxide (CO_2)	12.55
Oxygen (O_2)	4.87
Water vapor (H_2O)	7.76
Sulfur oxides (SO_x)	
Sulfur dioxide (SO_2)	0.22
Sulfur trioxide (SO_3)	0.001
Nitrogen oxides (NO_x)	0.04
Particulate matter	
Percent by weight	0.66
Grains per standard cu. ft. (scf)	3.59

* Composition, percent: C, 66.0; S, 3.3; ash, 12.0; N, 1.3.
** Except as indicated.

TABLE 1.6: EFFECT OF BOILER TYPE AND SULFUR CONTENT OF COAL ON STACK GAS COMPOSITION

Boiler type	Pulverized coal (horizontal, frontal-fired)			Cyclone-fired		
Sulfur content of coal, % by wt	2.0	3.5	5.0	2.0	3.5	5.0
Flue gas composition, % by volume						
Nitrogen	74.62	74.55	74.49	74.59	74.52	74.46
Carbon dioxide	12.57	12.55	12.54	12.57	12.55	12.54
Oxygen	4.86	4.86	4.85	4.83	4.83	4.82

(continued)

TABLE 1.6: (continued)

Boiler type	Pulverized coal (horizontal, frontal-fired)			Cyclone-fired		
Water	7.77	7.76	7.75	7.77	7.76	7.75
Sulfur dioxide	0.12	0.22	0.31	0.12	0.22	0.31
Sulfur trioxide	0.001	0.001	0.001	0.001	0.001	0.001
Nitrogen oxides	0.06	0.06	0.06	0.12	0.12	0.12
Fly ash loading						
Grains/SCF dry	4.11	4.11	4.11	1.37	1.37	1.37
Grains/SCF wet	3.79	3.79	3.79	1.26	1.26	1.26

Carbon dioxide is regarded as a pollutant by some because of the possible adverse effect on transmission of heat from the sun through the earth's atmosphere if the CO_2 content in the atmosphere increases significantly; however, little attention has been given to the possibility of removing it from stack gas, which would be an economic problem of major proportions. The content in the stack gas depends almost entirely on the carbon content of the coal. Assuming 65% carbon, the output of CO_2 from a 1000 Mw boiler is on the order of 17,000 tons per day, or over 6,000,000 tons per year (assuming an 80% operating factor).

The content of CO_2 in the gas does not appear to have any major effect on sulfur oxide removal processes. However, CO_2 must be considered because it enters into solution equilibria in such processes as wet scrubbing with alkali and alkaline earth solutions and slurries. Moreover, in some processes involving solid absorbents, the SO_2 and CO_2 compete in reaction with the absorbent, particularly at low temperatures.

The moisture content of the gas is a function of (1) the free water present in the coal, (2) moisture in the combustion air, and (3) hydrogen in the hydrocarbon constituents of the coal. Rough ranges of the contributions from these sources are given in Table 1.7.

TABLE 1.7: SOURCES OF WATER VAPOR IN STACK GAS FROM COAL-FIRED POWER PLANTS

Source	Moisture Content, %	Tons per Day of Moisture from 1000 Mw Plant
Coal		
Free water	5.4	541
Hydrocarbons*	40.5	4,050
Combustion air	1.32**	1,296
Total in combustion gas	4.71	5,887

*Hydrogen content expressed as equivalent moisture content.
**Lb. moisture/lb. dry air.

The water vapor content of the gas has two main effects in pollution control. Since the molecular weight of water is lower than that of most of the other gas constituents, change in moisture content changes the density of the gas and therefore affects plume rise from the stack. In addition, change in moisture content affects the wet bulb temperature, an important consideration in wet scrubbing because it determines the scrubbing temperature.

The oxygen content depends mainly on the amount of excess oxygen used in burning the coal and an air heater leakage. Some excess air is used to ensure complete combustion but the degree of excess varies widely with boiler type, type of coal, particle size of coal, and operator preference; a typical excess is about 20%. The objective is to get the best compromise between incomplete combustion on one extreme and undue heat loss because of increased stack gas volume on the other. (Since there is a lower limit for stack gas temperature, an increase in stack gas volume carries more heat out of the system to the atmosphere.) Incomplete combustion not only wastes coal but also makes the air pollution problem worse because of the larger amount of particulate matter, the black color of the unburned carbon, and the greater soiling potential of the escaping particulates.

Oxygen can have several types of effects on sulfur oxide removal processes, mainly in regard to oxidation of stack gas constituents and of absorbents. Examples are oxidation of SO_2 to SO_3, NO to NO_2, and of various sulfites to sulfates in absorption systems.

The sulfur oxide content of the gas (volume basis) can be estimated by multiplying the sulfur content of the coal (weight basis) by 0.067 (i.e., 3.55% sulfur in the coal gives about 0.23% SO_2 in the gas). Most of the sulfur in the coal leaves the boiler in the gas, although in some types of boilers a small proportion may become combined with the ash that falls out of the gas stream and is removed from the bottom of the boiler.

The SO_3 usually accounts for only a small proportion of the total sulfur oxides; the amount depends on the degree of SO_2 oxidation by oxygen, which in turn depends on boiler temperature, length of time in the boiler, and catalytic effects from metal surfaces in the boiler and from metal compounds in the ash. A rough range of SO_3 concentration is 0.5 to 5.0% of the total sulfur oxides present.

The relatively high dew point of SO_3 in the gas is important in operating both the boiler and a sulfur oxide removal unit. Condensation of sulfuric acid (from reaction of SO_3 with H_2O) begins typically at about 275°F, making it necessary to keep the gas temperature, in normal power plant operation, at a higher level until the gas leaves the stack. Otherwise, corrosion becomes a major problem.

In many sulfur oxide removal processes, the gas is cooled to the wet bulb temperature (about 125°F) prior to or during wet scrubbing, in which case resistant materials of construction may have to be employed to avoid corrosion. Moreover, the H_2SO_4 may under some circumstances condense as an aerosol that is extremely difficult to remove in a scrubber, and therefore may pass through the system and out the stack. The SO_3 also contributes to the problem of sulfate formation (rather than the preferred sulfite) in many absorption systems. Regeneration of sulfate to get the absorbent back for recycling is usually much more difficult than for sulfite.

The content of SO_2 in the gas affects the economics of recovery methods, the ease of bisulfite formation in processes that are favored by high bisulfite content in scrubber effluent liquor, and the degree of oxidation in absorption processes. The SO_2 concentration also affects efficiency of removal, of course, when the product of absorption has a vapor pressure of SO_2. Generally speaking, it is easier to achieve a given percentage removal with a high inlet concentration than with a low one.

Nitrogen oxides are formed in two ways, by oxidation of nitrogen compounds in the coal and by reaction of nitrogen and oxygen in the combustion air at the high boiler temperature. The oxides are pollutants themselves, as will be discussed later. In addition, they enter into sulfur oxide removal technology because some processes are aimed at removing both in one operation. It has also been postulated that nitrogen oxides promote oxidation of SO_2 and of sulfite. The amount of NO and the ratio between NO and NO_2 depend on several factors, including boiler type (as affecting boiler temperature), retention time of gases in the boiler, and rate of gas cooling after the boiler. For the concentration given in Table 1.5, a 1000 Mw boiler would typically emit 50 tons per day of NO and 5 tons of NO_2.

Hydrocarbons present in coal are volatilized at the high boiler temperature and are essentially burned completely. It is thought, however, that minute amounts, particularly of phenolic types, escape combustion and act as inhibitors to oxidation in sorption processes.

Particulate matter comes mainly from the inorganic compounds in the coal (unburned carbonaceous matter constitutes a minor source); burning of the coal releases the ash as fine dust particles that escape, at least partially, with the combustion gas. Again, the composition of the ash varies widely. A more-or-less typical composition is given in Table 1.8.

TABLE 1.8: COMPOSITION AND PARTICLE SIZE OF ASH FROM COAL COMBUSTION*

Constituent	Percent by Weight
CaO	6.1
MgO	1.1
Al_2O_3	20.5
Fe_2O_3	16.9
SiO_2	44.9
S	1.9
K_2O	0.8
Na_2O	0.3
Unknowns	7.5
	100.0

Particle Size, Microns	Percent of Total Number of Particles	Percent by Weight
0.5 and smaller	57.2	0.13
1	20.5	0.37
2	10.6	1.5
3	5.2	2.5
4	3.0	3.5
5 to 9	3.0	17.0
10 to 100	0.5	75.0
	100.0	100.0

* Coal from western Kentucky, burned in pulverized fuel boiler.

The amount of ash in the stack gas varies with the content in the coal and the type of boiler. Ash content of coal varies widely, from only a percent or so up to, in special cases, about half the weight of the coal. Boiler type affects the situation because in some boilers, for example, the cyclone type, much of the ash falls out of the gas and is removed at the bottom of the boiler. However, the most common type, pulverized fuel, carries over 75% of the ash out with the gas. A typical pulverized fuel boiler is shown in Figure 1.3.

FIGURE 1.3: TYPICAL COAL-BURNING POWER BOILER

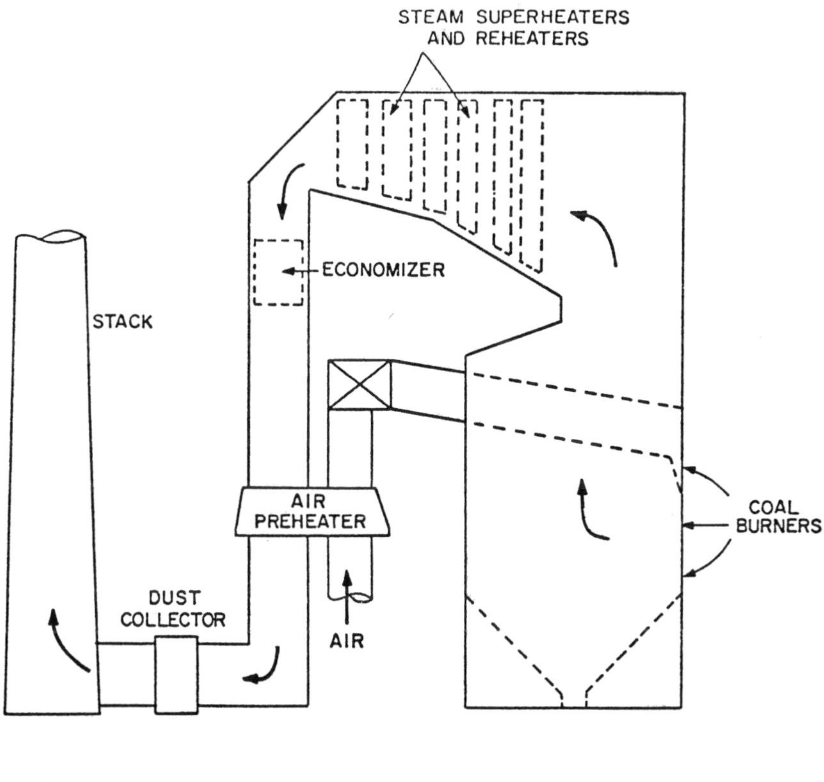

Emission of particulate matter has been considered in the past to be the major pollution problem in power production from coal. Public attention is drawn to it because of plume visibility and the visible evidence of deposition on objects in the area surrounding the power plant. As a result, devices for removing dust from the gas have been in use for a long time, whereas removal of other pollutants is just beginning.

The main connection of dust with sulfur oxide removal is that the latter may make it desirable to use a dust removal method different from those employed in the past. If a wet scrubbing system is used for SO_2, it is more economical to use a wet system for dust also than to install an electrostatic precipitator. If certain types of high temperature sorption are employed for SO_2, then a hot precipitator, operating at a temperature much higher than the usual type, is installed ahead of the SO_2 unit.

Residual dust passing through a wet scrubber or hot precipitator can affect operation of the following SO_2 unit adversely, both by deposition and obstruction of scrubber surfaces and by promoting oxidation. If the dust and SO_2 are removed in the same scrubber unit, as in some processes, there can be problems in separating and recovering scrubber liquor from the dust. Residual dust in the liquor and soluble compounds dissolved from the dust can affect process operation, particularly if all the liquor is recycled; in such closed loop operation, soluble materials build up and affect solution equilibria.

Most of the weight of the dust is in particles larger than 1 micron (μ) in diameter that can be removed in precipitators and wet scrubbers fairly completely. Most of the particles, however, are smaller than 1μ (Table 1.8) and tend to pass on through dust removal equipment. It has been considered that these very fine particles may be more injurious to health than are the larger ones, may contribute in a major way to smog and haze, and may promote obstruction of the sun's rays from the earth. The main effect of this on SO_2 removal is that new methods will be required to remove such fine dust and that SO_2 removal must be adapted to any new dust removal techniques.

The amount of stack gas is a major problem in design of SO_2 removal systems (Table 1.9). The quantity is so large that existing designs for sorption equipment must be extrapolated into unknown areas in most cases. Alternatively, multiples of existing sizes can be used, but this generally requires more space and fails to realize the reduction in investment usually obtainable from larger size.

TABLE 1.9: STACK GAS WEIGHT AND VOLUME*

Weight, tons/day	125,000
Volume, standard cu. ft. per minute (scfm)	2,100,000
actual cu. ft. per minute (acfm) at 300°F. (usual gas temp. after air heater)	3,100,000

* Approximate amount for a 1000 Mw boiler.

Power Production from Oil

In the U.S., consumption of oil in power production, on a megawatt-equivalent basis, is much less than that of coal. In some other parts of the world, however, the percentage is much higher. One of the main advantages is that the relatively low ash content, far lower than in coal, makes it a premium fuel in areas where disposal of a solid waste such as ash is a major problem.

The sulfur content of oil varies widely but generally is lower than in coal, even on a weight basis; on a heating value basis the effective content is even lower in comparison with coal. About 40% of the crude oil produced in the U.S. contains less than 0.25% sulfur and only about 20% contains more than 1%. In other parts of the world, Africa and Canada have somewhat comparable situations while South America and the Middle East generally produce high-sulfur oil (>1%).

During refining the sulfur is concentrated somewhat in the higher boiling, residual fractions, which, since they are the cheapest, are generally used as fuels. The very heaviest fraction, as might be expected, ends up mainly in power plants. Only about 25% in the U.S. residual oil contains less than 1% sulfur; the bulk (about 70%) is between 1 and 3%. Thus, although the sulfur becomes concentrated in the residual oil burned by power plants, the general average is still lower than in bituminous coal.

Except for the lower sulfur and ash contents, the composition of stack gas from oil-burning power plants is much like that from coal-burning units. One exception is that oil usually has a relatively high vanadium content, which can have some effect on oxidation of sulfur compounds during the removal process. Another difference is that the very low ash emission is quite beneficial in some sulfur oxide removal methods.

Other Uses of Coal and Oil

After power production, the next largest sulfur oxide producer is coal and oil combustion for industrial and residential purposes, mainly for process steam production and space heating. Although the total emission is large, unfortunately the individual unit size usually is quite small and therefore not economically suited to stack gas treatment for sulfur oxide removal. One of the alternative control methods discussed later would be more applicable in most cases.

Metal Ore Smelting

Several of the important metalliferous ores occur as sulfides (copper, lead, zinc, nickel, mercury, and molybdenum) and are oxidized (roasted) as the first step in metal production. The resulting SO_2 makes up about 12% of the total emitted in the U.S. Copper, zinc, and lead are the main contributors, in that order.

Unlike coal, the ores are almost always treated (beneficiated) before the roasting step to remove diluent materials. As a result, the actual material burned has a relatively high sulfur content and the SO_2 concentration in the stack gas is correspondingly high. In some cases the concentration is high enough (over 3% SO_2) for feeding the gas directly to a standard wet type sulfuric acid plant. About half the U.S. smelters are reported to be making by-product sulfuric acid in this way and there is also major production in Canada.

Much of the gas, however, is too dilute for direct acid production. It is still considerably more concentrated than stack gas from power plants, and thus can be treated for SO_2 removal by methods that would not be economical for the more dilute gases. Nevertheless, such methods (e.g., amine scrubbing) have not found much favor. Because of tightening restrictions on SO_2 emission, smelter companies are currently giving serious consideration to methods that are under development primarily for power plant use.

Although power plant gas cleaning methods are generally applicable to smelter gases, differences in composition of the two types of gas introduce complications. Moreover, a smelter may produce several gas streams, each varying in composition and temperature, and some of these may have a cyclic type of flow. The major composition differences are amount and composition of dust (which varies widely) and content of carbon dioxide and water vapor (since the ore serves as a major portion, or all, of the fuel). In some cases, the metal values give a return for dust recovery.

Petroleum Refining

The refining and processing of crude oil involve many operations and processing steps. Sulfur is usually a constituent because of its presence in the original raw material. However, the refining steps usually are of a reducing rather than oxidizing nature and H_2S rather than SO_2 is the usual sulfur end product. If H_2S concentration is high enough in a particular stream it is usually recovered. First is separation from the gas stream by absorption in a regenerable solvent such as an ethanolamine, followed by Claus process treatment to produce elemental sulfur. This process involves oxidation of part of the H_2S to SO_2 and reaction of this with the remainder of the H_2S, over a catalyst, to give sulfur.

$$H_2S + 1\ 1/2\ O_2 \longrightarrow SO_2 + H_2O$$

$$SO_2 + 2H_2S \longrightarrow 3S + 2H_2O$$

Many of the streams, however, present the problem that the H_2S concentration is too low for economical recovery and the H_2S cannot be released to the atmosphere because of the odor. The only practical course is to burn (flare) a waste stream to convert the H_2S to SO_2, a less undesirable form. Other sources are: (1) burning of oil as fuel in the various heaters and boilers needed in a refinery; (2) burning off of coke deposits on cracking catalyst to regenerate it; (3) burning of acid sludge for disposal; and (4) various air blowing operations. The growing use of hydrogen treating in refineries is reducing the sulfur emission problem.

The hydrogen, made in reformers or partial oxidation units, has the primary function of reacting chemically to reduce formation of heavy oil fractions and increase production of the more valuable light fractions such as gasoline. An incidental advantage is that most of the organic sulfur compounds are converted to H_2S, which can be separated and converted to sulfur. Up to 85% of the sulfur in crude oil can be recovered in this manner.

The waste gases from flares, boilers, and other combustion units are generally too dilute in SO_2 for feed to a sulfuric acid plant and are therefore vented without treatment. The range of concentration varies considerably, from as low as 300 ppm up to a percent or so. The sulfur oxide removal methods being developed for use in power plants should be applicable to such streams, but as far as is known, there are no current efforts in this direction.

Sulfuric Acid Production

The production of sulfuric acid in the U.S. is on the order of 30 million tons per year, mainly produced by burning elemental sulfur to SO_2, oxidizing over a catalyst to SO_3, and absorbing in water. The conversion of SO_2 to SO_3 is not complete, however, and the residual SO_2 is not absorbed to any significant extent in the SO_3 absorption step. The estimated emission is about 600,000 tons of SO_2 per year, or about 3% of the sulfur burned. The concentration of SO_2 in the acid plant stack gas is at about the same level as in power plants, from 2,000 to 3,500 ppm. The gas also contains, unlike the power plant gas, from 3 to 30 mg of sulfuric acid mist and about 0.5 mg of unabsorbed SO_3 per scf. These sometimes give rise to a white plume leaving the stack. One of the favored methods for reducing SO_2 emission is to improve SO_2 oxidation (conversion) by using two stages of catalyst with SO_3 absorption after each. A conversion of over 99.7% is claimed, which is equivalent to reducing the SO_2 concentration in the stack gas by about 90%.

ALTERNATIVES FOR CONTROL

Although removal of SO_2 from stack gases has received major emphasis, there are several other approaches to abatement of atmospheric pollution by SO_2. These will be reviewed to put removal from the gas in the proper perspective.

High Stacks

In the past, atmospheric pollution from stationary sources has been generally considered to be a local problem, of consequence only in the area surrounding and near to the emission source. At points farther away, the dilution of the pollutant with air reduces its concentration to levels so low that they have not been generally regarded as serious. The reason for the local high concentrations is that wind and other atmospheric conditions often are such that the gas plume bends back to the ground after leaving the stack, thus bringing the SO_2 to ground level before it has time for extensive dilution with air. The resulting content of SO_2 in the gas-air mixture at ground level is called the ambient concentration. An obvious way to reduce ambient concentration is to make the stack chimney tall so that the gas will be relatively high above the ground when it is released to the atmosphere, thereby giving it more chance for dilution if adverse meteorological conditions tend to bring it back to the ground.

High stacks have been quite successful over several decades as a means of dispersing pollutants and keeping ambient concentrations low enough to avoid significant damage from SO_2 emission. In the utility industry, for example, average stack height has increased steadily until the newer plants of today have stacks 1,000 feet or more in height. The same trend has taken place in the smelter field and to a lesser extent in other industries that emit SO_2. In England, use of high stacks is considered adequate for controlling the SO_2 problem; little work is being done on other approaches.

Because high stacks have been an effective and relatively economical solution to the problem, a large amount of research has been done in the past three decades or so on the path of the plume under various weather conditions. As a result, formulas have been developed that allow selection of stack height for a specific set of conditions with some confidence. The subject is a complicated one and will not be dealt with here in depth.

Although high stacks are effective in controlling ambient concentrations, there is increasing concern over the total amount of SO_2 emitted and the resulting general increase in ambient concentration over wide areas. Coupled with this is the particulate sulfate problem, since the SO_2 becomes oxidized in time to

SO_3 which reacts with water and with bases such as ammonia in the air to give liquid or solid sulfate particles. It is claimed by some that the sulfate is even more detrimental than SO_2 and that a hazardous concentration level will be reached unless SO_2 emission is reduced drastically.

For these reasons, EPA has set an emission limit of 1.2 pounds per 10^6 Btu for new power plants. Existing plants are more subject to state regulations, which vary widely but usually also have an emission limitation. In all cases, however, ambient concentration limits must also be met and therefore stack height is important.

In other countries, particularly in Japan, there are also emission limitations. These usually combine stack height with an empirical constant based on the local situation (overall emission in the area, population density) to give a formula from which the emission limitation for a given plant can be calculated.

Tall stacks are also of value, particularly in large power plants, for dispersing NO_x, CO_2, and submicron dust. Moreover, several of the other methods are subject to operational failure, in which case it is highly desirable to have available the dependable dispersion afforded by high stacks.

Low Sulfur Fuel

If raw materials that contained no sulfur, or not more than a tolerable amount, could be used, there would be no SO_2 problem. Most of the economically available fuels and nonferrous ores, however, contain too much sulfur for full release to the atmosphere. Where SO_2 emission standards have been imposed, many producers have been hard put to cope with the situation; methods for SO_2 removal from the gas are only partially proven and low sulfur fuel is scarce and expensive.

Although smelters and refineries might possibly have some choice regarding sulfur content of the raw material, the main possibilities and problems lie with the coal and oil burned in power production. The situation in regard to coal and oil in the natural state was reviewed earlier. There are large reserves that are low in sulfur content but they are generally much more expensive, delivered to the power plant, than are the high sulfur types. This penalty can be absorbed, of course, in the interest of abating pollution, and it is likely that use of low sulfur fuels will continue to be one of the solutions to the problem for some time. The extent will depend on the economic balance between the higher cost of the low sulfur fuel and the cost of fuel desulfurization, or of SO_2 removal from the combustion gas.

In the U.S., there is a major trend in the eastern part of the country to use of low-sulfur coal from the western areas (Montana, Wyoming, Utah) in new power plants being constructed. To meet the new source standard (1.2 pounds SO_2 per 10^6 Btu), it is necessary that the coal not contain more than about 0.6% sulfur but large supplies of such coal are available. The cost of shipping is so high that in many instances the use of scrubbers with high-sulfur eastern coal is considered preferable. Also, some of the eastern plants, particularly those east of the Allegheny Mountain range, have been able to acquire supplies of low-sulfur eastern coal.

In the West, low-sulfur coal is burned in both new and existing plants, of course, since it is the prevalent type. Nevertheless, scrubbers are generally required because state regulations, more rigorous than those in the East, require it. As in most other areas of the world, regulations grow increasingly restrictive as time goes on. It is conceivable that in time the western 0.6% sulfur coal will not be acceptable in the eastern U.S.

One consideration for existing coal-burning power plants is that modern, large boilers are designed for a particular type of coal and will not necessarily work well with another type. This is particularly true in going from a high-sulfur to a low-sulfur coal.

Fuel desulfurization is a major potential solution to the problem, second only in emphasis received and research carried out to combustion gas desulfurization. Numerous methods for removing sulfur from both coal and oil have been proposed and many have been tested. The technology is quite complex and will not be detailed here. The situation can be summarized by saying that (1) the technology for oil desulfurization is well developed, (2) methods for coal desulfurization are not as far advanced but appear feasible, (3) the sulfur is present as various compounds, some of which are relatively easy to remove, and (4) a high degree of desulfurization is quite expensive for either coal or oil because the more recalcitrant sulfur compounds must be attacked. Again, the situation has not crystallized enough for economic pressures to determine the status of fuel desulfurization.

Operational Control

Another method for preventing high ambient concentrations is merely to reduce production when meteorological conditions are such as to make high concentrations likely.

This has been developed and refined to a high degree. Since it is necessary to predict the adverse meteorological conditions if the production cutback is to be done in time to be of any help, meteorological data are collected and computerized calculations based on mathematical models are carried out to predict when and to what degree the boiler output should be reduced. TVA has pioneered in developing this technique and uses it at some of the stations.

Since it has been proved effective in meeting ambient standards, the utility industry has pushed such intermittent control [called Sulfur Dioxide Emission Limitation (SDEL) by TVA] as the most economical method for dealing with the problem. EPA has steadfastly resisted it, however, on the basis that overall emission would be reduced very little.

MAJOR CONSIDERATIONS IN REMOVAL FROM STACK GAS

The technology of SO_2 removal from stack gas has developed in different directions, each with its particular advantages and problems. New processes have been developed, old ones have been adapted, special equipment has evolved, and various approaches have been followed in regard to the many nonprocess problems encountered. In the past few years the development of processes and equipment has been under forced draft because of environmental pressures, with the result that an embryonic but complicated technology has evolved in a relatively short time.

Process Type

The general approach in process development has been to convert the sulfur to some form other than SO_2 that is easier to remove from the gas. This has taken different directions:

(1) Conversion of sulfur to H_2S in a partial combustion operation followed by sorption of the H_2S from the gas stream.
(2) Oxidation of SO_2 to SO_3 in the gas stream followed by condensation of sulfuric acid.
(3) Reduction of SO_2 to sulfur in the gas stream followed by condensation of the sulfur.
(4) Preferential sorption of the SO_2 by various liquid and solid materials to form sulfites or sulfates in the liquid or solid form.

The SO_2 sorption processes are the most numerous by far, followed by the SO_2 oxidation methods. Reduction of SO_2 to sulfur in the main gas stream appears to have serious technical problems, and conversion to H_2S is more a part of coal gasification technology (for preparation of low Btu or pipeline gas) than of combustion processes that produce SO_2.

Throwaway Versus Recovery

One of the major divisions in the technology is between: (1) recovery of the SO_2 in a useful form, and (2) formation of a product intended for discard as a waste solid. Although it is desirable to recover a usable product, all the recovery processes developed so far appear to have a higher production cost than the throwaway type. Since sale of a by-product is often a major problem, the producers involved—power producers, smelters, and refiners—would generally prefer a throwaway process unless the recovery approach showed promise of considerable economic advantage.

The throwaway methods have the major disadvantage, however, that the pollution problem is not completely solved; atmospheric pollution is merely converted to a solid waste problem. Coal-burning power plants and smelters have solid waste problems anyway—ash from the power plant and various solids from the smelter—so the disposal task is merely made larger. The main difficulty is with oil-burning power plants and other operations that normally produce little in the way of solid wastes. Solids disposal facilities must be installed and a disposal area found, which may be quite difficult in municipal and urban areas.

The recovery route has so many drawbacks and problems that a special chapter will be devoted to them. One of the major problems is the uncertainty regarding sale of the sulfur-containing by-product. If it cannot be sold, a very serious waste disposed problem is incurred.

Nevertheless, development of recovery processes has proceeded at an accelerating rate and several utilities and smelter companies have selected recovery over throwaway as the area of their expenditure of research funds. The usual reason is that the solid waste disposal problem in throwaway process operation is considered unacceptable.

Wet Versus Dry

Both the throwaway and recovery methods can be carried out either in wet or dry systems. Over the history of research and development on SO_2 removal, the two systems have varied in favor from time to time. At present the wet methods seem to be in the ascendancy.

One of the main advantages of the dry system is that the stack gas is not cooled in the SO_2 removal operation and therefore has the full thermal buoyancy leaving the stack as it does in normal operation. In contrast, wet methods involve the use of a scrubber installed just before the stack; the water solution of salts circulated through the scrubber cools the gas to its wet bulb temperature, which is considerably lower than the usual stack inlet temperature. In power plants the normal temperature is about 300°F whereas the wet bulb temperature is about 125°F.

Dry processes, particularly the absorbent type, also have some major disadvantages. For one thing, it is difficult to design for enough gas-solid contact time to allow permeation of SO_2 to the solid particle interior; as a result, sorbent utilization per pass normally is lower per unit of active sorbent than in a wet system.

Moreover, if cyclic operation is involved as it is in many processes, the regenerated particles tend to become inactivated and to deteriorate physically over several cycles of sorption and regeneration. This is especially true if a fluidized bed, moving bed, or solids-entrained reactor is used, because attrition is a major problem in such equipment. Parallel fixed beds can be used to avoid this; the beds operate in a cyclic manner, one absorbing while the other is being regenerated. The main drawback is that the very large tonnage of gas flow must be shifted from one vessel to the other at relatively short time intervals.

The wet system avoids these drawbacks; mass transfer is better in the liquid medium and physical deterioration of absorbent is not a problem. However, adverse chemical reactions are more likely to take place, and some cost is involved in separating water from the product, either for regeneration in recovery processes and for discard in throwaway methods. Because of the gas cooling in wet processes, the consensus is that the gas should be reheated before release, at least enough to prevent condensation in equipment after the scrubber. The degree of reheat, one of several factors affecting recovery economics, will be discussed in a later chapter.

Degree of Removal Needed

Regardless of process type or objective, the degree of SO_2 removal attempted is a major consideration. The removal requirement varies widely from place to place for several reasons. In the U.S., the Environmental Protection Agency (EPA) has set 1.2 pounds SO_2 per 10^6 Btu as the emission standard for new coal-fired power plants; for other SO_2 emitting industries there are other standards. For existing power plants, the rules vary widely since existing plants are under the direct jurisdiction of states and municipalities rather than EPA. In some states, regulations are still under negotiations; in others the allowable emission is as low as 0.15 pound per 10^6 Btu in some of the western areas and as high as 4.0 pounds in the East. The allowable emission for oil-fired boilers is generally lower than for coal-fired; the new source standard is 0.8 pound per 10^6 Btu for oil.

In some parts of the world, the setting of regulations has been approached on an area rather than plant basis. A study is made for a particular industrial area and emission regulations set for each plant in the area, based on parameters such as stack height, type of process from which the SO_2 is emitted, age of plant, sulfur content of raw material, and relative cost of an SO_2 removal system.

For particulate, the new source regulation is 0.1 pound per 10^6 Btu (about 0.05 grain/scf). There is also a stack opacity regulation of Ringelmann 1 (based on a crude method for estimating the opacity of the plume leaving the stack).

The degree of removal required thus depends on the initial SO_2 and particulate content in the gas. For a typical eastern coal (3.5% sulfur), which produces a concentration of about 2,500 ppm SO_2 in the gas (roughly 6.25 pounds SO_2 per 10^6 Btu), the degree of removal required to reach the 1.2 pound standard is about 80%.

New Plant Versus Retrofit

Some processes can be classified as the add-on type, that is, they can be installed on the gas stream after it has passed through all the equipment and is ready to enter the stack. Others require a higher temperature level and therefore must be located farther back toward the front end of the plant; in a power plant this usually means before the air heater (temperature of 600° to 900°F versus about 300°F after the heater). This affects process choice if the plant is an existing one. A high-temperature process involves the cost and time for cutting into the gas train, plus the uncertainties introduced regarding the effect on primary plant operation. Or the process can be installed after the air heater by reheating the gas to the required temperature, an operation that requires an expensive heat exchanger plus burning some additional fuel. Thus an add-on process is much more appropriate for a retrofit installation. Processes that must have high gas temperature are more suitable for new plants, where the design can be optimized for fitting in the removal unit.

SCOPE OF PRESENT EFFORT

For many years research on SO_2 removal from waste gases was carried on somewhat fitfully and was tied mainly to particular pollution situations. For example, the Consolidated Mining and Smelting Company developed ammonia scrubbing in the 1940s as a means of abating SO_2 emission that was causing an international pollution problem. The Battersea and Bankside lime scrubbing units in England were developed and installed mainly for the specific local problem, and other examples also could be given.

In the last decade or so, however, SO_2 removal research has become general in nature rather than aimed at some specific situation. The main impetus in this direction has come from governmental authorities, mainly in the U.S., Japan,

England, and Germany, who have promoted research in the interest of finding the best methods for general control of the problem throughout their countries. As a result, the volume of work has increased rapidly, as well as the number of research organizations participating. The general pressure for better environmental protection has also led many private companies to do research in the area at their own expense. A few of these have been companies operating emitting plants, but mainly the objective has been to develop process technology and equipment for sale.

In the U.S., the EPA is the leading organization in the field. A broad program of both in-house and sponsored research is under way, including surveys and analyses of various types, basic research, small-scale process research, pilot plant projects, and large prototype plant test programs. Other U.S. government organizations with major involvement in SO_2 research are the Bureau of Mines and the Tennessee Valley Authority (TVA). The Bureau of Mines is concerned with the proper use of coal and oil as natural resources and therefore has an interest in problems such as SO_2 emission that limit the use. Research on various methods of SO_2 removal and recovery are under way at Bureau research stations over the country.

TVA has a threefold interest in the problem. The agency operates one of the larger power systems in the world (about 40,000 Mw in operation or on order, of which about 18,000 Mw is from coal-fired units), in which high-sulfur coal is burned in many of the boilers. Emission of sulfur oxides therefore is a matter of concern. Secondly, TVA carries out a major research and development program in fertilizer technology, and one of the problems in this area is the recurring shortage of elemental sulfur. Since sulfuric acid (mainly made from sulfur) is a major raw material in phosphate fertilizer production, the potential of sulfur recovery from stack gases is an important consideration.

Finally, because one of the major TVA objectives is to promote conservation of natural resources, recovery of the sulfur from the coal burned in TVA power plants is a research goal. TVA carries out a major research and development program on sulfur oxide removal and recovery, both as a direct agency effort and in the form of projects carried out for EPA.

The Office of Coal Research also supports research on sulfur removal, but as part of the effort to make clean fuels and pipeline gas from coal. In most of the processes being tested, the sulfur is converted to hydrogen sulfide in the coal gasification or cleaning step and is never oxidized as it is in power production. Hence the removal processes differ considerably from those being developed for removing sulfur oxides from combustion gas.

In Germany, the state of Nordrhein-Westphalen, which includes the Ruhr Valley, has been quite active in the field through the Landesanstalt (Land Institute for the Protection of Air and Soil). In other countries the principal government agencies involved are: (1) the Central Electricity Generating Board in England, (2) the Resources Research Institute in Japan, (3) the Research Institute of Inorganic Chemistry and the Fuel Research Institute in Czechoslovakia, and (4) various groups in the USSR. All of these have either current or past research programs on sulfur oxide removal, either in-house or sponsored.

The work by private organizations varies over a wide range in type, scope, and stage of development. Proprietary processes of both the throwaway and recovery type are under development. Some of these are described in technical publications and others in news announcements. Still others have not been announced publicly but are known to those working in the field, and no doubt there is work under way known only within the particular organizations involved.

The SO_2 problem has attracted a remarkably large number of investigators. No effort has been made herein to cover all of the work under way, as much of it is not of much consequence as yet. However, all those processes on which a significant amount of work appears to have been done will be described in the subsequent chapters. For convenience, the principal investigating organizations, both public and private, are listed in Table 1.10. This is not presented as a complete listing; some important contributors to the technology may have been omitted because of insufficient information.

TABLE 10: ORGANIZATIONS ACTIVE IN RESEARCH ON SULFUR OXIDE REMOVAL FROM STACK GASES

Austria
- Osterreichische Stickstoffwerke AG

Bulgaria
- Energoproyekt

Canada
- Cominco Ltd.
- McGill University
- Ontario Hydro

Czechoslovakia
- Fuel Research Institute
- Institute of Chemical Equipment
- Research Institute of Inorganic Chemistry

France
- Electricite' de France
- Etablissements Kuhlmann
- Institute Francais de Petrole

Germany (Federal Republic)
- Bergbau-Forschung GmbH
- Chemiebau
- Gottfied Bischoff Bau Kompletter Gasreinigungs und Wasserruchkuhlanalger AG
- Grillo-Werke AG
- Hölter
- Land Institute for the Protection of Air and Soil (Landesanstalt)
- Lurgi Gesellschaft für Chemie und Hüttenwesen mbH
- Steinkohlen Elektrizitat AG (STEAG)

Hungary
- State Office for Technical Development

Italy
- Associazione Termotecnica Italiana

Japan
- Central Research Institute of Electric Power Industry
- Chisso Corporation
- Chiyoda Chemical Engineering and Construction Company
- Dowa Mining Company
- Fujikasui-Kogyo, Ltd.
- Hitachi, Ltd.
- Japan Engineering Consulting Company
- Kawasaki Heavy Industries, Ltd.
- Kobe Steel Company
- Kurabo Industries
- Kureha Chemica Industries
- Mitsubishi Heavy Industries, Ltd.
- Mitsubishi KK
- Mitsui Miike Machinery Company
- Nippon Kokan KK
- Resources Research Institute
- Showa Denko Company
- Sumitomo Chemical Company, Ltd.
- Sumitomo Metals Company
- Tokyo Institute of Technology
- Toyo Engineering Company
- Tsukishima Kikai Company

Netherlands
- Shell Oil Company

Poland
- Copernicus University
- Fedzerzhinskii Technology Institute
- Institute of Chemtechnology
- Institute of Scientific Research
- Politechniki Wroclawskiej

Romania
- Inorganic Chemistry and Fertilizer Engineering Institute (IPRAN)

Sweden
- AB Bahco

USSR
- All-Union Association on Gas Cleaning and Dust Collection
- State Scientific Research Institute of Gas Purification for Industry and Sanitation (NIIOGAZ)

United Kingdom
- BCURA Industrial Laboratories
- Central Electricity Generating Board
- Esso Research Centre
- National Coal Board

United States
- Aerojet-General Corporation
- Air Products
- Allied Chemical Corporation
- Ambient Purification Technology, Inc.
- Argonne Laboratory
- Arthur D. Little, Inc.
- Atomics International
- AVCO Corporation
- Babcock and Wilcox Company
- Battelle Memorial Institute
- Bechtel Corporation
- Bituminous Coal Research, Inc.
- Black, Sivalls, and Bryson, Inc.
- Callery Chemical Company
- Carnegie-Mellon University
- Chemfix
- Chemical Construction Corporation
- Chemico-Basic Corporation
- Chevron Research Company
- City College of The City University of New York
- Combustion Engineering, Inc.
- Commonwealth Edison Company
- Consolidation Coal Company
- Davy Powergas, Inc.
- Detroit Edison Company
- Dow Chemical Company
- Dravo
- Environmental Protection Agency (EPA)
- Envirotech
- Exxon Research and Engineering Company
- FMC
- Foster Wheeler Corporation
- General Technologies Corporation
- General Motors Corporation
- Institute of Gas Technology
- IU Conversion Systems
- M.W. Kellogg Company
- Kennecott Copper Corporation
- Monsanto Company
- Office of Coal Research, ERDA
- Peabody Engineering
- Pope, Evans, and Robbins
- Princeton Chemical Research, Inc.
- Radian Corporation
- Research-Cottrell, Inc.
- Riley Stoker
- Singmaster and Breyer
- Southern California Edison Company
- Stone and Webster-Ionics, Inc.
- Tennessee Valley Authority (TVA)
- Tracor, Inc.
- Uniroyal, Inc.
- United Aircraft Laboratories
- United International Research, Inc.
- Universal Oil Products Company
- University of California, Berkeley
- University of Washington
- U.S. Bureau of Mines
- Westinghouse Electric Corporation
- Westvaco Corporation
- West Virginia University
- Zurn Industries, Inc.

CURRENT STATUS OF PROCESS DEVELOPMENT

The limestone throwaway processes have been developed to a greater degree than the recovery methods, mainly because the throwaway route is simpler and the economics are more certain. In early 1975, wet scrubbing with lime or limestone was rapidly being adopted in the U.S. by the power industry as a standard method of SO_2 control. Ten boilers or more equipped with scrubbers were in operation and numerous new installations either had been announced or were known to be in the early stages of planning.

Major test programs on throwaway processes were also under way, particularly in Japan, Germany, and the U.S. The major effort was an EPA-TVA joint project at a TVA power plant, an extensive and in-depth project to optimize the process.

Several projects were underway in the U.S. which involved recovery facilities on full boilers, including catalytic oxidation, MgO scrubbing and sodium sulfite scrubbing. Similar projects were underway in Japan, for carbon absorption and sodium scrubbing. Although these are full boiler installations, the development aspect appears to have played a major role in justifications for the projects.

The many other development efforts, both throwaway and recovery, were in various stages of development, many of them in large pilot plants. The general consensus was changing, from the earlier position that no proven and reliable process for power plants was available, where proven and reliable is defined as extended, satisfactory operation on an operating boiler large enough to give an adequate evaluation of operating problems to the attitude that processes were maturing and becoming acceptable, although still crude, expensive, and in need of further development.

Sulfur oxide recovery from sulfuric acid plants and smelters is considerably farther advanced. Ammonia scrubbing has been used for several years in both industries, and lime scrubbing is used routinely on acid plants in Japan. Moreover, U.S. acid plants have been equipped with sodium and ammonia systems of wet scrubbing.

It is expected that stack gas cleaning to remove SO_2 will become a major technology over the next two or three decades. The regulations are very likely to require reduction of emission from most of the plants in the smelter, refinery, sulfuric acid, sour gas cleaning, and power industries. Although other ways of reducing emission may eventually be preferable, stack gas cleaning appears to be the most attractive at the moment and is being emphasized by the various industries in their plans.

The main need centers on existing plants, particularly in those industries in which new capacity is being added at a relatively slow rate. For the power industry, the principal emitter of sulfur oxides, it is expected that use of fuels containing sulfur will eventually decline and that the SO_2 problem therefore will be reduced. Nevertheless, equipping the present plants with SO_2 removal facilities will be a tremendous task, and it is expected that a large amount of new capacity will have to be added over the next several decades to bridge the gap until new energy sources can take care of the growth in power consumption. The heavy emphasis on fitting existing plants has resulted in the add-on type of system getting the most current attention. The main effect of this on process choice is that wet scrubbing methods, in general, are more amenable to adding on than are the dry processes. Installing equipment to treat the gas after it has passed through all of the power equipment (except the stack) gives several advantages, including fewer space problems, less possibility of reducing station reliability, and less downtime during the construction period.

It is expected that the nonregenerative, throwaway processes will be favored over the recovery type for some time. Although many of the SO_2-emitting organizations are carrying out research on recovery, the majority are looking to throwaway methods as the immediate solution to the problem. The lack of proven technology and the major problem of developing markets put recovery off into the future.

Thus it seems likely that a large number of lime-limestone slurry scrubbing plants will be built in the next few years. This is also a relatively undeveloped technology, although generally farther along than recovery. The current work is identifying some unexpected problems that will have to be solved before fully effective and dependable performance can be attained.

For the future, cleaning the fuel before it is burned may come into favor. One of the main problems in stack gas cleaning is that cost of cleaning per unit of energy produced rises rapidly as the plant size decreases. Thus it might be more economical to supply small boilers with a clean fuel rather than to install gas cleaning equipment. The coalplex approach may be the best way to accomplish this. In such a complex, the noncarbon constituents of the coal would be separated, by distillation, solvent extraction, or other methods, and the carbon supplied as a relatively sulfur-free fuel. Other products such as pipeline gas, chemicals, and of course, sulfur, might also be produced.

Another factor that may affect the future of stack gas cleaning is the possibility that the power industry may adopt advanced power cycles as a means of recovering more of the energy in the fuel. The first step in most such cycles is conversion of the coal or oil to a clean fuel gas by partial combustion. Thus the sulfur, as H_2S, would be removed from a fuel gas rather than as SO_2 from the combustion gas, and a different technology would be required.

THROWAWAY OPERATION—DRY PROCESSES

Most of the processes for removing sulfur oxides from waste gases involve introduction of some material into the gas stream to combine with the sulfur oxides and convert them to a removable liquid or solid. Either chemical reaction or adsorption may take place. In either case, if the product is to be discarded (throwaway process), the added material must be quite inexpensive. Only two materials have been seriously considered for throwaway processes, i.e., limestone and water. Of these, water has the disadvantage that SO_2 is not very soluble and therefore very large volumes are required (see discussion in Chapter 3). For this reason most of the emphasis in throwaway process development has been on use of limestone.

The term limestone as used herein is defined as a naturally occurring material made of calcium carbonate ($CaCO_3$) plus various amounts of other substances. Calcitic limestone is mainly $CaCO_3$ with relatively small amounts of impurities, mainly magnesium carbonate and silica. Dolomite ($CaCO_3 \cdot MgCO_3$) is only about half $CaCO_3$ and dolomitic limestones are in between. Limestone is a very widespread natural material, occurring in most parts of the world and usually available in the vicinity of plants that emit sulfur oxides. Surface mining is normally feasible. The combination of low-cost mining and proximity to plants makes the delivered cost quite low, on the order of \$2 to \$5 per ton in many instances.

The ways that have evolved for using limestone can be divided into two major classes, dry and wet systems. In the dry system, the limestone (as is or calcined) is injected into the gas and recovered after reaction with the sulfur oxides. Hence equipment both for injecting into the gas and for removal is required.

Limestone reacts well with SO_3 in the dry state but not with SO_2; even if the limestone is calcined to CaO, which is usually much more reactive than $CaCO_3$, little reaction takes place with SO_2 at ordinary temperature. At high temperature, on the order of 1800°F, both $CaCO_3$ and CaO react with SO_2. At this temperature, however, the $CaCO_3$ will become calcined rapidly to CaO so that the reacting species is CaO even if limestone is the material injected. The dry system has been considered mainly for power plants, where injection into the boiler furnace makes it possible to calcine the limestone at little cost. There is no equipment cost other than for grinding and injecting, and fuel and labor costs are relatively low as compared to separate calcination. From this it would appear that the low cost should make the limestone dry system a very promising solution to the SO_2 problem. There are, however, some serious disadvantages.

(1) Degree of SO_2 absorption is relatively low because of: (a) the very short retention time in an ordinary boiler; and (b) difficulty in getting SO_2 penetration into the limestone particles beyond the surface layer. Excess limestone can be used to increase SO_2 removal, but even with a large excess a high degree of removal is difficult to attain.

(2) Reaction of lime with ash in the boiler may cause fouling (deposits on tubes with resulting increase in gas pressure drop).

(3) Unreacted lime in the gas stream reduces efficiency of the electrostatic precipitators normally used in power plants to remove fly ash from the gas. Therefore, precipitator capcity must be increased considerably more than would be necessary merely to handle the additional solids. For example, although the limestone addition may only double the solids content of the gas, it may reduce dust collection efficiency to the point that four times or more increase in precipitator capacity is required.

Because of these drawbacks, it is now generally concluded that injection of lime or limestone into a standard boiler is not an adequate way to control SO_2 emission. The absorbent utilization is too low, in most instances, to give the high degrees of removal now required by regulations. A much more promising approach is to use the method in a fluidized-bed boiler, a new type now being developed rapidly both in the U.S. and England. In the fluidized-bed boiler, limestone and coal are fluidized together in the combustion zone. Because of the extended retention time, the lime resulting from calcination of the limestone has time to react much more completely with SO_2 than in a boiler of the standard type. Because of the emphasis placed on the method in the past, the history of both the studies with standard and fluidized-bed boilers will be given.

HISTORY OF THE PROCESS

The limestone dry system is one of the oldest methods for SO_2 removal and has been studied by many investigators, mainly in Germany, Czechoslovakia, Japan and the United States. The more active organizations have been Bergbau-Forschung, Landesanstalt, Steinkohlen-Elektrizitat (STEAG) and Technical University of Stuttgart in Germany; Central Research Institute of Electric Power Industry (CRIEPI) and Resources Research Institute in Japan; Fuel Research Institute in Czechoslovakia; Energopomiar in Poland; EPA, TVA, OCR, Argonne, Exxon, Westinghouse, and Pope, Evans, and Robbins in the United States, and BCURA, NCB, and Exxon in England.

Because of the difficulty in studying limestone injection on a small scale, much of this work has been at the pilot plant or prototype level. Possibly the most promising results at this scale of operation were obtained by the Fuel Research

Institute in Czechoslovakia. Limestone injected into a 5 feet i.d. pulverized coal boiler gave SO_2 removal of about 50% for some limestones at 1.5 times stoichiometric feed. Up to 76% removal resulted at higher feed ratios (about 2.2).

There have also been at least four full-scale test programs in operating boilers. In tests by STEAG, 20 to 30% SO_2 removal was obtained in a 20 Mw boiler, with 1.0 to 1.3 times the stoichiometric amount of limestone. The relatively low sulfur content of the coal (1.5%), however, probably limited the percentage removal because of the lower driving force of SO_2 in the gas. In another German plant (BASF, in Ludwigshafen), tests were carried out by the University of Stuttgart in a small boiler. Over 50% removal was accomplished with 2.8 times the stoichiometric amount of hydrated lime injected at a point in the boiler where the temperature was 1000°C. In these tests the boiler fuel was coal; when oil was burned, 77% removal was obtained under the same conditions.

Plant-scale tests have been made by CRIEPI in Japan. Limestone injected at 1100° to 1200°C into an oil-fired boiler removed 40 to 50% of the SO_2 when two to three times stoichiometric was used. The most extensive plant-scale test program was a joint EPA-TVA project carried out at TVA's Shawnee power station near Paducah, Kentucky. This was preceded by a conceptual design and cost study made by TVA for EPA (then NAPCA) in 1968, which indicated that capital cost for limestone injection is lower than for other methods and that operating cost is lower for small units.

These considerations led EPA to sponsor the Shawnee test program. It was not expected that the method would be acceptable for genral use in large, basic power plants. However, for small boilers (particularly those operated intermittently) limestone injection, even though less effective, could be preferable to the more complicated systems because of their very high cost in small sizes. The objective of the Shawnee test program was to optimize the system and determine its potential under the best operating conditions.

Equipment for limestone receiving, grinding and injecting into the boiler at several points was installed in 1969 (see Figure 2.1). The boiler, rated at 175 Mw, is front-fired and burns pulverized coal that averages about 2.7% sulfur content. The cross section is 46 ft by 24 ft, and gas velocity is on the order of 33 ft/sec. The distance from the injection point to point **4** in the superheater section, where the temperature has dropped to 1200°F and therefore little further reaction is to be expected, was such that the retention time in the reaction zone was only about 1.5 seconds. The system was capable of feeding at a rate of three to four times stoichiometric. Both the injection velocity and the injection angle could be varied to improve uniformity of limestone distribution in the boiler.

The main independent variables in the test program (partial factorial design) were: (1) limestone distribution, (2) temperature at injection point, (3) stoichiometry, (4) limestone type, and (5) limestone particle size. The unit was started up in early 1970 and operated for about three years. The results of the tests were generally unpromising. It was difficult to get good limestone distribution and SO_2 concentration was also found to vary over the boiler cross section. As a result of these and other limitations, getting more than about 40% removal was difficult even with three times the stoichiometric amount of limestone. There was also a major loss in precipitator efficiency.

FIGURE 2.1: DRY SYSTEM FOR REMOVING SO_2 FROM STACK GAS BY SORPTION IN LIMESTONE

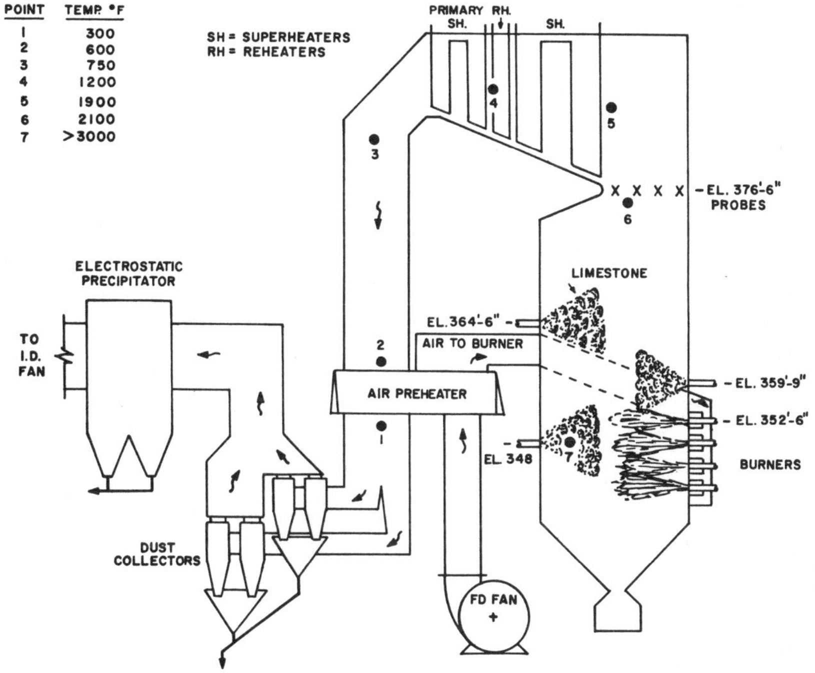

Flowsheet shown is for the EPA-TVA test facility in TVA's Shawnee power station near Paducah, Kentucky.

Much of the work on fluidized combustion, with accompanying absorption of SO_2 by lime, has been done in England by the Central Electricity Generating Board (CEGB), the British Coal Utilization Research Association (BCURA), the National Coal Board (at the Stoke Orchard station), and Exxon Research. CEGB is not active in the field at present. NCB has emphasized operation at atmospheric pressure and BCURA pressurized operation, while Exxon has worked with oil combustion.

The results have been fairly promising although lime utilization is still not as high as desired, especially in the pressurized work. Tests on a larger scale, which are badly needed, are to be carried out in a unit installed on a small industrial boiler.

In the United States, small-scale and pilot-plant work has been carried out by various organizations, including Argonne, Exxon, Westinghouse, and Pope, Evans and Robbins (PER). OCR is funding construction of a 30-Mw test unit based on the PER work, with start-up planned for late 1975. EPA is also planning a test unit (probably about 20-Mw) to be located at TVA's Watts Bar station and based on the Argonne-Exxon-Westinghouse work.

THERMOCHEMISTRY

The limestone injection process is complicated by the fact that three different types of reactions must take place, i.e., calcination, sorption and oxidation. There are several possible reactions involved.

$$CaCO_3 \longrightarrow CaO + CO_2$$

$$CaO + SO_2 \longrightarrow CaSO_3$$

$$CaCO_3 + SO_2 \longrightarrow CaSO_3 + CO_2$$

$$CaSO_3 + 0.5O_2 \longrightarrow CaSO_4$$

$$CaO + SO_2 + 0.5O_2 \longrightarrow CaSO_4$$

$$CaCO_3 + SO_2 + 0.5O_2 \longrightarrow CaSO_4 + CO_2$$

$$4CaSO_3 \longrightarrow 3CaSO_4 + CaS$$

$$CaSO_3 \longrightarrow CaO + SO_2$$

$$CaSO_4 \longrightarrow CaO + SO_3$$

For the very short retention time available in a power plant boiler, the rates of the various mechanisms involved, both chemical and physical, are far more important than the thermochemistry of the chemical reactions taking place. The chemical driving forces and reaction equilibria are of interest, however, because they set the limits of performance attainable. Moreover, for the extended reaction time obtainable in fluidized combustion, the thermochemistry gains in importance as contrasted to the kinetics.

Calcination

Experimental data indicate that calcination of the limestone in the boiler is necessary for optimum sorption. Direct sulfation (or substitution) is possible but the driving force for the reaction (negative free energy change) varies inversely with temperature and is quite low at boiler temperature. For an equilibrium concentration of SO_2 in the gas low enough to represent adequate removal, the temperature would have to be well below 1000°F, a level at which the rate is much too low to be acceptable.

Direct sulfation without calcination can also occur, which involves oxidation by the oxygen present in the boiler gases. The thermochemistry is quite favorable; although the negative free energy change decreases with rise in temperature, it is still quite high even at 3000°F. The equilibrium is such that adequate absorption is indicated at temperatures well over 2000°F. Nevertheless, experiments have shown that for the situation in standard boilers, the substitution reaction is far slower than calcination.

The thermochemistry of calcination is such that the reaction reverses, in a typical boiler atmosphere, at about 1400°F. Because the gas typically contains about 14% CO_2 (from combustion of the fuel), CaO introduced into the boiler at a temperature below 1400°F would absorb CO_2. $CaCO_3$, of course, would not be affected. Above 1400°F, $CaCO_3$ loses CO_2 because the equilibrium vapor pressure of CO_2 is higher than that corresponding to a 14% content, and the reaction proceeds because the CO_2 produced is swept out of the boiler. In fluidized boilers, which operate at 1500° to 1600°C, the temperature is closer to the reversal point, especially in pressure operation. Dolomite or dolomitic limestone behaves somewhat differently. The reversing point for $MgCO_3$ calcination is only about 650°F as compared to the 1400°F for $CaCO_3$. Hence the driving force for $MgCO_3$ calcination is relatively high at boiler temperature. This may be a disadvantage because it can lead to overburning or calcination to a degree that absorption efficiency is reduced.

Sulfation and Oxidation

Sulfation of CaO appears to be the second step when limestone is injected into the high-temperature environment in the boiler. Sulfation can proceed either by formation of $CaSO_3$ or $CaSO_4$. The $CaSO_3$ reaction, however, reverses at about 1900°F and the temperature must be as low as about 1200°F to give an equilibrium SO_2 concentration in the gas low enough for adequate desulfurization. Formation of $CaSO_4$, which involves oxidation, is much more favorable. Assuming a level of about 3% oxygen in the gas, the temperature can be as high as 2000°F with an equilibrium SO_2 concentration of less than 100 ppm. Above this temperature, however, the equilibrium is displaced rapidly to higher SO_2 vapor pressure.

Disproportionation of Sulfite

Disproportionation of calcium sulfite could give calcium sulfide in the product, a result that would be quite undesirable because of likely hydrogen sulfide evolution from reaction with water in the disposal pond or area. However, with oxygen present in the gas, the thermochemistry strongly favors formation of sulfate rather than sulfide at boiler temperature.

Decomposition of Sulfite and Sulfate

Decomposition of sulfite or sulfate defeats the purpose of limestone injection. As noted earlier, sulfite decomposition prevents attainment of adequate desulfurization above about 1200°F in the absence of oxygen. Because of the presence of oxygen, sulfate decomposition is of more concern. However, appreciable SO_3 formation above about 1500°F is not indicated because decomposition to SO_2 ($SO_3 \rightarrow SO_2 + \frac{1}{2}O_2$) is favored. This reaction inhibits SO_2 absorption above 2000°F.

KINETICS

The paramount importance of SO_2 sorption rate has led to a large amount of research on the kinetics of the dry limestone process. Unfortunately, the system is so complex that development of meaningful data is extremely difficult. The system is polythermal in that the limestone particles are heated rapidly to high temperature upon entry into the boiler furnace, are cooled to some extent by the endothermic calcination reaction, are heated by the exothermic reaction with SO_2, and are cooled rapidly as the heat in the gas stream is extracted by the heat exchange surfaces in the convection section of the boiler. The short retention time, very high temperature, turbulence in the boiler gases, and high velocity of the gases leaving the boiler furnace all make up a system that is very difficult either to simulate on a small scale or to test on a large scale. The situation is easier in fluidized combustion but still relatively complicated. One of the difficulties is that most workers have found it convenient to work with a synthesized stack gas, either by mixing gases or by operating an oil or gas-fired burner. The nature of coal and the complexities of boiler operation, however, give the actual stack gas from a coal-burning power plant somewhat different characteristics. Experience has shown that data obtained with actual stack gas are more likely to be valid.

Further, limestones vary widely in effectiveness and no method other than actual sorption testing has been available to determine the differences. Hence, results obtained by one worker may vary considerably from those obtained by another who used the same test conditions but a different limestone. Finally, the physical conditions in the boiler cannot be fully duplicated in a smaller scale test equipment. Although a set of consistent and interpretable data can be obtained in a given test unit, there is no way of knowing whether the same results would have been obtained in the boiler.

Conversely, tests made in full-scale boilers are usually difficult to reconcile. The disparities very likely result from the fact that a boiler is a very poor test unit, subject to many types of operating variation that affect sorption but cannot be controlled. Laboratory simulations can be controlled better but they are usually of the fixed-bed type and therefore do not achieve mass transfer conditions comparable to those in the boiler; retention time is much longer in such small-scale tests and the effects of particle dispersion are not accounted for.

Notwithstanding these experimental difficulties, some progress has been made in elucidating the kinetics and some general conclusions can be set down. It cannot be said, however, that any unanimity of opinion has been reached; the differing scale and conditions of the various test programs appear to have led to a considerable divergence in view on the basic mechanisms involved. In this discussion, the objective will be to present as good a consensus as possible. No detailed analysis and reconciliation of the individual findings will be attempted; because of the wide variance in test conditions, results, and conclusions, this does not appear feasible.

Chemical Considerations

It is generally considered that a major limiting factor in SO_2 sorption by CaO, assuming an adequate system for getting the SO_2 to the CaO particle surface, is a first-order chemical reaction with respect to the concentration of SO_2 in the gas phase. After some degree of reaction has taken place, internal diffusional resistance becomes limiting. Chemical considerations are also important in calcination because chemical factors affect the porosity of the resulting calcine, an important variable since high porosity increases the surface area available for the chemical reaction.

Calcination: The calcination reaction takes place at phase boundaries, proceeding from the outside surface into the particle interior. The rate increases with temperature and decreases as particle size and CO_2 partial pressure are increased.

Other important factors are the crystallite size of the carbonate mineral in the stone, the content of magnesium carbonate, and the SO_2 content of the gas. The rate can be represented by the equation shown on the following page.

$$(1 - \alpha)^{1/3} = 1 - \frac{K_0}{3q} \int_{T_0}^{T} \exp(-E/RT)\, dT$$

dT	=	qdt
$d\alpha/dt$	=	$K(1 - \alpha)^{2/3}$
K	=	$K_0 \exp(-E/RT)$
α	=	fraction of limestone calcined in time t
K	=	specific rate constant
T_0	=	temperature at start of calcination (t = 0), °K.
T	=	temperature of time t, °K.
q	=	heating rate, °K./min.
E, K_0	=	Arrhenius parameters for K
R	=	gas constant

E, the apparent activation energy, ranges from 30 to 75 kcal/mol for various limestones (in a nitrogen atmosphere and for a particle size of -24+28 mesh). The frequency factor, K_0, is related to the activation energy by the expression: $\log K_0 = 1.233 + 0.2144\, E$. Because of the wide variation in activation energy among limestones, it is obvious that rate of calcination varies over a wide range. Actually, however, the temperature in standard boilers is so high that the small limestone particles calcine almost instantaneously even at the higher levels of activation energy. The problem is not rate of calcination but rather the rate of dead-burning, which is a collapsing of pores and recrystallization of calcium oxide resulting from exposure to high temperature. Such effects are highly undesirable becuase they close pores and reduce permeability to SO_2.

Dead-burning is a common problem in the lime industry. It is difficult to calcine a particle all the way to the center without dead-burning the particle surface. The usual practice is to use a relatively low uniform temperature and somewhat extended calcination time. Since degree of calcination is a function both of time and temperature, extending the time allows operation at a temperature level low enough to avoid excessive dead-burning.

Such a procedure is not feasible in a standard boiler as the retention time is fixed, and quite short. The limestone must be injected at a relatively high temperature to accomplish both calcination and SO_2 sorption in the short time available. Limestones vary widely in their resistance to dead-burning. They can be grouped into four categories.

(1) Those that fracture and decrepitate readily during preheating and at low calcination temperature.

(2) Those that give a porous, reactive calcine under most calcination conditions and are difficult to dead-burn.

(3) Those that produce a dense, unreactive calcine under the mildest calcination conditions.

(4) Those that give a porous calcine under mild temperature conditions but a dense and unreactive lime at higher temperature.

Most of the readily available limestones appear to fall in the last group, which makes selection of injection temperature quite important. Magnesium carbonate is less subject to dead-burning than is calcium carbonate because magnesium oxide crystals grow much more slowly during recrystallization. Dolomite therefore usually gives a more porous calcine than does limestone when each is calcined at its optimum calcination temperature. However, the optimum temperature is lower for magnesium carbonate than for calcium carbonate. Hence, it is difficult to calcine dolomite at the high temperature required in the boiler to calcine the calcium carbonate in the short retention time available without dead-burning the magnesium carbonate portion. Nevertheless, inferiority of dolomite has not been clearly demonstrated.

The situation in fluidized boilers is somewhat more favorable because of the longer retention time and lower combustion temperature (1500° to 1600°F). Dead burning is much less likely in this temperature range. The main question is the back pressure of CO_2 when the combustion is carried out under pressure, which, as noted earlier, can make calcination difficult.

Even if the limestone is calcined properly and no dead-burning occurs, there can still be wide differences between limestones in regard to pore size and pore volume in the calcined material. Many investigations have been made with the objective of determining the limestone characteristics that affect porosity of the calcine but the situation is still not clear. A major factor seems to be the original crystallite size of the limestone.

For example, Iceland spar, which is a pure calcitic limestone having very large crystal size in the natural state, gives relatively small pore volume after calcination. A typical porosity is 0.04 cc of pore volume per gram after calcination at 1100°C. In contrast, for a normal calcitic limestone of intermediate initial crystal size, 0.49 cc per gram is typical. Moreover, the pores in the calcined spar are quite small, mainly less than 0.2 microns in diameter, whereas limestone calcines have a large part of the pore volume in the 0.2 to 17.5 micron range, which is much better for SO_2 absorption. Empirical equations to correlate these properties to SO_2 sorption efficiency (which depends on porosity) have been developed from laboratory data in an EPA-TVA research program.

$$w_3 = -1.571z_4 + 5.685z_5 - 32.14z_6 + 3.607z_7 - 0.3480z_9 + 2.818z_{10}$$

$$\log w_3 = 0.05114z_2 - 0.06688z_4 + 0.2942z_5 + 0.1834z_7 - 0.01403z_9 + 0.1200z_{10}$$

w_3 = SO_2 sorbed in a given time (mg. SO_3/50 mg. limestone)

z_2 = % rhombic carbonates/%CaO

z_4 = microscopic grain size (expressed as code number in statistical experimental design)

z_5 = unit cell length ($\underline{a}$) of the carbonate mineral (expressed in angstroms)

z_6 = x-ray crystallite size of the carbonate mineral (microns)

z_7 = Na_2O content of limestone (log of percentage)

z_9 = content of illite, montmorillonite, tremolite, limonite, feldspar, and muscovite in the limestone (total %)

z_{10} = % limonite (iron oxide mineral)/% total Fe_2O_3

The best estimate of w_3 is the arithmetic average of the two equations. The relative importance of these parameters depends not only on the magnitude of the coefficients and the measurement units but also on the range available in which changes to improve sorption can be made. The comparison given in Table 2.1 shows the importance of small grain and crystallite size and the large adverse effect of impurities.

It should be noted that some of the factors are closely related to the choice between dolomitic and calcitic limestones. Dolomites usually have higher rhombic carbonate/CaO ratio and smaller grain and crystallite size, but contain more total impurity, have more iron in the crystal lattice (not as limonite), and have small lattice constants. Thus although the MgO content of dolomitic limestone absorbs little or no SO_2, other properties could improve porosity of the calcine (and possibly other properties) to the extent that sorption of SO_2 (weight of SO_2/weight of limestone) might be as good as for calcitic limestone.

Some of the test work has indicated this to be true; apparently the presence of the MgO, although nonreactive, promoted calcine porosity and perhaps reduced blockage of pore surfaces by reaction product. The concensus of experimenters, however, seems to favor high calcium limestone.

TABLE 2.1: RELATIVE EFFECT OF VARIOUS LIMESTONE PROPERTIES ON RATE OF SO_2 SORPTION

Property	Range[a]	Change Required to Increase Sorption 10%	Significance Factor[b]
% Rhombic carbonates per % CaO	1.76-3.42	+0.81	21
Microscopic grain size (coded)[c]	1-3	-0.62	32
Lattice constant, angstroms	4.805-4.989	+0.14	13
X-ray crystallite size, microns	0.070-0.211	-0.031	46
Na_2O content, %	0.005-0.25	(68.2)[d]	73
Clays, feldspar, and limonite, %	0.1-31	-2.95	105
Fraction of Fe as limonite	0-1	+0.34	30

[a] For 35 limestones tested.

[b] Maximum improvement in sorption in range available, %.

[c] 1 = $<63\mu$; 2 = $>63\mu, <250\mu$; 3 = $>250\mu$.

[d] Percentage increase in Na_2O.

It should be noted in this respect that in some instances dolomite has been reported superior on the basis of Ca/S ratio. However, since the material is bought on a weight basis rather than Ca content, the important consideration is gross weight of sorbent per unit of SO_2 sorbed. It should be remembered that the data given here were obtained in laboratory tests, with extended reaction time, and do not necessarily reflect the situation in actual boilers. Full-scale tests have been too limited to allow such a correlation.

Sulfur Oxide Sorption: The sulfation reaction is favored thermodynamically by low temperature but, as discussed earlier, the temperature must be below 1200°F for adequate equilibrium conversion; at this temperature the rate is so low as to be of little consequence. The use of hydrated lime, $Ca(OH)_2$, has been proposed as a method for increasing sorption efficiency. Any significant absorption by $Ca(OH)_2$ must be at a temperature below about 700°F, as decomposition to CaO at higher temperatures makes the system the same as with CaO. The equilibrium is favorable at lower temperatures but, as for CaO, rate becomes the ruling consideration. Tests on a boiler in the U.S. in which $Ca(OH)_2$ was injected have not been promising.

The conclusion that sulfite formation at low temperatures is very slow is supported by the fact that little or no $CaSO_3$ is ever found in the sorption product from limestone addition under boiler conditions. Because of the presence of unreacted CaO in the solids, there is ample opportunity for sulfite formation (if the rate were fast enough) as the gas passes through the lower temperature regions of the boiler system. In laboratory tests, where longer reaction time is feasible, $CaSO_3$ can be obtained by extended exposure of CaO to SO_2 at relatively low temperature, on the order of 900° to 1100°F. At this temperature level the oxidation reaction is relatively slow.

Sulfite has also been found in products from pilot plant tests made at high temperature and very short retention time (0.2 to 0.6 second). When the retention time is extended to 1.0 second or so the sulfite decreases and then disappears. A possible explantion is that sulfite is formed at relatively low temperature while the injected particle is heating up and that when the particle reaches gas temperature decomposition and oxidation promote formation of sulfate. From this it appears that the presence of oxygen in the boiler gas is quite advantageous because, by oxidizing sulfite to the more thermally stable sulfate, it allows operation at the high temperatures that are required for an adequate reaction rate.

The sulfite and sulfate decomposition reactions are important both because they limit the degree of sorption at high temperatures and because the decomposition might make it possible to regenerate the CaO (for recycling) and to recover the evolved SO_2. As noted earlier, these reactions make it necessary for the temperature to be below about 2000°F for adequate SO_2 sorption. If decomposition were desired, as in a regeneration operation, the temperature would have to be considerably higher, on the order of 2600°F for substantially complete decomposition of $CaSO_4$ in a reasonable length of heating period. Calcium sulfite is less stable, decomposing in a short time at 1900°F.

Mass Transfer and Overall Reaction Models

Several reaction models have been proposed for the combined sulfation-oxidation reaction which from the foregoing appears to be the significant one involved.

These are mainly based on small-scale work, either isothermal conditions or with a slow heating up of the stone. Parameters included in most of the models are (1) retention time, (2) temperature, (3) SO_2 concentration in the gas, (4) particle diameter, (5) sulfate loading at a given rate, and (6) the Arrhenius parameters (frequency factor and activation energy). Others used in some of the models are (1) stoichiometry, (2) degree of dead-burning, (3) heating and cooling rates of the limestone after injection, (4) surface area of calcine, (5) bulk density of calcine, (6) oxygen and moisture contents of the gas, and (7) diameter of the pores in the calcine.

General Mechanism: From a consideration of all the proposed models and postulated effects of variables, the following conclusions can be set down as to what probably happens when limestone is injected into a boiler.

(1) Calcination produces a characteristic pore structure within each particle, depending mainly on limestone type and temperature.

(2) Assuming that transport rate of SO_2 to the particle surface is adequate and that most of the pores are large enough for ready penetration of SO_2 into the particle, reaction takes place fairly rapidly on the exposed CaO surface within the pores. During this period the rate of the chemical reaction is the controlling resistance.

(3) After the initial reaction, further sorption proceeds in two ways: (a) by diffusion of SO_2 into those pores so small that diffusion rate is slower than reaction rate, and (b) diffusion of SO_2 through the surface layer of $CaSO_4$ into unreacted CaO underneath (or possibly diffusion of Ca^{++} ions to the surface).

The importance of mass transfer in this is obvious. For maximum sorption in the time available, it is important that the chemical reaction be allowed to proceed to the fullest extent possible without having to wait for SO_2 to reach the CaO reaction sites. First, the limestone must be calcined in such a way as to give a large number of pores of the proper diameter range (not too small or too large). Then the SO_2 must move from the gas to the CaO particle surface at a rate as fast as the CaO can accept it. Diffusion throughout the pore system must also be fast enough to keep the reaction supplied with SO_2, and finally, there must be some diffusion into the smaller pores and through the reacted layer on the pore surfaces if an acceptable degree of CaO utilization is to be obtained.

Obviously such diffusion into small pores and through the reactant layer is relatively slow and is helped by the extended reaction time that a fluidized bed affords. However, a considerable degree of such diffusion apparently occurs even in a standard boiler; calculations have shown that only about 4% of the limestone would be reacted if all available surface area in a typical calcine were covered with a layer of $CaSO_4$ one molecule deep, whereas in practice up to 25% or more reaction of limestone has been obtained.

All of this, of course, should take place at the optimum temperature, as high as possible in order to get high chemical reaction rate but not so high that equilibrium considerations (reversing of the sulfation-oxidation reaction) prevents adequate SO_2 removal from the gas. Again, this is much more feasible in a fluidized bed than in a standard boiler; in the latter the temperature drops rapidly in a second or so to a level at which the reaction rate is too slow to give much benefit.

Pore Size: If pore diameter is too small, the SO_2 cannot reach the internal reactive surfaces fast enough. The size can even be so small that the initial reaction products block the pore openings and effectively prevent penetration into the interior. The resulting shell formation is considered by some to be the principal limiting mechanism (unreacted-core model). Others conclude that such small pore size does not occur very often and that in most cases the pores are large enough to allow surface reaction throughout the internal structure; they consider that pores small enough to cause shell formation can result from dead-burning and that a few limestones give very small pore diameter even under good calcination conditions.

Very large pore size also appears to be undesirable as it reduces the reactive surface area and thus lowers the initial reaction rate. At extended reaction time, however, large pore size gives higher capacity, where capacity is defined as the amount of SO_2 taken up before sorption rate becomes so slow as to be insignificant. Preseumably this is the point at which unreacted CaO deep in the particle interior becomes well nigh inaccessible because of blockage by reaction products. Large pore size increases capacity because of the generally more open internal structure and smaller distances between pores.

Ideal pore size appears to be 0.2 to 0.3μ. Pores smaller than 0.1μ become rapidly blocked by reaction product and as pore diameter increases above 0.3μ the reactive surface area decreases rapidly. Since pore size depends both on the properties of the original limestone and on the conditions of calcination, it is difficult to generalize as to the range of pore size to be expected. Mean pore size from 0.07 to 4μ has been found for limestones of various types; surface areas ranged from 0.6 to 10 m^2/gram. The net effect of limestone properties and calcination conditions on SO_2 sorption can be expressed by a model that includes activation energy for the limestone and effect of dead-burning during calcination:

$$\ln \frac{1 - \alpha/S}{1 - \alpha} = -K_{20}\, a\, (S - 1)\, I/p$$

$$I = \int_{T_1}^{T_2} \exp(-E_2/RT)\, dT$$

$S = S_0 (1 - y)$

K_{20} = frequency factor

E_2 = activation energy

p = cooling rate of limestone after injection, °K./sec.

T_1 = injection temperature, °K.

T_2 = temperature at end of reaction, °K.

S_0 = effective stoichiometry in absence of dead-burning

y = degree of dead-burning

a = SO_2 concentration

α = degree of SO_2 removal from gas

Use of this model indicates that, for what appear to be typical values for some of the terms, activation energy should not be much higher than 10 kcal/mol if a reasonable degree of desulfurization is to be obtained (about 65% at 10 kcal, 3,000 ppm SO_2, and twice stoichiometric; only 25% under same conditions at 16.7 kcal). Activation energy for various limestones has been found to range from 8 to 18 kcal.

Particle Size: The particle size of the CaO is considered to be important in various ways. (It should be noted first that particle size of the calcine may differ from that of the injected limestone, either because of decrepitation or agglomeration.) First, increase in number of particles by reduction in particle size would be expected to increase the reactive surface area.

However, since most of the surface is inside the particle, reduction in size has little effect. A more important consideration is that the smaller the particle diameter the less is the distance from the outside surface to the center and therefore the SO_2 does not have to diffuse so far through pores. It has been shown in small-scale tests that SO_2 diffusion through the pores is not fast enough to avoid slowing the reaction when the particle size is larger than 0.01 cm unless pore size is so large that the reactive area is unduly diminished. The smaller the particle size the smaller can be the pores and still have chemical reaction rate controlling (unless they are so small as to become blocked).

In fluidized boilers there is the additional consideration that the particle size must be such as to fluidize properly, whereas in standard boilers the lower limit is the smallest that is economical to produce. The particle sizes that have been used in fluidized beds vary over a wide range, some as large as ⅛ inch.

Residence Time: The effect of contact time between gas and solid is one of the more difficult parameters in regard to reconciling results of various investigators. Most report that sorption rate in injection tests decreases almost to zero after a certain time (usually about a second), presumably because reactive surfaces become blocked. Others report a continuing pickup of SO_2 even with total retention times of 4 seconds or more. The point is of little importance in standard boilers because contact time is more or less fixed, but is significant for fluidized boiler operation. It has also been proposed that a packed bed type of sorption be used, based on various designs for moving the limestone continuously through the packed bed (e.g., a panel filter) and out of the boiler. The main drawback appears to be difficulty in maintaining such a structure in the high-temperature section of the boiler.

SO_2 Concentration: The general consensus from small-scale batch tests is that sorption rate is first order with respect to SO_2 concentration in the gas during the period when chemical reaction rate is controlling. There are some, however, who do not necessarily agree with this, particularly those who have made pilot-plant injection tests. Moreover, it has been pointed out that high SO_2 concentration and consequent high initial reaction rate may give such rapid formation of $CaSO_4$ that pores are blocked which otherwise would remain open. In contrast to the reaction rate theory, one investigator postulated that the rate is diffusion-controlled at SO_2 concentrations below 2,000 rpm.

Oxygen Content of Gas: The effect of oxygen in the gas (from use of excess air in combustion) is another point on which there is not full agreement. Some of the small-scale work has indicated no effect from variation in oxygen content. In most of the pilot plant tests, however, increase in oxygen has had a fairly significant beneficial effect. This would seem reasonable because the extra oxygen should push the combined sulfation-oxidation reaction farther to the right. The practicality of improving sorption by increasing excess air is questionable. The increased heat loss in the stack gas could be a significant operation cost.

Catalysts: If oxidation rate is important, then catalysts to promote oxidation should be helpful. Again, there is not good agreement. Limestones with a high content of iron have been reported to be superior sorbents by some investigators; in one case, limestones containing 1% or higher iron oxide were found to be 50% more effective than others of lower iron content. Addition of iron or vanadium salts to the limestone, however, has not been found very effective by those who have tested it. In the work on limestone properties discussed earlier, the total content of iron was not significant; in fact, the sorption was improved if any iron present was in the form of limonite (mixture of oxides as a separate mineral) rather than a constituent in the CaO lattice.

It has also been claimed, in connection with fluidized combustion work, that sodium chloride is a good catalyst. This would probably be infeasible in a standard boiler because of chloride corrosion.

Effect of Water Vapor: Only a few investigators have studied the effect of water vapor on sorption. One found that water vapor was generally harmful, although decrease in limestone particle size greatly reduced the degree of adverse effect. Another reported no effect with CaO but a major beneficial effect with MgO. In any event, it would be difficult to vary the water vapor content of the gas in power plant operation.

PROCESS CONSIDERATIONS

In operational use of the dry limestone method there are several process choices that must be made, some of which do not involve chemical factors. These differ widely between standard boilers and the fluidized-bed type. As noted earlier, the important considerations in standard boiler operation are point of injection, limestone and SO_2 distribution, stoichiometry, limestone type, limestone particle size, and precipitator performance. Because of the relatively low promise in this approach, these factors will not be discussed here in detail in regard to standard boilers; reference can be made to the first edition if details are desired.

Fluidized Bed Combustion

For fluidized boilers, the point of injection is not pertinent and the limestone and SO_2 distribution presumably would be uniform. The bed in which the coal burns is composed of CaO and ash. Since the technology is not yet well developed, it is not clear whether the product calcium sulfate will be bled off as fines in the outlet gas or as larger particles removed through an outlet from the bed; in any event there will be some fines escaping with the gas and the questions of stoichiometry, limestone type, and particle size will be significant.

Adoption of the fluidized bed for SO_2 removal, however, depends first on its adoption for combustion. The following advantages are claimed.

(1) Combustion intensity of up to 50 times that allowable in a conventional boiler.

(2) Very fast heat transfer to boiler tubes immersed in the bed, on the order of five times as fast as the conventional boilers.

(3) No fouling by molten slag as the bed is operated below the ash fusion temperature.

(4) No erosion by solidified particles of slag.

(5) Greatly reduced volatilization of corrosive alkali salts.

(6) Smooth operation at very high ash content in the fuel and with varying ash content. Some coals that cannot be used in a conventional boiler burn satisfactorily in the fluidized-bed type. Moreover, it may be possible to burn waste materials such as garbage along with the regular fuel.

(7) No need to pulverize coal to small particle size.

In operation (see Figure 2.2), an upward flow of air expands the bed of inert material (granular ash, refractory particles, or coarse limestone) to a dense suspension. Crushed coal is injected pneumatically into the bed at several points near the bottom. The coal, which constitutes only a small portion of the total solids in the bed, disperses through the bed and burns at 1500° to 1650°F. Steam tubes immersed in the bed remove the heat. Typical bed depth is 0.6 to 1.0 meter, with inlet air velocity ranging from 1 to 4 meters/second. Combustion is usually not as complete as in conventional boilers so it is desirable to collect most of the ash in cyclones and recycle it to the combustor or to an auxiliary fluid bed to complete combustion.

FIGURE 2.2: FLUIDIZED BED BOILER

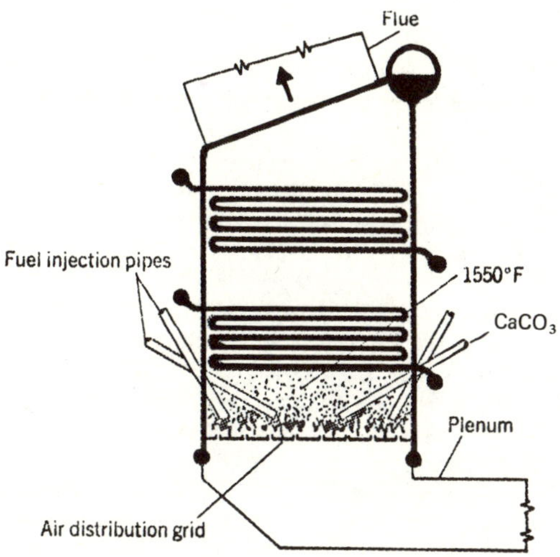

Operation with oil is much the same, except that the bed must be something other than ash, either limestone or a material such as alumina. The oil is sprayed in, and it is considered by some that spraying the oil directly on limestone particles promotes desulfurization.

For SO_2 removal, limestone is introduced into the bed, usually with the coal in coal-burning plants. Two design concepts are being developed: in one, coarse limestone (up to ⅛ inch) serves both as SO_2 sorbent and as the inert combustion medium; in the other, fine limestone (similar in size range to that used in injection tests on conventional boilers) sorbs SO_2 and is blown out of the bed. The fine limestone has a higher sorption rate than the coarse limestone but retention time is shorter. Much of the work with coarse limestone has been aimed at development of a recovery process, and therefore will be described later.

The choice between the two operating modes may depend in part on the nature of the coal ash. Ash from some coals tends to retain about the same particle size as the coal from which it was formed and therefore is amenable to the coarse ash-fine limestone combination. Other coals give a fine ash even if coarse coal is burned; for this the fine ash-coarse limestone concept is more appropriate.

In work with fine limestone (25μ average), SO_2 removal was 40 to 50% with 1.0 to 2.0 stoichiometric. At 2.2 to 3.0 stoichiometric, almost 80% removal was obtained. In other work results have not been as good. The data have varied widely for coarse limestone also, in one instance from 65 to 99% removal (at twice stoichiometric) depending on type of coal and type of limestone. The effect of coal type is not well understood; it has been postulated that this may be associated with caking tendency, as the sulfur in coals which agglomerate on heating may not burn as rapidly as that in noncaking coal.

No conclusions can be made yet as to fluid bed performance for SO_2 removal and as to the relative merits of fine and coarse limestone. The work under way should give the answers eventually. The approach is a promising one because: (1) retention time is longer; (2) limestone distribution is no longer a problem; (3) there is little danger of dead-burning; (4) simultaneous calcination and sulfation, which small-scale tests have shown is desirable, can take place; and (5) since much of the solid product is removed from the gas by the fluid bed, there is less dust to collect later.

However, even with less dust in the gas it will be more difficult to collect because of the reduced conductivity. In England, tests have been made with cyclones to avoid this problem but such equipment is not likely to meet particulate emission standards in the U.S. With the SO_3 absent, it may be possible to operate a precipitator at low temperature, say 200°F, and get an adequate increase in conductivity by this means. The feasibility of this remains to be demonstrated.

There are two trends in the fluidized bed combustion development, however, that could make the method less effective for SO_2 removal. HIgher gas velocities than those in current tests may be necessary to make fluidized bed combustion economical for power plants. The reduced contact time would probably reduce SO_2 removal efficiency. Secondly, fluidized bed combustion under pressure (with a turbine to recover the compression energy) is being studied, again to improve economics by reducing equipment size. At the pressure level contemplated (5 atm or higher), calcination of high-calcium limestone is difficult and the internal pore structure does not develop well. It has been reported, however, that dolomite is clearly superior for pressurized sorption; apparently the beneficial properties of MgO are accentuated in operation under pressure.

Limestone Availability

The very large and continuing supply of limestone that would be required for SO_2 sorption in a large power plant makes limestone availability an important consideration. At twice stoichiometric for 3.5% sulfur in the coal, about one ton of limestone is needed per 4 tons coal, or about 2,200 tons per day (700,000 tons per year) for a 1,000-Mw boiler.

Fortunately, limestone is a very plentiful material. Limestone deposits are prevalent in practically all of the states of the United States and most of the countries of the world. Economically minable limestone does not occur so frequently, however, because many of the deposits are too small, too impure, or too deeply embedded in the earth. Athough there are no good data on the tonnage in deposits that are economical from the standpoint of size and accessibility, it appears to be quite large.

Purity is another matter. It is estimated that only about 2% of the deposits are chemical grade, that is, containing 95% or more $CaCO_3$. Because of the adverse effect of impurities on development of porosity during calcination, chemical grade may be required for SO_2 sorption. If dolomitic limestones should also be found unacceptable because of water pollution (see later discussion), the supply would be further curtailed. In some areas there is already evidence of a shortage of chemical grade, high-calcium limestone.

The U.S. deposits being mined are mainly in the central and northeastern sections of the country, presumably because these are the areas of heaviest demand. The developed deposits are concentrated in areas of heavy power production, again probably because these industrialized regions require large amounts of both limestone and power. This gives the advantage that limestone should be readily available, without the problem of quarry development, if limestone injection were widely adopted for SO_2 control.

Limestone is a major commodity in the U.S., ranking second after sand and gravel. The annual tonnage appears to be on the order of 600 to 700 million tons, but over half is produced as crushed or broken stone used as aggregate in concrete or road beds. Since this is enough for about 1,000 boilers of the 1,000-Mw size, the limestone industry should be able to handle any power plant demand without difficulty, assuming that the quality needed can be produced at acceptable cost in the quantity required.

One possibility for reducing cost is to use limestone that is a by-product of other operations. Examples are limestone overburden from ore mining and fines produced in sizing limestone to specific requirements. Sometimes very large amounts of such by-products are available at low cost.

A special quarrying operation may be economical for very large power plants. For example, the 700,000 tons per year needed for a 1,000-Mw boiler (3.5% S, twice stoichiometric) would exceed the annual capacity of 95% of the crushed stone plants in the United States. A special quarrying operation, set up at the nearest suitable deposit to the power plant, should be more economical than buying from several existing sources at varying distances from the plant.

Fuel Type

The difficulty and effectiveness of limestone injection differs depending on whether coal or oil is used as the fuel, mainly because of the ash associated with the coal. A major disadvantage for oil as fuel is that because of the low ash content many oil-burning boilers are not equipped with electrostatic precipitators. If limestone is injected, precipitators must be installed. Cost would probably be higher than for increasing existing precipitator capacity in coal-burning plants, and space for a new precipitator could be difficult to find. Moreover, oil-burning plants would have the problem of suddenly being required to expand the dust removal and disposal facilities, to a much greater extent than for a coal-fired installation.

Fuel type is also a consideration in use of limestone to absorb SO_2 in fluidized bed combustion. The version of the method in which the inert bed is made up of ash, with fine limestone injected into the bed, cannot be used of course if oil is the fuel. Rather, the bed must consist of limestone particles large enough to avoid entrainment with the gas. (Otherwise a material such as refractory alumina must be used.) The large particles are not as effective in sorbing SO_2 as the smaller ones; however, there is some evidence that more intimate contact can be achieved between oil and limestone, as compared to coal and limestone, and that this promotes sorption efficiency.

Solids Disposal and Water Pollution

The limestone injection method for controlling SO_2 emission has the drawback that it converts the atmospheric pollution problem to a solid waste problem. For a typical ash content of coal (12%), use of twice the stoichiometric amount of limestone almost triples the output of waste solids [about 2.7 times, assuming 50% removal of SO_2 and hydration of unreacted CaO to $Ca(OH)_2$]. For coal-burning power plants, the problem is that of expanding the waste solids transport and disposal system over what would otherwise be needed.

For oil-fired plants, which produce so little ash that removal is not always practiced, installation of waste solids disposal facilities may represent a cost that otherwise would not have been involved at all. Disposal of ash from coal-burning power plants is already a major problem. The production is on the order of 30 million tons per year, of which only a small part is used, as fill, as aggregate, and in cement. The ash-$CaSO_4$-CaO mixture possibly may have superior properties for use in cement and concrete; research on this is under way. However, most of it must be discarded.

Ash is conveyed to the disposal point at power plants in various ways. A major practice is to sluice through pipelines to a settling pond; when the pond becomes full of solids, another is constructed. In some cases, the material is transported in dry form, usually by truck, to some point where the topography affords a good disposal area. This is more expensive than ponding, but sometimes preferable because of land scarcity near the power plant or because of esthetic considerations. For the ash-$CaSO_4$-CaO mixture, the ultimate solution would appear to be transporting it back to the quarry from which the limestone came in the first place. This may actually be economically feasible for quarries located near power plants, since the waste solids could be carried to the quarry on the back haul. In most cases, however, ponding will probably be the method used. For this, about 250 annual acre-feet will be required for a 1,000-Mw power plant (twice stoichiometric limestone and 3.5% sulfur).

In regard to water pollution, one favorable factor is that the reaction product is $CaSO_4$ rather than $CaSO_3$. Dissolved $CaSO_4$ is not generally considered harmful unless the concentration is high; watercourses often contain appreciable amounts of calcium sulfate from natural sources, and there are major natural deposits of gypsum ($CaSO_4 \cdot 2H_2O$) that cause no harm.

In some plants the sluice water is overflowed to an adjoining watercourse. Acceptability of this, under current regulations in most areas, depends on the content of Ca^{++} and $SO_4^=$ in the watercourse, which in turn depends on the quantities of Ca^{++} and $SO_4^=$ involved, the content in the watercourse before influx of the sluice water, and the volume flow in the watercourse. The nature and quantity of other dissolved constituents (Mg, Na, Cl, NO_3) is also important. If overflow of sluice water is unacceptable, it can be recycled. The problem then is whether leaching from the ponded solids by ground water would be excessive, which seems unlikely for Ca salts but could be a problem if dolomitic limestone were injected into the boiler. Although the MgO does not appear to absorb SO_2 under boiler conditions, it has been shown that it can react in the pond with $CaSO_4$ and atmospheric CO_2 to form $MgSO_4$. The high solubility of $MgSO_4$ might well make leaching from the ponded solids a problem.

ECONOMICS

As compared to other ways of removing SO_2 from stack gas, limestone injection has a low investment and a relatively low ratio of fixed charges (investment-related) to total operating cost.

Investment

The main investment costs for the dry limestone method are for (1) limestone receiving, crushing (or grinding), and handling, (2) waste disposal, and (3) possible increase in cost of particulate removal from the gas. Investment for a fluidized bed system cannot be estimated with any accuracy because there is no good information yet on limestone type, particle size, and stoichiometry required (which affects the size of the handling and disposal equipment). The investment is likely to be much lower than for other systems, however, a prospect that is bringing about growing emphasis on the process.

Operating Costs

The operating cost depends mainly on the purchase price of the limestone, which is determined by the volume, location, and quality. Open-market prices in the U.S. vary widely, prices ranging from \$0.73 to over \$10.00 per ton, fob. Much would depend on the annual volume and the type of purchase agreement. If a long-term contract for a large quantity were entered into, the supplier might make special provisions for supplying the material and sell at a relatively low price. Shipping cost would also be a major consideration, particularly if the nearest suitable supply of limestone were some distance away. From \$0.50 to \$1.50 per ton might be expected for distances up to 50 miles. Barging would be a relatively low cost way of shipping and might be widely applicable because the bulk of the power producing capacity in the U.S. is located on navigable waterways. A large quarrying operation supplying several power plants by barge could deliver limestone at low cost.

Because of the relatively low ratio of capital charges to total cost, operating time is especially important in cost of limestone injection, mainly because no limestone is needed when the plant is not operating. This makes the method attractive for the older plants in a power system that are operated only when the newer and more efficient ones cannot supply the demand.

THROWAWAY OPERATION
DIRECT LIME/LIMESTONE SCRUBBING

Since use of limestone in an injection-type dry sorption system has proven generally inefficient for standard boilers both in regard to degree of SO_2 removal and sorbent utilization, emphasis has shifted to wet systems. The dry system may make a comeback in the fluidized boiler area but considerable further development will be required. The mobility of liquid absorbents is a major advantage because the reacting surface can be constantly and rapidly renewed by turbulence in the liquid, a very large area of reactive surface can be developed, recycling can be used to advantage in getting high absorbent utilization, and absorbent solubility in aqueous systems makes the absorbent molecules more accessible to the SO_2.

There are drawbacks to wet systems, however, including corrosion, pressure drop through scrubbers, growth of crystal deposits in the equipment, and cooling of the stack gas. These are serious problems; nevertheless, the promise of better scrubbing efficiency has directed the bulk of the current research to wet systems. For throwaway processes of the wet type, lime and limestone are the absorbents that have received the most attention, for the same reason as in the dry throwaway system, namely, low cost. Scrubbing with water has been studied and is still considered promising by some. However, very large quantities are required because most of the systems are based on once-through use of the water. Moreover, the acidity of the scrubber effluent makes disposal a problem.

In Japan there are numerous small operations of the throwaway type in which an alkali (usually sodium carbonate) is the absorbent and the product alkali sulfite-sulfate is discarded. There is one example of this in the U.S. (Reid Gardner station, Nevada Power). The main drawback is the water pollution problem.

Several systems have been developed for using limestone in a wet scrubbing process. The major ones are shown in Figure 3.1.

(1) Scrubber Addition of Limestone – The simplest approach is to use a slurry of ground limestone in the scrubber.

Major drawbacks are need for grinding and a somewhat slow reaction between SO_2 and the raw limestone.

(2) Scrubber Addition of Lime – Scrubbing efficiency can be improved by using lime (CaO) rather than limestone in the scrubber. The cost of calcining limestone to produce the lime, however, is a major disadvantage.

(3) Boiler Injection of Limestone – Cost of calcining can be reduced by injecting the limestone into the boiler as in the dry system (Chapter 2). The resulting lime then is carried by the gas into the scrubber. Problems include danger of increased scaling in the scrubber (as compared to scrubber addition of lime or limestone), possible fouling of the boiler, danger of dead-burning the lime, and no opportunity for countercurrent scrubbing.

(4) Alkali-Lime/Limestone – Superior scrubbing efficiency can be obtained by using an alkali salt (NH_4^+, Na^+, K^+) as the sorbent and regenerating the alkali outside the scrubber by reaction of the scrubber effluent solution with lime or limestone. The discarded solids ($CaSO_3$ and $CaSO_4$) may or may not be washed to avoid alkali loss.

FIGURE 3.1: VARIATIONS IN USE OF LIME OR LIMESTONE FOR REMOVAL OF SO_2 FROM STACK GASES

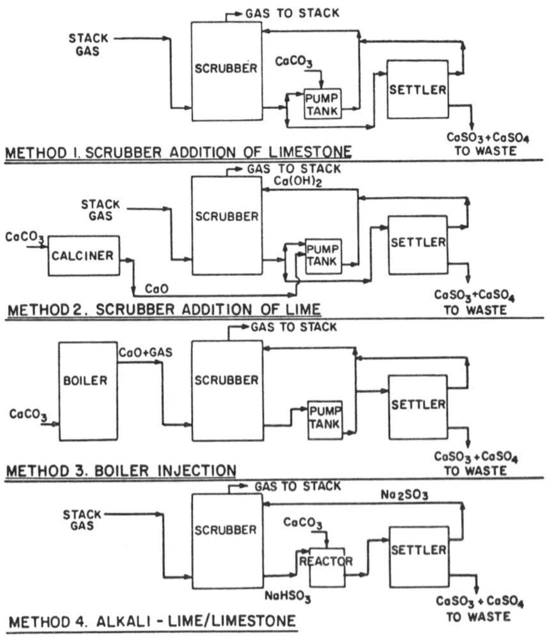

Of these, boiler injection was the first method tried (at Union Electric's Meramec station and Kansas Power and Light's Lawrence station) but results were poor. The several scrubbers now being installed in the U.S. all involve scrubber addition of lime or limestone. The alkali-lime/limestone type (usually called "double-alkali") has not yet been installed on a full-scale in the U.S. but several are operating or under construction in Japan (see Chapter 4).

HISTORY OF THE PROCESS

The history of limestone scrubbing is tied in closely with that of water scrubbing, mainly because the tests of water scrubbing have been carried out with natural water and therefore much of the SO_2 absorption obtained was due to dissolved alkaline carbonates ($CaCO_3$ and $MgCO_3$) present in the water; as will be discussed later, pure water has a low affinity for SO_2. Moreover, in many of the water scrubbing tests a small amount of lime or limestone was added.

One of the earlier examples of the use of water was the work of Schroeder and Hanisch in 1884. There was other similar work but most of it was aimed at absorbing from relatively rich streams of SO_2, where the high vapor pressure of SO_2 increases solubility. One of the first efforts to use water for scrubbing flue gases was in England shortly after the turn of the century (about 1906), when at the works of Messrs. Roundtree at York an experimental unit was operated. Other such efforts are reported, but the most thorough study of water scrubbing appears to have been made by the London Power Co. around 1930.

A fairly large test unit (4.5 tons coal per hour; once-through water flow) was built and a test program carried out, which resulted in construction of a large gas washing installation at the Battersea Power Station on the south bank of the Thames in London. The Battersea scrubbers went through various stages of revision, first to the addition of a chalk ($CaCO_3$) slurry through the upper sprays of the countercurrent scrubber (with water introduced through the lower ones) and then to the use of the chalk slurry only.

Since only a small amount of chalk was used, the pH of the effluent slurry was too low for release to the river; it was increased to an acceptable level by dilution with condenser water (river water containing 150 to 200 parts per million of $CaCO_3$ alkalinity). Thus the scrubbing was, in effect, done with $CaCO_3$.

The Battersea process was later installed at another London power plant (Bankside). This was about the limit, since it was not considered that further scrubber effluent should be introduced into the Thames. When the Fulham Power station was authorized, therefore, attention was turned to development of a noneffluent process which would produce a waste solid that could be put somewhere other than in the river. Imperial Chemical Industries and James Howden & Co. collaborated in the development of what came to be called the ICI-Howden process. A slurry of lime or chalk was circulated through a grid-packed scrubber and a sidestream of the slurry was filtered to remove the product as a solid. There was thus no liquid effluent at all.

The ICI-Howden method was first applied on a full scale at the Sir John Power Station in Swansea, England. However, the high ash in the coal being burned overloaded the scrubber and this, plus other troubles, caused the scrubbing unit to be abandoned. The process was then installed at the Fulham station, where it operated efficiently but maintenance cost was reported to be high. The unit was shut down during World War II because steam plumes from the stacks were thought to be guiding enemy aircraft and was not put back into use after the war.

The Battersea operation (coal-fired) has been partially abandoned but the B unit is still in service because the scrubber is needed to remove particulate. The entire station will probably be shut down before long. The Bankside SO_2 scrubbers (oil-fired station) are still in service; although the scrubbers are expensive to operate, the operation is being continued because the stacks are not very high and the scrubbers can be operated with an acceptable degree of reliability.

The next large scale uses of lime/limestone scrubbing were in Russia and Japan. A scrubber installation has been in operation in Russia since 1964, removing SO_2 from about 3 million m^3/hr of waste gas from an iron ore sintering plant; the gas contains about 0.5% SO_2. In Japan, the first lime/limestone scrubbing was on a large sulfuric acid plant, where a lime scrubber was started up in 1966. Since then, lime/limestone processes have proliferated at a rapid rate in Japan until today the country leads the world in number of installations.

In the U.S., TVA carried out small scale and limited pilot plant studies in the 1950's. The first major pilot plant work appears to be that of Universal Oil Products (at a Wisconsin electric power plant) beginning in 1965. A limestone slurry circulated through a mobile-bed scrubber (2,000 cfm of gas) gave good SO_2 removal. Injection of limestone into the boiler was tested in a separate study, but the combination of injection followed by scrubbing, although proposed at the time, was not studied.

Injection followed by scrubbing was tested by Combustion Engineering in a pilot unit (3,000 cfm of gas) at a Detroit Edison power plant beginning in 1966. At 110% of the stoichiometric amount of limestone (dolomite in this case), SO_2 removal was 98%. On the basis of the pilot plant work, CE offered the process to the utility industry. Two installations have been made, by the Union Electric Co. in St. Louis, Mo. (140 Mw) and the Kansas Power and Light Co. in Lawrence, Kansas (125 Mw). Kansas Power and Light Co. has also used the method in a 420 Mw unit started up in early 1971.

In the past several years, research and development work on lime/limestone scrubbing has been accelerated, mainly under EPA contracts. The EPA effort can be divided roughly as follows.

(1) Basic chemistry and kinetics–EPA (in-house), Battelle, Radian, TVA.
(2) Pilot plant–EPA (in-house), TVA.
(3) Prototype testing (single project)–Bechtel, TVA.
(4) Conceptual design and cost study–TVA.

In addition, many other organizations have carried out their own research programs, each including pilot-plant or large-scale tests. Most of the work is aimed at solving the problems in scrubber addition of lime or limestone.

The EPA prototype testing program (item 3 above) is by far the most extensive effort yet attempted in development of lime/limestone scrubbing. Bechtel designed a versatile and flexible test facility located at TVA's Shawnee power plant near Paducah, Kentucky (the same location as for the dry system test unit). The unit consists of three parallel and independent scrubber trains, each with a capacity of about 30,000 cfm at 300°F and all taking gas from the same boiler. Three types of scrubbers are being tested: fluidized bed of plastic balls, flooded marble bed, and venturi followed by a packed bed. The primary purpose is to optimize the process in all respects for the three types of scrubbers.

In mid-1975, the test scrubbers had been operating for almost three years. Many problems have been encountered of both the mechanical and chemical type but steady progress has been made. The emphasis in current work is on eliminating scaling in the mist eliminators and on finding an acceptable solution for the sludge disposal problem.

The early efforts on limestone injection were unsuccessful. The Union Electric installation has been abandoned and the Kansas Power and Light system is being changed to one in which limestone is introduced into the scrubber. Many other scrubbers based on scrubber addition are now operating, under construction, or planned. A partial list is given in Table 3.1; several others are planned but have not been announced.

TABLE 3.1: LIME/LIMESTONE SCRUBBING INSTALLATIONS IN THE UNITED STATES

Organization	Plant	Size, Mw	Scrubbers, No.	Status[a]
	Lime Scrubbing			
Kansas City Power and Light	Hawthorne	2 x 140	4	O
Louisville Gas and Electric	Paddy's Run	70	2	O
Southern California Edison	Mohave	160	1	O
Kentucky Utilities	Green River	70	1	C
Duquesne Light	Phillips	100	1	O
Ohio Edison	Mansfield	2 x 880	12	C
Louisville Gas and Electric		1650	-	P
Arizona Public Service	Four Corners	575	6	P
Southern California Edison	Mohave	2 x 790	-	O
Columbus and Southern Ohio	Conesville	2 x 375	-	P
Allegheny Power	Pleasants	2 x 625	-	P
	Limestone Scrubbing			
Commonwealth Edison	Will County	163	2	O
Kansas City Power and Light	La Cygne	820	7	O
Arizona Public Service	Cholla	115	2	O
City of Key West		37	-	O
Northern States Power	Sherburne County	2 x 680	-	C
Southern California Edison	Mohave	160	1	C

(continued)

TABLE 3.1: (continued)

Organization	Plant	Size, mw	Scrubbers, No.	Status[a]
	Limestone Scrubbing			
Tennessee Valley Authority	Widows Creek	550	4	C
Detroit Edison	St. Clair	180	1	C
Texas Utilities	Martin Lake	2 x 750	12	P
Texas Utilities	Monticello	575	6	P
Central Illinois Light	Duck Creek	100	1	P
Salt River Project	Navajo	3 x 750	-	P
Arizona Public Service	Four Corners	1500	-	P
Public Service of Indiana	Gibson	75	1	P
Springfield Utility Board	Southwest	200	-	P
Southwest Public Service	Nichols	350	-	C

[a]O = operating; C = under construction; P = planned.

In Japan the use of lime/limestone scrubbing has also grown rapidly. A list of the installations is given in Table 3.2. Of the 56 installations listed, 26 are of the indirect, or double-alkali type.

The present trend is strongly in favor of lime/limestone scrubbing as the practicable solution to the SO_2 problem for the next decade or so. Although not fully developed and optimized, the process appears to be usable without major risk. Recovery of the SO_2 in a useful form is obviously preferable, but evaluations indicate that the commercial risk is much less with the throwaway processes and that several years will elapse before recovery methods are developed enough for general adoption.

CHEMISTRY

The chemistry of lime/limestone scrubbing is quite complex and not yet fully understood. The overall reaction is that of SO_2 with CaO or $CaCO_3$ to form calcium sulfite, with some oxidation of the sulfite to sulfate. Any MgO or $MgCO_3$ present in the feed materials may react in a similar manner or may precipitate as $Mg(OH)_2$. The actual reaction path to these end products, however, appears to be complex, with gas-liquid reactions, both ionic and nonionic reactions in the liquid phase, and liquid-solid reactions all taking place.

An exhaustive study of lime scrubbing chemistry has been carried out by the Radian Corp. for EPA. Radian started by assuming that the key species involved are Ca and Mg from the lime; CO_2, SO_2, SO_3, O_2 and NO_x from the gas; and Na, K, Cl and others from the ash. It was further assumed that these can interact to give over 41 neutral and ionic species in solution and 7 solid species. The ionic reactions (22 in all) were then considered that represent

TABLE 3.2: LIME/LIMESTONE SCRUBBING INSTALLATIONS IN JAPAN

Process developer	User	Plant site	Absorbent	MW[a]	Type of plant[b]	Year of completion	Gypsum (tons/day)
Mitsubishi-JECCO	Kansai Electric	Amagasaki	$Ca(OH)_2$	30	Utility boiler	1972	20
"	Onahama Smelter	Onahama	"	29	Copper smelter	"	450
"	Kawasaki Steel	Chiba	"	37	Sintering plant	1973	16
"	Kansai Electric	Kainan	"	150	Utility boiler	1974	22
"	Tohoku Electric	Hachinoe	"	125	"	"	53
"	Tokyo Electric	Yokosuka	$CaCO_3$	130	"	"	20
"	Kyushu Electric	Shinkanda	$Ca(OH)_2$	175	"	"	77
"	Kawasaki Steel	Mizushima	"	232	Sintering plant	"	130
"	"	"	"	279	"	1975	100
"	"	Chiba	"	130	"	"	30
"	Kansai Electric	Amagasaki	"	125	"	"	42
"	Teijin	Ehime	"	83	Industrial boiler	"	93
"	Mizushima Power	Mizushima	"	192	Utility boiler	"	105
"	Niigata Power	Niigata	$CaCO_3$	117	"	"	47
"	Kyushu Electric	Katsura	"	250	"	"	75
"	"	"	"	175	"	1976	50
"	"	Ainoura	"	250	"	"	105
"	"	"	"	250	"	"	105
"	Chugoka Electric	Oase	$Ca(OH)_2$	375	"	"	365
"	"	"	"	375	"	"	365
Chemico-Mitsui	Mitsui Aluminum	Omuta	"	119[c]	Industrial boiler	1972	d
Mitsui Miike	"	"	$CaCO_3$	25[c]	"	1974	22
"	"	"	"	175[c]	"	1975	220
"	Elec. Pow. Dev.	Takasago	"	250[c]	Utility boiler	1974	220
Bahco-Tsukishima	Yahagi Iron	Nagoya	$Ca(OH)_2$	26	Sintering plant	1971	28
Babcock-Hitachi	Chugoku Electric	Mizushima	$CaCO_3$	104	Utility boiler	1974	28
"	"	Tamashima	"	500	"	1975	385
Ishikawajima-TCA	Chichibu Cement	Kumagaya	$Ca(OH)_2$	62	Diesel engine	1972	24
Sumitomo-Fuji Kasui	Sumitomo Metal	Kokura	"	32	Heating furnace	1974	28
Chubu-MKK	Ishihara Sangyo	Yokkaichi	$CaCO_3$	77	Industrial boiler	"	55
"	Mitsubishi Gas	"	$Ca(OH)_2$	22	"	"	13

a Actual for boilers and equivalent gas flow for others (at 1900scfm per MW).
b Boilers are oil-fired unless noted.
c Coal-fired.
d Waste sludge of calcium sulfite.

formation of ion pairs from the various individual ions; examples are:

$$Ca^{++} + HCO_3^- \longrightarrow CaHCO_3^+$$

$$Ca^{++} + CO_3^= \longrightarrow CaCO_3^o$$

$$Mg^{++} + SO_4^= \longrightarrow MgSO_4^o$$

$$H^+ + HSO_3^- \longrightarrow H_2SO_3^o$$

A critical review of the literature was made to determine equilibrium constants for these reactions. Some good values were found, some were perhaps questionable, and no data were found for the neutral ion pairs $CaSO_3^\circ$ and $MgSO_3^\circ$. An experimental study was made to determine the dissociation constants for $CaSO_3$ and $MgSO_3$. The liquid-solid and gas-liquid equilibria were also considered:

$$CaCO_3\ (s) \rightleftharpoons Ca^{++} + CO_3^=$$

$$CaSO_4\ (s) \rightleftharpoons Ca^{++} + SO_4^=$$

$$CaSO_3\ (s) \rightleftharpoons Ca^{++} + SO_3^=$$

$$Ca(OH)_2\ (s) \rightleftharpoons Ca^{++} + 2OH^-$$

$$Mg(OH)_2\ (s) \rightleftharpoons Mg^{++} + 2OH^-$$

$$MgCO_3\ (s) \rightleftharpoons Mg^{++} + CO_3^=$$

$$SO_2\ (g) + H_2O \rightleftharpoons H_2SO_3$$

$$CO_2\ (g) + H_2O \rightleftharpoons H_2CO_3$$

With published and experimentally determined data, the solubility products were determined for the liquid-solid reactions. Solubility data for SO_2 and CO_2 were available from the literature. Since lime/limestone scrubbing will likely involve closed loop operation (no outflow of liquid from the system) in order to minimize water pollution, the content of dissolved salts in the liquor or slurry circulating through the scrubber will build up.

Thus activities of the dissolved species must be considered rather than the concentrations. Activity coefficients for the various species were calculated. With the resulting set of equilibrium constants, solubility products, solubilities and activities, a set of 28 equations was set up covering both the ionic reactions and the solid-liquid equilibria.

The equality in these equations holds for a saturated solution. In addition, seven mass balance equations were set up for the major input species: SO_2, CO_2, $SO_4^=$, NO_3^-, Ca^{++} and Na^+. An example of this type of equation, where m is molarity, is shown on the following page.

$$\text{Total } SO_2 = mHSO_3^- + mSO_3^= + mH_2SO_3 + mCaSO_3 + mCaSO_3\,(s) + mMgSO_3$$

Some input species were not included in these equations; K was combined with Na, and O_2 and Cl were assumed to have no effect on equilibria. The partial pressure of SO_2 or CO_2 can be used as the input value in place of the amounts of these species. Finally, the electroneutrality equation (sum of positive ion concentrations equals sum of negative ion concentrations) was set up, giving a set of 36 equations and 36 unknowns. A computer program was developed, written in Fortran V for the UNIVAC 1108 digital computer, for solving these equations at any given input of the major species. The program was checked by comparing computed values with published data for systems SO_2-H_2O, Ca-CO_2-H_2O, Ca-SO_2-H_2O, Mg-SO_2-SO_4-H_2O, and others. Generally satisfactory agreement was obtained.

The Radian program is an extremely useful tool not only for elucidating the chemistry of the system but also for use in calculating solution or slurry composition in a given scrubber situation. For the latter, certain assumptions must be made, mainly in regard to type of liquid flow in the scrubber, degree of CaO hydration and sulfite oxidation, and amount of water leaving the system. Moreover, the program gives equilibrium data only and therefore it must be assumed that sufficient time has elapsed before or at the point of measurement (for example, scrubber exit, surge tank or scrubber feed) for equilibrium to have been obtained.

This may or may not be true; further data on rates must be obtained before dependable predictions can be made regarding solution or slurry compositions. Certain general conclusions as to the process chemistry have resulted from the Radian study.

(1) For typical solution composition, the equilibrium pressure of SO_2 is low enough for good SO_2 absorption even at a pH as low as 4.0.

(2) Ionic strength (concentration of ionized fraction of dissolved salts) is important in regard to concentration of Ca and Mg into solution. The optimum ionic strength appears to be about 0.7 (mols/kg H_2O).

(3) Degree of sulfite oxidation affects the SO_2 equilibrium vapor pressure. An optimum level appears to exist from the standpoint of SO_2 absorption.

(4) The solid Mg species formed in the scrubbing system under some conditions is likely to be $Mg(OH)_2$ rather than $MgSO_3$ or $MgSO_4$.

In TVA pilot-plant work on limestone scrubbing, data were obtained that were fed into the Radian equilibrium program to calculate relative saturation levels. For simplicity, the relative saturations for sulfite, sulfate, and carbonate can be denoted as follows:

$$R_1 = \frac{(A_{Ca^{++}})(A_{SO_3^=})}{K_{sp\ CaSO_3\cdot\frac{1}{2}H_2O}}$$

$$R_2 = \frac{(A_{Ca^{++}})(A_{SO_4^=})}{K_{sp\ CaSO_4 \cdot 2H_2O}}$$

$$R_3 = \frac{(A_{Ca^{++}})(A_{CO_3^=})}{K_{sp\ CaCO_3}}$$

where A = activity of the individual species, calculated from the measured concentration (mol/l) and its associated activity coefficient (Y). Ksp = solubility product.

The results indicated that calcium sulfite remained highly supersaturated in the scrubber circuit. Calcium sulfate was also supersaturated in the scrubber circuit but to a lesser extent. Calcium sulfate supersaturation increased in the settling tank (clarifier) whereas sulfite liquor desupersaturated.

Other results indicate that for calcium sulfate there is indeed a critical value of relative supersaturation beyond which nucleation and a greatly increased rate of precipitation (or scaling) take place. The results show that, beyond a value of 1.3 to 1.4, the rate of calcium sulfate dehydrate precipitation markedly increases. Photographs of the crystals formed at values of the ratio greater than this critical value clearly indicate the formation of small new crystals where nucleation has occurred.

The calcium carbonate concentration was roughly twice as high in the settling pond bottom as in the scrubber loop. This indicates precipitation of limestone in the settling pond, which is reasonable since the clarified recycle liquid is saturated with respect to calcium carbonate whereas the scrubbing liquor is unsaturated in carbonate. Also, some oxidation occurs in the settling tank.

In order to scrub a specified amount of acidic species from the gas entering the scrubber, sufficient basic species must be available in the scrubber feed. The L/G (gallons liquid per 1,000 cu ft of gas) required to scrub a specified amount of SO_2 is dependent upon the amount of available basic material in the scrubber feed.

Normally, a liquid is considered basic (or alkaline) when the pH is high (denoting a high concentration of OH^-). For aqueous lime/limestone scrubbing of SO_2 (the acid anhydride of sulfurous acid), a number of other liquid species should be considered basic. By this, it is meant that these species are basic with respect to the first ionization step of sulfurous acid, i.e.,

$$H_2SO_3(\ell) \rightleftarrows HSO_3^- + H^+$$

Of the chemical species considered in the Radian equilibrium program, the following chemical species are basic with respect to SO_2 sorption:

OH^-	$CaOH^+$	$MgOH^+$	$NaOH(\ell)$
$SO_3^=$	$CaSO_3(\ell)$	$MgSO_3(\ell)$	$NaCO_3^-$
HCO_3^-	$CaCO_3(\ell)$	$MgHCO_3^+$	$NaHCO_3(\ell)$
$CO_3^=$	$CaHCO_3^+$	$MgCO_3(\ell)$	

Basic species available in the scrubber feed come from two sources. All of the basic species in the aqueous phase that were given above are considered available for reaction. In addition, more basic species are made available by limestone and calcium sulfite dissolution in the scrubber according to the following equations:

$$CaCO_3(S) \rightarrow Ca^{++} + CO_3^{=}$$

$$CO_3^{=} + H_2O \rightarrow HCO_3^{-} + OH^{-}$$

$$CaSO_3 \cdot 1/2H_2O(S) \rightarrow Ca^{++} + SO_3^{=} + 1/2\ H_2O$$

$$SO_3^{=} + H_2O \rightarrow HSO_3^{-} + OH^{-}$$

Generally, for limestone systems, it is the calcium carbonate dissolution in the scrubber that adds additional basic species whereas for lime systems it is the calcium sulfite. In summing the basic species available from the scrubber feed, most of the species in solution are given a weighting factor of 1.0. The carbonate species [$CO_3^{=}$, $CaCO_3(\ell)$, $MgCO_3(\ell)$, $NaCO_3^{-}$] are given a weighting factor of 2.0 since they can react with two hydrogen ions. The limestone dissolving in the scrubber is given a weighting factor of 2.0 since the $CO_3^{=}$ is equivalent to two mols of the other species such as HCO_3^{-} and OH^{-}. Using the same analogy, calcium sulfite dissolution in the scrubber is given a weighting factor of 1.0.

In a TVA limestone pilot plant run, analysis indicated that the following basic species were available in the scrubber feed for removing SO_2:

Liquid species	gmol/kg H_2O	Weighting factor	gmol/min (basic species)
OH^-	6.655E-08	1	2.519E-05
$SO_3^=$	9.972E-05	1	3.775E-02
HCO_3^-	2.262E-03	1	0.856E-00
$CO_3^=$	2.102E-07	2	1.592E-04
$CaOH^+$	1.013E-08	1	3.834E-06
$CaSO_3(\ell)$	1.818E-03	1	0.688E-00
$CaCO_3(\ell)$	2.395E-06	2	1.813E-03
$CaHCO_3^+$	2.686E-04	1	0.102E-00
$MgOH^+$	2.187E-07	1	8.278E-05
$MgSO_3(\ell)$	6.780E-04	1	0.257E-00
$MgHCO_3^+$	1.507E-04	1	5.704E-02
$MgCO_3(\ell)$	4.459E-06	2	3.376E-03
$NaOH(\ell)$	3.234E-11	1	1.224E-08
$NaCO_3^-$	8.284E-09	2	6.271E-06
$NaHCO_3(\ell)$	2.97E-06	1	8.694E-04
		Total liquid phase basic species:	2.00E-00

A rough value for the rate of $CaCO_3$ dissolution in the scrubber was calculated. Assuming no precipitation in the scrubber, the average rate of dissolution was 2.5 ± 0.4 mols/min which with the weighting factor converts to 5.0 gmol/min.

The total basic species available to react and remove SO_2 from the inlet gas stream thus was 7.00 mols/min. Under the conditions of the pilot-plant run, approximately 5.10 mols/min SO_2 entered the scrubber and 4.70 mols/min SO_2 was absorbed. Hence, more than enough basic species were available to remove all of the entering SO_2.

It should be noted that far more than 7.00 mols/min basic species could have been available if more limestone dissolved in the scrubber. For the case presented, the 2.5 mols/min of limestone dissolution represented only 1.6% of the total limestone in the scrubber feed. Any additive or absorber configuration which would increase limestone dissolution would necessarily either increase SO_2 removal efficiency or allow a lower liquid pumping rate for the same efficiency or possibly both–increase SO_2 removal efficiency at a lower pumping rate.

A similar pilot-plant test was made with lime. The following basic species were available in the scrubber feed:

Liquid species	gmol/kg H_2O	Weighting factor	gmol/min (basic species)
OH^-	1.923E-05	1	8.735E-03
$SO_3^=$	9.085E-06	1	4.633E-03
HCO_3^-	3.593E-04	1	0.163E-00
$CO_3^=$	8.580E-06	2	7.794E-03
$CaOH^+$	2.500E-06	1	1.135E-03
$CaSO_3(\ell)$	1.431E-04	1	6.500E-02
$CaCO_3(\ell)$	8.433E-05	2	7.661E-02
$CaHCO_3^+$	3.648E-05	1	1.657E-02
$MgOH^+$	1.358E-04	1	6.168E-02
$MgSO_3(\ell)$	1.321E-04	1	6.000E-02
$MgHCO_3^+$	5.069E-05	1	2.302E-02
$MgCO_3(\ell)$	3.908E-04	2	0.355E-00
$NaOH(\ell)$	6.774E-09	1	3.077E-06
$NaCO_3^-$	2.490E-07	2	2.262E-04
$NaHCO_3(\ell)$	2.644E-07	1	1.201E-04
	Total liquid phase basic species:		0.82 gmol/min

In comparing basic species in the liquid phase for the lime and limestone systems, it is evident that additional dissolution of calcium carbonate and/or calcium sulfite is required in the lime system. However, chemical analysis showed very little solid calcium carbonate available in the scrubber inlet liquor. Therefore, the increase must come from calcium sulfite. If one assumes that 7.0 mols per minute basic species are required in the scrubber feed (based on the limestone data), then 6.18 mols/min of calcium sulfite must dissolve in the scrubber. This would account for approximately 3% of the total calcium sulfite in the scrubber feed.

In more recent work, organizations such as M.W. Kellogg, Dravo, and EPA have stressed the importance of soluble magnesium salts in the chemistry of the system. In the Kellogg work it was found that maintenance of a fairly high concentration of a dissolved magnesium compound in the scrubber slurry improves SO_2 absorption and reduces or prevents sulfate scaling. The magnesium can be added as MgO or $MgSO_4$; in either case $MgSO_4$ is the active compound because the MgO rapidly converts to $MgSO_3$ and then to $MgSO_4$. The beneficial effect of the $MgSO_4$ is based on the reaction:

$$MgSO_4 + CaSO_3 \longrightarrow MgSO_3 + CaSO_4$$

Since the $MgSO_3$ is about 300 times as soluble as $CaSO_3$, the reaction increases the concentration of dissolved basic species (species such as $SO_3^=$ that are basic to H_2SO_3 and therefore can absorb SO_2 as bisulfite). The absorption then is improved because of the reaction:

$$MgSO_3 + SO_2 + H_2O \longrightarrow Mg(HSO_3)_2$$

In the hold tank addition of limestone gives the reaction:

$$Mg(HSO_3)_2 + CaCO_3 \longrightarrow CaSO_3 + MgSO_3 + CO_2 + H_2O$$

Since the $CaSO_3$ is relatively insoluble most of it precipitates and is removed as product.

Any soluble sulfate will increase the $SO_3^=$ concentration and thus improve absorption. For example,

$$Na_2SO_4 + CaSO_3 \longrightarrow Na_2SO_3 + CaSO_4$$

However, $MgSO_4$ has the advantage that the $MgSO_3$ remains largely undissociated in solution whereas Na_2SO_3 dissociates to Na^+ and $SO_3^=$. Thus with magnesium both $SO_3^=$ and $MgSO_3^\circ$ are provided and both will absorb SO_2.

$$SO_3^= + SO_2 + H_2O \longrightarrow 2HSO_3^-$$
$$MgSO_3^\circ + SO_2 + H_2O \longrightarrow Mg(HSO_3)_2$$

Calculations indicate that most of the limestone (over 90%) dissolves in the hold tank in such a process. Most of the SO_2 absorption in the scrubber is by $MgSO_3^\circ$, which accounts for about 90% (on the molar basis) of the total SO_2 absorbing species in the slurry returned to the scrubber. The relative supersaturation of $CaSO_4$ and $CaSO_3$ in the scrubber effluent are 1.1 to 1.2 and 0.90

to 0.94, respectively, which minimizes any tendency to scaling. The very small increase in Ca^{++} ion concentration through the scrubber also contributes to low scaling potential.

In a further exploration into the chemistry of the process, EPA has investigated the phenomenon of sulfate unsaturated operation, in which the scrubber solution remains unsaturated in calcium sulfate even though oxidation in the scrubber oxidizes calcium sulfite continuously. For the solution to remain unsaturated, the sulfate formed must be purged from the system at such a rate that saturation does not develop at steady state.

This is difficult to visualize because sulfate solubility is so low that even a low degree of oxidation will produce more sulfate than the solution can hold and crystallization therefore must occur unless an extreme of open-loop, once-through type of operation is used. Up until recently it was generally assumed that calcium sulfate crystals must of necessity be formed in the scrubber and that therefore the solution is, by definition, either saturated or supersaturated.

In very early work by ICI (1951), a mechanism had been proposed by which sulfate can be incorporated within calcium sulfite crystals. Examination of solids from the Bankside scrubber showed no gypsum or anhydrite crystals even though chemical analysis indicated about 10% calcium sulfate. It was postulated that the sulfate was present within the sulfite as a solid solution.

EPA carried out various tests to explore this hypothesis. The scrubber solids, like those of ICI, were found to contain 10 to 15 mol percent sulfate by chemical analysis, but X-ray diffraction showed no crystalline calcium sulfate. Stirring with water did not extract sulfate, confirming that it was not present as a separate phase. Dissolution in HCl followed by water extraction and recrystallization gave pure gypsum in amounts to be expected from the chemical analysis. It was concluded, therefore, that the sulfate was tied up with the sulfite in such a manner that, although precipitated, it did not saturate the solution with gypsum ($CaSO_4 \cdot 2H_2O$).

Thus it appears that sulfate has been leaving the scrubber loop within the sulfite crystals in all the systems operated in the past. Oxidation was so high, that not all the sulfate formed could be purged this way. As a result the concentration increased to saturation and beyond with consequent eventual crystallization of gypsum without scaling by using the conditions worked out by ICI–high L/G, high sulfate crystal content in the scrubber slurry, and extended delay time outside the scrubber–but it was not realized that reducing oxidation could make these requirements less rigorous or perhaps not necessary at all.

Oxidation can be reduced by (1) keeping excess air in the gas as low as possible, (2) reducing the amount of sulfite in solution (where oxidation occurs) by keeping pH as high as possible throughout the scrubber loop, and (3) minimizing entry of air into the reaction tank. Maximum allowable oxidation is about 20 mol % of total sulfur. More oxidation can be tolerated if Mg is present since $MgSO_4$ accumulates in the solution and drives more sulfate into the sulfite. Chloride has the opposite effect. Lime is better than limestone because the generally higher pH throughout the loop makes it easier to keep oxidation below the critical level. However, EPA has demonstrated unsaturated operation with limestone; Mg is especially helpful by offsetting the effect of higher oxidation.

KINETICS

Very little has been done in the way of a systematic study of the factors that affect mass transfer and reaction kinetics. In the early English work, the experimental emphasis was on solving the scaling problem (deposition of an adherent crystal mass on equipment surfaces) rather than absorption kinetics. One of the requirements worked out for preventing scaling was a very high slurry recirculation rate, so high that good absorption of SO_2 was obtained without need to study or optimize other variables affecting absorption. The main work on kinetics has been done in recent years, most of it under EPA contracts. Other work has been concerned with testing of various scrubber types, on bench and pilot scale; most of this has been limited in scope and not carried out in a very systematic manner.

Limestone slurry scrubbing can be looked on as a system for making a weak acid (H_2SO_3) and reacting it with a relatively unreactive solid ($CaCO_3$); the same analogy can be made for lime slurry scrubbing if it is assumed that the bulk of the dissolved or slurried $Ca(OH)_2$ reacts with CO_2 before it has a chance to react with SO_2 which may or may not occur. Both the formation of H_2SO_3 by water scrubbing of SO_2 and the reaction of $CaCO_3$ with weak acids have been studied extensively. Both are known to be relatively slow; the question is which is the slowest and whether the difference is large enough for the slowest to become rate controlling. Various resistances to mass transfer exist in the system.

(1) Diffusion to and through the gas film at the liquid surface.
(2) Rate of SO_2 dissolution and hydration to H_2SO_3, H^+, and HSO_3.
(3) Diffusion of H_2SO_3 and ions through the liquid film and into the droplet interior.
(4) Dissolution rate of $CaCO_3$.
(5) Reaction rate of Ca^{++} with acidic ions to precipitate $CaSO_3$, thereby allowing further dissolution of $CaCO_3$.

Other resistances may also be important; absorption of CO_2, for example, may promote dissolution of $CaCO_3$, in which case steps 1 through 4 would be involved for CO_2. Rate of oxygen absorption may also enter, as degree of sulfite oxidation affects SO_2 partial pressure over the slurry.

It is known that SO_2 is a moderately soluble gas. Therefore both gas film and liquid film resistance would be expected to affect the rate of absorption by water to form H_2SO_3; this is supported by published data. When $CaCO_3$ is added to the water and reaction of $CaCO_3$ and H_2SO_3 becomes a part of the system, some additional driving force is established for moving H_2SO_3 through the liquid film and into the liquid, which should reduce the liquid resistance and make the gas film transfer even more important.

However, there may be some question regarding the data published on water scrubbing. Most of it was apparently obtained with natural water rather than distilled water, in which case dissolved alkaline compounds no doubt contributed to the absorption and reduced the liquid film resistance in the once-through scrubbing tests. In limestone scrubbing, the original charge of water becomes exhausted of its alkaline content quickly as it is recycled and thereafter contains only that supplied by the dissolving limestone. Therefore, the overall resistance in the liquid phase, including limestone dissolution, could be higher than that reported for once-through water absorption. Some have postulated that it is so high as to make the gas phase resistance noncontrolling.

Rate Factors

There are several operating factors that can affect both the gas-liquid transfer and the $CaCO_3$-SO_2 reaction rate.

Inlet Gas Temperature: One of the most significant variables in the TVA small-scale work has been inlet gas temperature; data illustrating the effect are shown in Figure 3.2. At the relatively low pH (4 to 5) of the slurry leaving the countercurrent scrubber, the solution contains a significant amount of SO_2 species (H_2SO_3 and HSO_3^-) that have not had time to react with $CaCO_3$. The slurry then flows into the pump tank where the reaction is completed. However, if gas entering the bottom of the scrubber is hot, the unreacted H_2SO_3 and HSO_3^- may be distilled out and reabsorbed higher in the scrubber after the gas has reached the wet bulb temperature.

FIGURE 3.2: EFFECT OF INLET GAS TEMPERATURE ON SO_2 REMOVAL IN SMALL-SCALE CONTINUOUS SCRUBBING TESTS

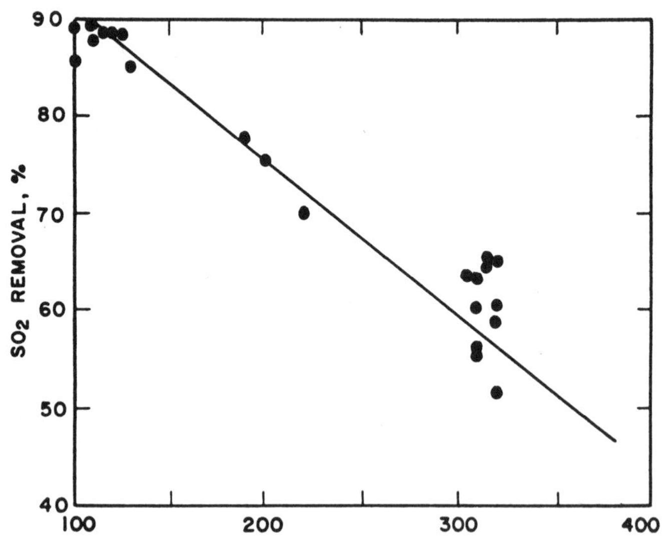

Scrubber Temperature: The temperature of the scrubber liquid is normally not considered to be an independent variable as it depends on the wet-bulb temperature of the incoming gas, which in most power plants is 120° to 130°F. However, in some of the reported tests, the liquid temperature was lower, either because of once-through water use or because the gas came from a sulfuric acid plant and therefore had a relatively low wet bulb temperature. The lower temperature in these tests helped gas-liquid transfer because of the lower equilibrium vapor pressure of SO_2.

Inlet SO_2 Concentration: The SO_2 concentration in the inlet gas should affect gas-liquid transfer because of vapor pressure considerations. Moreover, high inlet concentration allows a higher equilibrium SO_2 partial pressure at the outlet of a countercurrent scrubber (that is, a more acidic scrubber effluent), which has been shown to have a first order effect on the chemical reaction.

Limestone Type: Variation in limestone properties may or may not affect absorption rate. In batch tests chalk has given better results than ordinary limestone and practically as good as lime, which might be expected since chalk is a soft, fine-grained material that breaks down easily to very small particle size. Tests in continuous equipment, however, have indicated that performance comparable to that of chalk can be obtained from ordinary limestone if it is ground to small particle size, comparable to the grain size of chalk resulting from attrition to a scrubber system.

The same general results have been obtained in comparing calcitic limestones (ordinary types) from various sources. At the same particle size, coarse or fine, performance is comparable. There has been no indication that impurities in calcitic limestones affect absorption rate, which is quite different from the situation in boiler injection. There are limestones that are somewhat more porous than the usual types and therefore might be more effective because of the internal as well as external surface area.

The effect of Mg content in the limestone is not clear. It is known that $MgCO_3$ is attacked by weak acids at a slower rate than is $CaCO_3$, so it might be expected that absorption rate would be lower for dolomite and dolomitic limestone. Small-scale work has also shown slightly lower efficiency for dolomite but not enough to be conclusive. Once the $MgCO_3$ is converted to $MgSO_3$ in the scrubber circuit, it should be an excellent absorbent. And as noted earlier, the MgO in lime is quite beneficial.

Stoichiometry: The amount of lime or limestone used per unit of SO_2 should be significant in regard to rate of the $CaCO_3$-SO_2 reaction because of its effect on the concentration of absorbent surface area per unit volume of liquid. This is borne out by operating experience. A large excess of absorbent can produce a problem, however, by pushing the pH above the optimum range.

Particle Size: The dissolution rate of $CaCO_3$ should be affected strongly by particle size because of the effect on absorbent surface area. This applies mainly to use of limestone, since hydration of lime produces very fine particles.

Solubility of Absorbent: When CaO is the absorbent, either formed in the boiler and brought into the scrubber with the gas or produced in a separate calciner and introduced into the scrubber liquid feed, there is the advantage that hydration gives $Ca(OH)_2$, which is relatively soluble as compared to $CaCO_3$.

Moreover, CaO is much more porous than natural $CaCO_3$ and if some degree of this higher internal area should survive the conversion to $Ca(OH)_2$ followed by precipitation of $CaCO_3$, the effective surface area would be increased. Finally, the $Ca(OH)_2$ is on the favorable side of the equilibrium, that is, the Ca is already in solution whereas the Ca from the natural $CaCO_3$ is not. Any delay in coming to equilibrium should favor $Ca(OH)_2$. In the early United States work it was generally considered, on the basis of pilot plant tests, that lime was a superior absorbent. The mass transfer data discussed later indicate little resistance in the liquid phase when lime is the absorbent, whereas significant resistance for solid-liquid transfer has been found in work with $CaCO_3$.

Some of the more recent pilot plant work also indicates superiority for lime, particularly in scrubbers such as the venturi type where gas-slurry contact time is relatively short. It appears that as contact time and liquid holdup increase, finely ground $CaCO_3$ approaches lime in efficiency. In other words, $CaCO_3$ can be about as effective as lime but at the expense of high recirculation rate, restriction to certain scrubber types, or other means to get high contact time.

Hydration Rate of CaO: Since CaO must hydrate to $Ca(OH)_2$ before it becomes effective, hydration rate is important when lime is used as the absorbent. Hydration rate depends on several factors, the main one being the porosity developed during calcination. If the limestone is not calcined properly (as discussed in Chapter 2), it will be hard-burned or dead-burned and the reduced porosity will make the material more difficult to hydrate. Moreover, use of slaking water containing dissolved salts interferes with hydration.

Shell Formation: It has been postulated by several that $CaSO_3$ (or $CaSO_4$) precipitates on the surface of $CaCO_3$ particles and inhibits further reaction. In TVA pilot plant tests with limestone slurry, crystallization of calcium sulfite on $CaCO_3$ surfaces apparently occurred. The difficulty was associated with closed loop operation, in which clear liquid from the solids disposal stream was returned to the system. This liquid, which had relatively low pH because of the countercurrent scrubbing, contained a relatively high amount of dissolved sulfite as a result of higher sulfite solubility at low pH.

The return of this sulfite to the system, plus the generally low pH in the scrubber circuit at the time, apparently loaded the liquid phase with sulfite to the extent that crystallization on the limestone surfaces occurred. It is postulated that the binding mechanism involves formation of a liquid film around a $CaCO_3$ particle in which the dissolving $CaCO_3$ raises the pH and the resulting decrease in sulfite solubility causes crystallization of sulfite on the $CaCO_3$ surface. The $CaSO_3 \cdot 0.5\,H_2O$ crystals formed were quite small and plate-like, square or rectangular and very thin. Examination by electron microscope showed that the crystals tend to grow on surfaces such as limestone in an edgewise fashion.

Oxidation Rate: Previously it was noted that oxidation has offsetting effects on SO_2 solubility for a given SO_2 partial pressure, a decrease because of H^+ ions formed and an increase because of sulfite removal by sulfate formation. For limited oxidation, the H^+ ion effect overrides the sulfite removal, but as it progresses the sulfite removal becomes more significant until finally a net increase in solubility is attained. The point at which sulfite removal moves ahead depends on the magnitude of the partial pressure over solution (Figure 3.3). Up to about 0.065 mm Hg, even 90% oxidation gives a net decrease in solubility. At 1.0 mm, about 25% is the break point, and at 0.23 mm, corresponding to roughly 90% removal from gas containing 3,000 ppm of SO_2, about 75% oxidation is required to make any gain.

FIGURE 3.3: EFFECT OF SULFITE OXIDATION ON SOLUBILITY OF SO_2 IN ALKALINE WATER

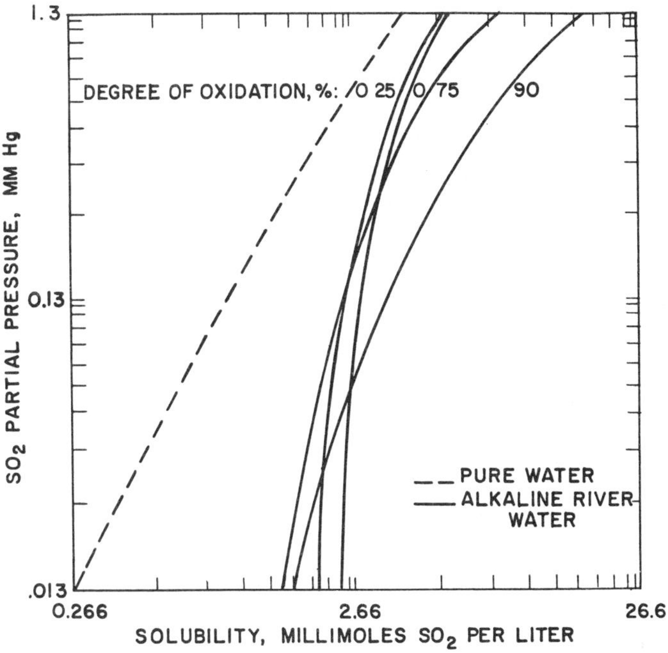

Source: Rees, *Journal of the Institute of Fuel,* Vol XXV, No. 148, 350 (March 1953)

Oxidation of sulfite solutions has been studied widely, both in work on sulfur dioxide removal from gases and in other fields. For sulfur dioxide removal, the principal studies appear to be those of NIIOGAZ (USSR) on the ammonium and calcium systems and Aerojet-General Corporation on the sodium system, but several other research organizations have worked on the problem. There appears to be a considerable amount of disagreement in the results reported, although the differing scrubbing conditions used make comparison somewhat difficult. Factors affecting degree of sulfite oxidation in scrubbing waste gases have been reported to be as follows:

(1) Rate of oxygen dissolution per unit of sulfite. Among the factors affecting this are (a) oxygen:sulfur dioxide ratio in the gas, (b) liquor circulation rate, (c) type of scrubber, (d) scrubber temperature and (e) solution viscosity and density.

(2) Amount of nitrogen dioxide in the gas and rate of dissolution.

(3) Presence of oxidation inhibitors in the gas or in raw materials.

(4) Presence of oxidation catalysts such as iron and manganese in the ash or in raw materials.

(5) Bisulfite:sulfite ratio.

The major work appears to have been done by Chertkov and coworkers at NIIOGAZ in Russia. A generalized expression, based on data from studies of both calcium sulfite and alkali sulfite systems, was developed.

$$G_o = \frac{(0.8\ Q^{0.7})\ (a)\ (S/C)^6}{(\gamma)\ (\mu)}$$

G_o = g. O_2 absorbed/(hr.) (m.^{2})

Q = liquid flow rate, m.^{3}/(m.^{2}) (hr.)

a = t/50 where t is the average solution temperature, °C.

S/C = molar ratio of sulfur to "effective base"

γ = solution density, kg./m.^{3}

μ = solution viscosity, (kg.) (sec.)/m.^{2}

This equation covers all the known effects except that of inhibitors or catalysts. Since it deals only with amount of oxygen absorbed, it does not include, of course, the effect of O_2/SO_2 ratio, and G_0 will also vary with scrubber type. The important effect of S/C has been noted in small scale lime/limestone scrubbing tests. At the lower pH range covered in the tests, where S/C is higher, the highest degree of oxidation was obtained.

It should be noted, however, that low pH does not necessarily indicate higher oxidation even though S/C is higher at low pH and the $(S/C)^6$ term in the equation has a major effect. In some of the tests reported by Chertkov, the oxidation was lower in low pH tests because the change in other terms more than offset the effect of $(S/C)^6$.

Another factor affecting oxidation in the calcium system, and possibly in alkali systems, is the increase in sulfite solubility as the pH is decreased. The fact that there is more sulfite in solution to be oxidized has an important effect on rate of oxidation. Again, this may be outweighed by other factors such as the increase in density and viscosity resulting from the higher amount of dissolved solids at low pH.

The general theory of sulfite oxidation has also been studied in a general way, without reference to the problem of SO_2 removal from gases. It is reported that the mechanism is based on the interaction of free radicals or intermediates which contain atoms with an unshared electron; examples are O_2^-, OH°, $HSO_3^°$, and SO_3^-.

These radicals are presumably formed and propagated by a complicated set of reactions, including the chain type, between ions in solution and between radicals. Inhibitors and promoters may affect the rate of oxidation by affecting the production of radicals or in some manner deactivating or enhancing the activity of intermediates. A general expression, consistent with the results of many investigators, is as follows.

$$-\frac{d(SO_3^=)}{dt} = \frac{g(HSO_3^-)}{[H]^{1/2}}$$

$SO_3^=$ and HSO_3^- = concentrations of these species

g = term including effects of free radicals and other effects characteristic of the system under study

[H] = activity of the hydrogen ion

t = time

From this it appears that bisulfite concentration and pH have opposite effects on oxidation. The lack of agreement or consistency among the reports of the various investigators indicates that more study is needed, particularly on the lime/limestone scrubbing system. The recent findings indicate that rate of oxidation may be a more important variable than previously assumed.

Gas-Liquid-Solid Flow Relationships: There are several interrelated variables that involve gas, liquid and solid flows, including liquid holdup in the scrubber, liquid:gas flow ratio, gas:liquid contact time, gas velocity, ratio of solids recycle to solids bleed, and concentration of solids in the recycled slurry.

(1) Liquid holdup in the scrubber is quite important in regard to the $CaCO_3$-H_2SO_3 reaction because it affects the amount of limestone surface to which the SO_2 is exposed. The sequence in the scrubber can be visualized as a progression of (a) $CaCO_3$ going into solution and saturating a film around the particle, (b) the dissolved species reacting with and exhausting H_2SO_3 (by precipitation of $CaSO_3$), (c) SO_2 dissolving to replace the H_2SO_3, and (d) $CaCO_3$ dissolving again as in (a). Hence the holdup must be relatively large in order for the scrubber to perform both absorption and reaction.

If CaO is the absorbent and it hydrates adequately, there will be more calcium in solution in the scrubber feed slurry; however, liquid holdup is still an important consideration, although probably less so than for $CaCO_3$.

(2) The liquid:gas ratio should be an important variable in regard to gas-liquid transfer (and effect on liquid holdup in many scrubber types). In tests where liquid flow variation was tested with other variables held constant, increase in liquid:gas ratio has been quite beneficial.

(3) Gas-liquid contact time is a factor in the interdependent absorption-reaction mechanism described in (1). It is also important in gas-liquid transfer; in the English work on water scrubbing, for example, several seconds contact time were considered necessary for adequate absorption.

(4) The concentration of solids in the recycling slurry is tied in with the above flow relationships. High solids content gives high $CaCO_3$ surface concentration.

(5) The gas velocity is an important consideration when gas-film transfer resistance is significant, as it has been shown to be in lime/limestone scrubbing. The velocity is also a factor in gas-liquid contact time.

These flow relationships are quite important in absorption rate. Very few data have been reported, however, that can be used in relating the various flows and their effect.

Transfer Rates

A limited amount of data has been reported on overall SO_2 transfer rate in pilot-plant scrubbers. In Russian tests with a grid-packed scrubber, using lime as the absorbent, the mass transfer coefficient varied directly with the gas velocity at a pH of 6.1 to 6.2. The data indicated an almost complete lack of liquid phase resistance at this pH; at lower pH, however, transfer rate in the liquid became important. At the higher pH, K, the overall transfer coefficient, was 2,000 to 7,300 kg/(m^3)(hr)(atm) over a range of 0.38 to 1.37 m/sec. K decreased somewhat as the pH was reduced. In a grid-packed tower tested in Japanese work (with lime as absorbent), K was 120 kg (mols)/(hr)(m^3)(atm) at 2.5 m/sec and a liquid-gas ratio of 1.4 l/m^3. The transfer rate was proportional to the gas velocity.

PROCESS AND DESIGN CONSIDERATIONS

The work on the chemistry and kinetics of lime/limestone scrubbing has done much toward making it possible to select design and process conditions on a rational basis. There is still much to be done, however, and work is proceeding in laboratories and pilot plants of several organizations. Process selection cannot yet be done on a sound theoretical basis.

In the following discussion, the advantages and disadvantages of the various process routes are presented along with an analysis aimed at selecting the optimum flowsheet. It should be emphasized that this is based on the current level of knowledge on the subject, and that new information developed in the several current programs may change the outlook. The rate studies underway in various pilot plants, the large-scale test program at Shawnee, and the various full-scale operational units should fill the gaps in current knowledge and make it possible to reduce cost and improve performance.

Choice of Absorbent

One of the more difficult decisions in the initial phases of lime/limestone scrubber design is which absorbent to use. There are several factors affecting the choice, some of which have not been elucidated well enough to allow adequate evaluation.

Basic Cost: Lime is usually the most expensive because of the calcining cost, which usually adds about $10/ton (of lime) to the raw material cost. However, if the limestone quarry is some distance from the power plant, the lighter weight of the lime, because of CO_2 and water elimination during calcining, can produce a substantial freight saving. Thus the cost per unit weight of available calcium delivered to the power plant depends mainly on how far the lime or limestone must be shipped.

Limestone has one shipping cost factor in its favor, however, because in lump form it can be shipped in open vehicles without any particular precautions. Because of this it is usually more economical to grind limestone at the point of use, with crushing at the quarry only to lumps on the order of 1 inch diameter or smaller. Lime, on the other hand, is usually shipped in closed vehicles, even though produced in lump form, in order to avoid undue hydration and carbonation during transit. Shipping and handling cost is likely to be somewhat higher, per ton, than for limestone. At least one company is considering grinding limestone at the quarry and pumping the slurry through a pipeline over 100 miles long. Presumably slaked lime could be transported in the same way.

Absorbent Preparation: A further cost differential is incurred in preparing the absorbent for introduction into the scrubber. The limestone must be ground and the lime must be slaked. The cost of grinding depends on several factors, including fineness of grind, limestone hardness, type of mill, and whether dry or wet grinding is used. Slaking cost of lime is not as variable. The main consideration appears to be rate of slaking as affected by the dissolved salt content of the slaking water. There has not been enough experience in this area to produce much useful data, but in one plant at least it has been considered advisable to use fresh water rather than recycled liquor for slaking; this has an adverse effect on the water balance, whereas limestone can be ground with recycled liquor. There has also been some difficulty in slaking lime with blowdown water from cooling towers because of the high salt content.

Another cost that enters in lime preparation results from incomplete calcination of the limestone. In making lump ("pebble") lime the lumps usually have an uncalcined center because it is difficult to calcine the lump completely without deadburning the outside. The resulting grits accumulate in the slaking equipment and must be either purged or carried along with the slaked lime into the scrubber. If the former, cost is incurred in handling and disposal; if the latter, the larger lumps may plug equipment. In any event, the grits are mainly ineffective in absorption and must be subtracted from the useful calcium content.

Effect of Impurities: The effective calcium content of limestone is reduced if it contains magnesium. Dolomite ($MgCO_3 \cdot CaCO_3$), the form in which the magnesium is usually present, reacts very slowly with weak acids—to the extent that not more than about 10% is likely to be dissolved in a normal scrubbing operation. This is a double disadvantage because not only is the magnesium ineffective but also the calcium associated with it in the dolomite.

If the limestone is calcined, however, the magnesium not only becomes available for absorption but also helps the operation in other ways, as discussed earlier. Although $Mg(OH)_2$ is quite insoluble, the fine particle size resulting from calcination and slaking apparently makes it quite reactive with bisulfite species in the delay tank, giving $MgSO_3$ which is relatively soluble. Thus lime has another factor in its favor if the original limestone is dolomitic.

Other than magnesium, silica is usually the main impurity in limestone. As between lime and limestone, the silica grains are more likely to become separated from the matrix when lime is slaked as compared with when limestone is ground. No data appear available as to whether or not this has any effect on erosion by the silica.

Reactivity: Limestone reacts more slowly than lime in the scrubber loops, both because the particle size is usually larger and because the reaction rate with H_2SO_3 or bisulfite species is lower. Therefore if the same degree of SO_2 removal is to be attained, the limestone system must be adjusted in some way to get more surface area in contact with the solution so that the amount of calcium dissolved per unit time will be comparable to that in the lime system. This can be done by finer grinding, use of a scrubber with larger hold-up, higher slurry recirculation rate, larger reaction tank, or use of excess limestone—or some combination of these parameters. Practice varies in the means used to offset the lower reactivity but in most cases, particularly for high inlet SO_2, the lower reactivity of limestone increases cost.

Dust Removal

Since the gas contains dust as well as SO_2 they must be considered together in designing a control system. The main situations encountered in practice follow.

Existing Plant with Inadequate Precipitator: Because of the more rigorous particulate regulations promulgated in recent years, many plants find themselves not only lacking in SO_2 removal capability but partially so in regard to particulate. Rather than expand or replace the existing precipitator the logical course is to install a scrubber capable of removing both the SO_2 and the residual particulate as a means of reducing overall cost. In some areas, however, there is little pressure to install SO_2 scrubbers on existing plants. In such cases the tendency is to upgrade the precipitator in some way to an efficiency adequate to meet regulations.

This is done either by adding capacity, building a new unit (so that it can be cut in quickly to avoid boiler downtime, as opposed to precipitator enlargement which can require considerable downtime), or adding something to the gas to improve precipitator efficiency. As an example, TVA has a major program on upgrading precipitators, even though no SO_2 scrubbing is contemplated except on an experimental basis. TVA also has systems in which SO_3 or NH_3 is added to the gas before the precipitator to increase conductivity of particles (by adsorption of the additive) and thus improve precipitator efficiency. There are also examples of upgrading dry collection systems by adding scrubbers even though SO_2 removal was not required.

Existing Plant with Adequate Precipitator: There are very few of these as yet because most of the existing precipitators were installed in an era when regulations were either nonexistent or not as strict. There should be several in the future, however, because many are upgrading their precipitators, like TVA, without any reference to SO_2 removal. If and when SO_2 scrubbing becomes necessary in these plants there presumably will be no necessity to design particulate removal capability into the SO_2 system.

New Plants: The problem of dust removal comes into focus mainly for new plants, where the planning can include many options and where, in most cases, removal of both SO_2 and dust is a requirement. If a wet lime/limestone system is adopted for SO_2 removal, then the most economical course—by a considerable margin—is to remove particulate also by scrubbing. Adapting the SO_2 scrubber to do the double job of SO_2 and particulate removal is almost always less expensive than installing a precipitator ahead of the SO_2 scrubber. In most double-duty wet systems, a venturi scrubber for removing particulate precedes the SO_2 scrubber; however, the venturi is part of the SO_2 system because the absorbent slurry passes from the second scrubber to the venturi.

A major drawback of this arrangement is the awkward bypass situation. It is highly desirable to have a bypass around the SO_2 scrubber so that scrubber breakdown will not shut down the power plant (if regulations allow boiler operation without SO_2 removal. However, the particulate removal equipment should continue to function so that the boiler can be operated without emitting a heavy particulate plume.

If the dust is removed by a precipitator ahead of the SO_2 scrubber this is easy to arrange; the duct after the precipitator can be split with one branch supplying the SO_2 scrubber and the other, with appropriate dampers serving as the bypass. However, if a particulate scrubber is used instead of the precipitator then bypassing the SO_2 scrubber becomes difficult; a second mist eliminator system must be placed in the bypass and all the auxiliary facilities for mist elimination provided. The damper must also be placed in a very wet area, which can cause corrosion and plugging problems.

If a single scrubber is used both for particulate and SO_2, as has been proposed by some, then, of course, bypassing particulate removal is impossible. Thus for new plants there is a trend to selecting the high cost alternative of precipitator plus scrubber to avoid these problems. This has advantages aside from allowing bypassing; keeping the dust out of the SO_2 scrubbing system reduces wet-dry interface problems, minimizes mist eliminator plugging by reducing slurry solids concentration, and probably decreases sulfite oxidation by avoiding any catalytic effect of the ash.

Double-function scrubbing is still cheaper, however, and is likely to be used in the future if the problems can be worked out. Particulate scrubbers may be involved when the fuel sulfur content is low enough to make SO_2 removal unnecessary (at least at the time of planning) because the same low sulfur makes precipitator cost high (because of the low dust conductivity). Such scrubbers do not absorb much SO_2 from the gas because the SO_2 content of the recycled water soon comes to equilibrium with that in the gas and no more is absorbed.

The resulting low pH, however, causes both corrosion and scaling—the latter because calcium dissolved from the ash reacts with dissolved SO_2 to form calcium sulfite which oxidizes rapidly because the low pH holds it in solution and thus available for oxidation. The resulting calcium sulfate then crystallizes on scrubber surfaces. It is now the general practice to raise the pH in the scrubber by adding lime to the scrubber loop. Since the lime absorbs SO_2 a relatively large amount must be added to get the pH high enough, with the result that SO_2 removal is obtained incidentally.

Gas Precooling

If it is assumed that precooling of the gas is helpful in countercurrent scrubbing, as small-scale tests indicate, there are various methods or precooling that can be used.

Indirect Cooling in a Low-Level Economizer: The gas can be cooled by a heat exchanger after the air heater. Corrosion from condensation of H_2SO_4 then becomes a problem and use of corrosion-resistant material is indicated for the heat exchanger (economizer). Even with corrosion inhibition, however, deposit formation from interaction of H_2SO_4 and ash or lime could be a problem. Indirect cooling has been considered mainly as a way of getting heat for reheating the gas. Cooling in this way has a secondary advantage in that the wet bulb temperature of the gas is reduced by a few degrees and thus the scrubbing temperature is lower.

Direct Humidification: Spraying water directly into the gas has the drawbacks of possible corrosion of ducts or formation of deposits if the water strikes the wall of the duct. It is not easy to humidify a gas completely without using excess water beyond that which will evaporate. It is not essential, of course, that the gas be cooled all the way to the wet bulb temperature before it reaches the scrubber; the choice would be between giving up some of the precooling advantage and accepting the wetted duct problem.

Prescrubber for Dust Removal and Humidification: Another solution is to use two stages of scrubbing and design the first one primarily for dust removal and humidification. A venturi is known to be a good humidifier, although other types should be effective also. Such an arrangement has the problems discussed earlier.

Cooling below the wet bulb temperature, to a temperature approaching that of the cooling water, would improve scrubbing but it appears that in most cases the cost would be too high. Indirect cooling from air heater exit temperature to below the wet bulb would require very large heat exchangers, and direct cooling with water would involve a very large cooling load for condensing moisture from the saturated gas in going downward from the wet bulb temperature.

Deposition of Soft Solids

One of the major problems in operating lime/limestone scrubbers is the tendency of slurry solids to stick to surfaces. The resulting mud deposition can produce large accumulations of solids in a relatively short time. One of the main trouble spots is the point at which the gas passage walls (duct or scrubber) change from a dry to wet condition. Spattered solids tend to dry out at that point and adhere to the surface. If a venturi is used for the first scrubbing stage the wet-dry interface is at the venturi entrance and may not cause trouble. However, there can be spattering within the unit and resulting buildup on the venturi ceiling.

If a single scrubber is used, the wet-dry interface is likely to be in the duct entering the scrubber because of slurry drops bouncing back into the duct against the gas stream. The usual remedy is to use a sootblower type of operation (jets of air, steam, water, or slurry) to dislodge any deposits and blow or wash them back into the scrubber.

Mud-type deposits can also form on any equipment such as supporting members or scrubbing elements within the scrubber, usually on the underside from slurry drops carried upward with the gas and particularly when the gas velocity in the scrubber is high. Also, if the scrubber design is such that the gas makes a change in direction between the scrubbing section and the mist eliminator, deposits are likely to form in the bend. For all such situations use of washing jets is indicated.

Scaling

Crystallization of calcium sulfite or calcium sulfate on scrubber surfaces was the main problem in early lime/limestone work and continues to be a major consideration although considerable improvement has been made. Scaling is a very complex phenomenon with many interrelated factors affecting it. The total system is not yet fully understood but considerable progress has been made in the past 2 or 3 years, both in elucidating the basic mechanisms and in identifying the critical process variables.

Basic Factors: In a typical lime/limestone scrubbing system, a large amount of slurry enters the scrubber at the top, flows downward in contact with the gas, passes to a reaction tank where fresh absorbent is added, and returns to the scrubber. A bleed stream flows from the reaction tank to a thickener or waste pond where the product solids settle out; the supernatant liquor is returned to the scrubber system.

In such an arrangement the slurry circulating through the scrubber normally contains crystals of both calcium sulfite and calcium sulfate. Sulfite is formed in the scrubber and goes in and out of solution as the pH changes at various points in the circuit. Sulfate, which is not affected much by pH, forms both in the scrubber and the reaction tank by oxidation of sulfite and crystallizes whenever the supersaturation gets so high that the solution can no longer hold it. The crystallization occurs preferably in the reaction tank, where it ordinarily does no harm, but can occur in the scrubber and either plug gas flow openings or form masses that eventually drop off and plug the liquor circulation system.

Scaling can be caused either by calcium sulfite or sulfate, or by calcium carbonate, for that matter, if lime is the absorbent. Of these, sulfate scaling is normally the most difficult to control but sulfite and carbonate must also be considered in system design.

Sulfite Scaling: In lime systems, sulfite scaling can occur if the pH of the slurry returned to the scrubber is unusually high, say above 10. There is usually some back mixing in scrubbers, especially in a tray type such as the marble bed used at Paddy's Run; when the sulfite-laden solution already in the scrubber back mixes with the entering high-pH slurry, the sharp rise in pH throws sulfite out of solution as crystals that may attach to surfaces and form scale. The remedy is to keep the entering pH at a level on the order of 9 or less. The actual critical level is not known very well and probably depends on the type of scrubber, which affects the degree of back mixing. This area of lime scrubbing technology does not appear to be understood very well.

There are some limitations as to how much the return pH can be controlled. The elevation of pH in the reaction tank is due to addition of the makeup lime, a quantity that cannot be varied very much if it is desired to keep the addition near the stoichiometric amount. The actual pH depends on the following.

SO_2 Content of the Inlet Gas – For high inlet SO_2 concentrations more lime must be added than for low concentrations at a given ratio of liquor circulated to gas throughput (usually stated as L/G, gal/Mcf).

L/G – At high L/G there will be more liquor for the amount of lime added, thus reducing the pH rise. For very high inlet SO_2 concentrations such as encountered in smelters or in power plants burning very high sulfur coal it may be necessary to use high L/G because otherwise the stoichiometric feed would raise the return pH to an intolerable level. This would reduce the attractiveness of lime in comparison to limestone because one of the major advantages usually attributed to lime is the low L/G required.

Delay Time – The return pH can be reduced to a minimum by providing adequate retention time in the reaction tank for the lime to react completely with the scrubber effluent solution. In the scrubber, as noted earlier, bisulfite species form as well as calcium sulfite; if there is adequate time in the reaction tank the added lime will convert bisulfite completely to calcium sulfite and, at the high pH, most of the sulfite will precipitate. Low retention time is more of a problem with limestone because it reacts more slowly. However, even with lime the delay time can be too short for complete reaction and attainment of minimum pH.

Absorbent Feed Ratio – If the system is so designed that excess absorbent must be used to get the desired SO_2 removal then return pH will be higher and scaling may result. Conversely, less than the stoichiometric amount can be used to reduce pH but at the expense of lower SO_2 removal.

Stirred vs Plug-Flow Reactor – In the usual stirred reaction tank there is some bypassing of the added lime to the tank outlet. As a result some of the solution does not get reacted adequately with the lime and the pH will be higher in the solution entering the scrubber. In contrast, if the lime is mixed with the solution and the mix then passed in plug flow through a reactor without back mixing the bypassing is eliminated.

In limestone scrubbing, high return pH is not a problem because even a high excess added in the reaction tank does not increase the pH very much. The main problem is that either an excess or insufficient limestone dissolution in the scrubber can cause too high a pH at the scrubber outlet. As the slurry flows down through the scrubber the dissolved sulfite concentration increases rapidly but is held in solution because the pH is decreasing. If the pH does not decrease enough then sulfite can precipitate in the lower part of the scrubber. The L/G, of course, is also an important factor in this.

This is not so much of a problem with lime because of the smaller amount of sulfite returned to the scrubber. At a pH of 8 to 9 there is very little sulfite in solution in the return slurry, whereas with limestone the return pH of 6.0 to 6.5 gives a considerable amount that increases the total concentration at the bottom of the scrubber.

The remedy for sulfite scaling in limestone scrubbing is to keep the excess at as low a level as possible and to adjust other variables in the proper direction. Fortunately, sulfite supersaturates to such a degree that sulfite formation in the scrubber can exceed solubility by severalfold without crystallization occurring until the solution reaches the reaction tank. If it were not for this, sulfite scaling could be a major problem. In practice it seldom occurs in limestone scrubbing.

Carbonate Scaling: It has been shown in small-scale tests that a high-pH lime slurry fed to the scrubber can react with CO_2 in the gas with resulting scaling of calcium carbonate on scrubber surfaces. The net situation is obscured by the fact that high pH can also cause sulfite crystallization. Because of this, it is not clear whether carbonate scaling is a significant problem or not. If it is, then keeping the return pH below some critical level appears to be an adequate solution to the problem.

Sulfate Scaling: Unlike sulfite, the decrease in pH down through the scrubber does not help hold sulfate in solution. Thus if there is a considerable amount of sulfite oxidation in the scrubber sulfate crystallization can occur, especially in the lower portion. Moreover, the increase in calcium content as the lime or limestone dissolves tends to throw calcium sulfate out of solution.

Another difference, in comparison to the lime system, is that high pH in the return liquor does not help as it does for sulfite, by decreasing solubility and throwing the dissolved salt out of solution. In the case of sulfate, the return solution is more likely to be supersaturated because sulfate tends to supersaturate to a high degree (although not as much as for sulfite) and complete dissipation of the supersaturation may not occur in the reaction tank. Moreover, it has been shown that a major part of the total sulfite oxidation can occur in the reaction tank.

The traditional method for preventing sulfate scaling, as prescribed by Imperial Chemical Industries in the early English work, involves three operating variables. (These are also effective for preventing sulfite scaling but as noted earlier sulfite is seldom a problem if the return pH is not too high.)

L/G Increasing the slurry pumping rate reduces the amount of sulfate picked up per unit volume of solution in the scrubber for a given oxidation rate. ICI used an L/G of over 100 for an inlet SO_2 concentration of about 2,000 ppm. In the U.S. somewhat lower L/G's are normally used. EPA pilot-plant results have indicated an L/G of 65 as the minimum for limestone scrubbing. Although there are not yet enough data for cost optimization, use of high L/G appears to be a good way both to reduce scaling and to get high SO_2 removal efficiency.

Crystal Content in Slurry – If the circulating slurry contains calcium sulfate crystals in suspension any crystallization of sulfate in the scrubber is more likely to occur on the existing crystals than on scrubber surfaces. In the ICI work tests showed that, for a retention time of 5 minutes, even 0.5% (by weight) of $CaSO_4 \cdot 2H_2O$ (gypsum) crystals would reduce the sulfate content in solution by about half—as compared with no crystals. A higher crystal content was helpful up to 5%, beyond which no further advantage was gained. At 5%, the $CaSO_4$ remaining in solution was only about 1,800 ppm as compared with about 5,900 ppm without crystals. ICI specified 5% gypsum in suspension for the Fulham plant operation.

In the U.S. plants, practice has varied widely in regard to sulfate crystal concentration. For plants in which part or all of the dust is collected in the scrubber there is some limit to the salt crystal content because the dust makes up part of the total allowable solids. The upper limit for solids is usually considered to be about 15% because of the increasing viscosity and stickiness at higher levels. However, EPA studies have shown that the presence of fly ash is also helpful in preventing scaling and that the total solids content should be 10% or higher.

Retention Time in Reaction Tank – For the usual type of operation the solution leaving the scrubber is supersaturated with gypsum even at high L/G and solids content. If the degree of supersaturation does not exceed a critical level, the magnitude of which depends on several considerations, scaling will be absent or at a minimum. In the reaction tank outside the scrubber, however, enough time must be allowed for the supersaturation to dissipate; otherwise the high sulfate content returned to the scrubber will soon cause a steady-state situation in which the critical degree of supersaturation will be exceeded in the lower part of the scrubber and scaling will occur.

ICI test work indicated that 5 minutes would be a safe retention time. EPA recommends about 10 minutes. As more is learned about the cost-benefit relationships it should be possible to optimize the reaction tank size against the other factors involved.

By proper adjustment of these parameters it has been possible to minimize sulfate scaling and get reasonably acceptable operation. However, some scaling still occurs; in TVA pilot-plant work with limestone, for example, a thin layer of scale usually forms on the scrubber walls, reaching a steady-state thickness. It is difficult to analyze the situation in the large-scale systems because of the many factors involved and the difficulty in sorting sulfate scaling out of the other problems encountered. It appears likely, however, that there will always be a potential problem when high levels of L/G, solids content, and reaction time are depended on to prevent scaling.

A major factor in this is the amount of fresh water that can be added to the system. It has been well demonstrated, in the TVA pilot plant and several others, that open-loop operation can prevent scaling that would otherwise occur. By open loop is meant discharge of liquor from the system, where the system is defined as the scrubber loop plus the discarded solids. If, for example, the liquor accompanying the solids to a settling pond can be overflowed from the pond to a watercourse rather than returned to the scrubber, the additional amount of fresh water that can be used is a major factor in preventing scaling.

Most of the full-scale lime/limestone systems operating both in this country and abroad are fully or partially open loop because the water pollution regulations allow it. This situation is not likely to continue, however, in which case these installations, and others that are being built like them, are likely to have more trouble with scaling. As noted earlier, recent EPA work has pointed the way to a new procedure for avoiding sulfate scaling–namely, the sulfate unsaturated mode of operation. It appears that in some large-scale operations this method of preventing sulfate scaling has occurred by chance. Data on dissolved sulfate concentration from such installations indicates unsaturation. Apparently the equipment design and other factors produced a degree of oxidation low enough to stay within the unsaturated sulfate operating mode.

In TVA pilot-plant tests good results were obtained by keeping oxidation low enough to stay in the unsaturated area. It appears that this is a major step forward in managing sulfate scaling, especially for plants burning high-sulfur fuel.

For low sulfur the relatively high oxygen:SO_2 ratio in the gas will make it difficult to keep oxidation low enough unless a combination of low excess air, use of lime, proper reactor design, and a high concentration of magnesium keeps the system within bounds. The situation will be aggravated at plants where blowdown from cooling towers is used as makeup water and therefore relatively large amounts of chloride are brought into the system.

Scrubber Design and Performance

Scrubber selection is an important consideration in designing lime/limestone systems because a compromise must be made between absorption efficiency on the one hand and operating reliability on the other. Several other factors also enter, giving a group of tradeoffs that have not yet been optimized. Some of the general considerations in scrubber selection are as follows:

Mass Transfer: The usual fixed type of packing gives good mass transfer but is more subject to plugging and scaling than other types. High gas velocity is helpful also but in vertical scrubbers is likely to carry too much mist up into the mist eliminator.

Slurry Holdup: Since limestone dissolves at a relatively slow rate, it is helpful to have a large holdup of slurry in the scrubber so as to provide a large limestone surface area in contact with the solution. Scrubbers differ in the amount of holdup provided.

Dust Removal: In some installations the scrubbers must remove part or all of the dust in addition to the SO_2. This requires a scrubber with some pressure drop and even more resistance to plugging.

Turndown: Unlike most scrubber installations, those attached to SO_2 emitters such as power plants and smelters are subjected to widely varying gas flow. This can be taken care of by installing several scrubbers on each boiler and shutting off one or more if the gas flow drops. The operation is simpler, however, if all the scrubbers can be left on the line and the flow through each reduced, which requires a type that can operate efficiently at a gas velocity less than that for which it was designed.

Gas and Liquor Distribution: For lowest investment it is desirable to make the scrubbers quite large, but the cross section may become so large as to give problems in getting uniform gas and liquor flow. Some scrubber types do much better in this respect than others.

No one type of scrubber gives the best performance for each of these criteria. Each design group in the field has made a judgment as to which gives the best overall performance and has worked at removing or minimizing its faults. In the following analysis an attempt is made to evaluate the current status of the various types, but it should be noted that more experience and operating data must be obtained before definitive judgements can be made.

Fixed Packing: Several fixed packing scrubbers are known to be in service particularly in Japan. The only one known to be operating in the U.S. is the one at Cholla (Arizona Public Service) built by Research-Cottrell (Figure 3.4).

FIGURE 3.4: VENTURI-FIXED PACKING COMBINATION TESTED BY RESEARCH-COTTRELL

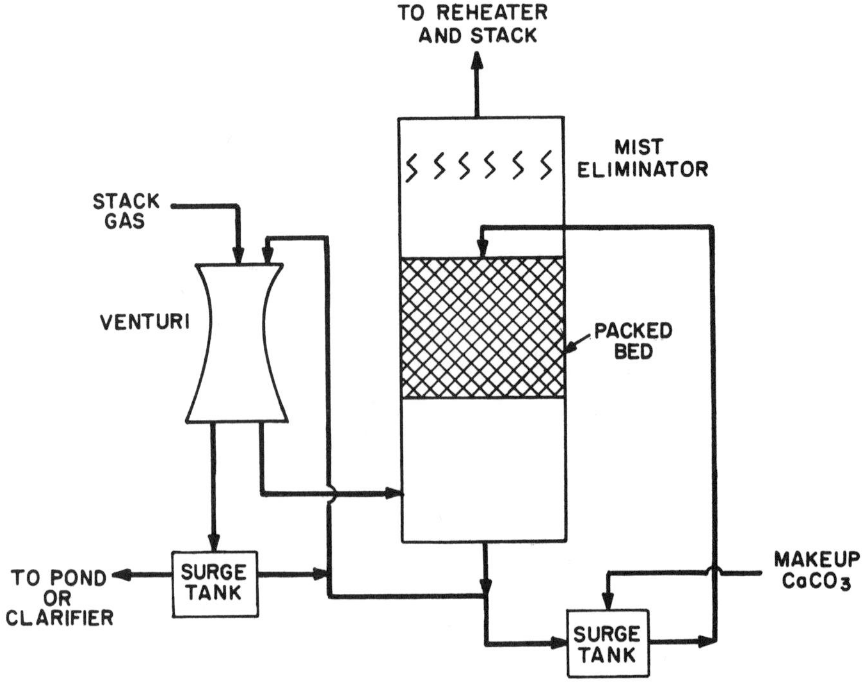

The main functions of the venturi and the fixed packing section are to remove particulate and sulfur oxides, respectively. The fixed packing is of the open type.

An open packing (cooling tower type) is used. Absorption efficiency is good but even with low-sulfur coal and open-loop operation scaling in the packing has been a problem. It may be possible to avoid this by the methods described earlier.

Mobile-Bed Packing (Hollow Spheres): Mobile packing has the advantage of good mass transfer with little chance for plugging because the movement of the spheres keeps solids from adhering. UOP uses the TCA type (ping pong balls) (Figure 3.5) and has installed several units for Public Service of Colorado. These are for particulate only; units for SO_2 have been installed at smelters in the United States and Japan, and UOP has several systems under construction at utilities in the United States.

Good removal of both SO_2 and particulate has been obtained in various pilot-plant tests and in the EPA-TVA test facility at Shawnee. The main drawbacks are excessive ball wear and limited gas velocity. Unless a ball with acceptable life can be developed other scrubber types may be preferable.

FIGURE 3.5: MOBILE BED SCRUBBER OF THE TCA TYPE

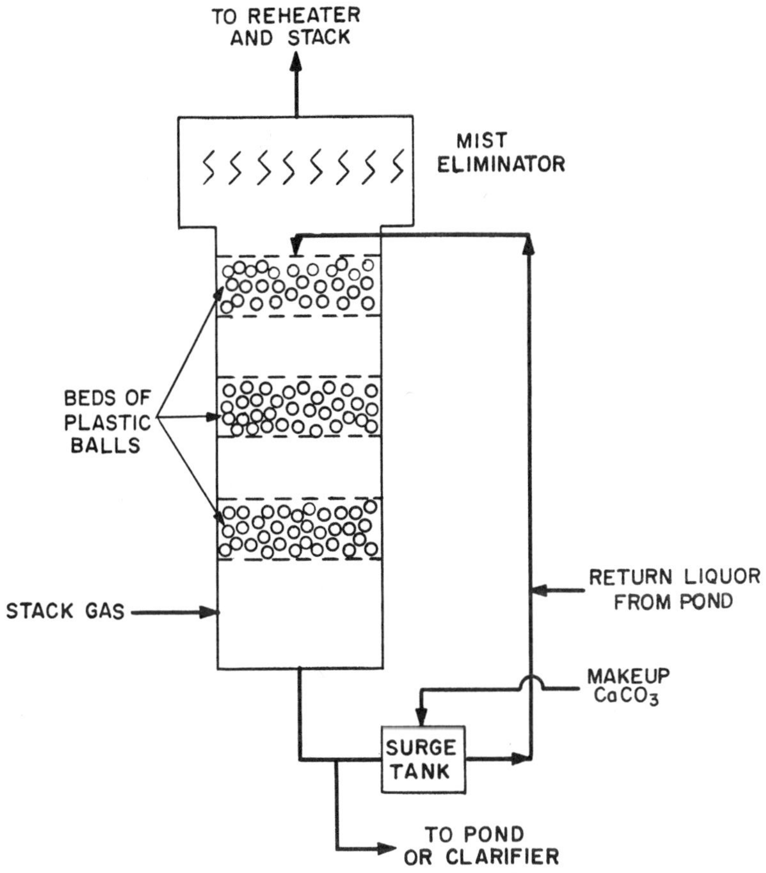

Both sulfur oxides and particulates are removed in the three beds. Use of such a scrubber in limestone-slurry scrubbing has been proposed by Universal Oil Products.

Mobile-Bed Packing (Solid Spheres): Combustion Engineering has used the marble-bed scrubber (originally developed by Environeering) in several installations, including Lawrence, Hawthorne, Paddy's Run, and several under construction (Figure 3.6). The glass spheres do not move as vigorously as the TCA hollow balls but nevertheless move enough to keep themselves clean under most conditions. Mass transfer is good (if two or more beds are used) and particulate removal seems to be adequate. Problems are the very limited turndown, difficulty in maintaining the bed at uniform depth, and some tendency to plugging at high-slurry solids or under scaling conditions. The pressure drop is also somewhat high unless particulate removal is required.

FIGURE 3.6: TRAY TYPE SCRUBBER WITH BED OF GLASS SPHERES (MARBLES) CARRIED ON THE TRAY

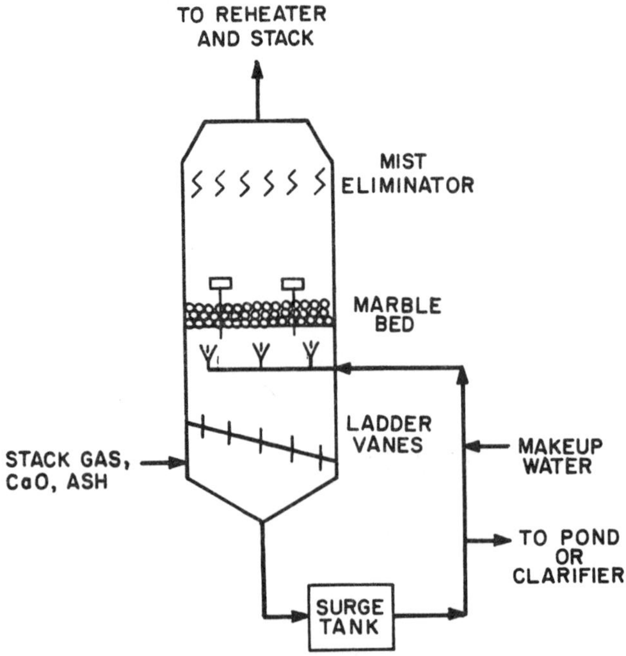

This type was developed by Environeering and is used in the Combustion Engineering process. Both sulfur oxides and particulates are removed in the same scrubber.

Entrained Packing (Spheres): American Air Filter has developed a scrubber in which spheres (solid balls) are entrained in the gas stream, passed through the scrubber, and disengaged for recycling. Little is known about the performance since it has been tested only on a pilot-plant scale. A unit is under construction at Green River (Kentucky Utilities) and Louisville Gas and Electric will use this design.

Sieve Tray: Babcock and Wilcox uses a perforated plate scrubber that seems to fit into the sieve tray class. Several plates spaced in the tower provide a body of slurry on each plate that is violently agitated by the gas coming up through the holes. B and W has installed this design at Will County, LaCygne, and a plant in Japan. Mass transfer appears to be good but turndown is questionable; gas velocity is limited, and the internals increase both cost and the possibility of plugging.

Environeering and TVA have designs that probably fit into the sieve tray classification, although there appears to be less of an agitated foam layer than in the B and W type. Environeering uses closely spaced rods and TVA a grid or screen.

There is some question whether either does as well in regard to mass transfer as the perforated plate.

Spray Tower (Vertical): Ontario Hydro and Peabody Engineering use a simple spray tower. The scrubber being installed at St. Clair (Detroit Edison) will be of this type. Maximum protection against plugging is provided and turndown is good, but mass transfer is relatively poor. Nozzle wear may also be a problem because there is more need to break the slurry up into fine drops than in scrubbers that depend on internal elements to get good gas-slurry contact.

Spray Scrubber (Cross Flow): One of the scrubber modules at Mohave is a cross-flow (horizontal) type in which the slurry is sprayed downward through the horizontally flowing gas. This is the only type that is staged, that is, the slurry goes in and out of the scrubber several times during its progress from the inlet to the exit. The cross-flow arrangement apparently allows much higher gas velocity without undue mist development than in vertical scrubbers.

At least the Mohave scrubber is operated at over 20 feet per second whereas the vertical types generally must be operated at not more than 8 fps. The higher velocity not only reduces scrubber size and cost but also apparently increases mass transfer to the point that the lack of packing or other internals is offset. Nozzle wear is a problem and more area is required than for vertical scrubbers. On balance, the cross flow appears superior to all the other types but further operation, particularly on high-sulfur coal, is needed to confirm this conclusion.

Venturi: Chemico uses the venturi type for SO_2 absorption and other designers may have a venturi ahead of a vertical scrubber when particulate removal is required. In most such cases the slurry passes from the main absorber to the venturi and thus the latter does part of the SO_2 removal. The main advantage is in situations where high particulate removal is required in which case the venturi may be the only type capable of the high efficiency needed. It should be noted that this is subject to some question because tests at Shawnee and elsewhere have indicated a very high particulate removal by the TCA.

The venturi has no internals, velocity is high, and there is no nozzle wear. However, turndown requires a variable throat mechanism and retention time is so short that two venturis in series have been required for adequate SO_2 removal [Mitsui Aluminum, Phillips (Duquesne Light), and Mansfield (Ohio Edison)]. Tests are underway with additives and sprays in the mist knockout section to determine if one venturi can do the job. If these are successful then the position of the venturi would be greatly improved.

It should be noted that the development work almost continually underway on most of these scrubber types may remove some of the shortcomings noted.

Absorption Efficiency

Ideally, the absorption efficiency should be high enough for essentially complete SO_2 removal with the stoichiometric amount of lime or limestone at high gas velocity and low pumping rate. This ideal can be approached in alkali scrubbing (soluble sodium or ammonium compounds) but in lime/limestone scrubbing the low solubility of the absorbent presents a problem. It is necessary to get enough Ca^{++} in solution, by dissolving $CaCO_3$, $Ca(OH)_2$, or $CaSO_3$, to react with the bisulfite species formed in the scrubber. The scrubber design is a factor in this, as discussed earlier, but there are other considerations.

Excess Absorbent: In installations where excess absorbent is used, there is a significant amount of unreacted absorbent in the slurry leaving the scrubber. To avoid this and approach full utilization the amount fed is cut until there is a considerable excess of SO_2 over absorbent as compared with the stoichiometric ratio, where the degree of SO_2 removal is likely to be unacceptably low. Experience with this in the boiler-injection systems where the limestone feed was cut back to below the stoichiometric ratio (quantity of Inlet SO_2) in order to prevent scaling by reducing scrubber pH; very good limestone utilization was obtained but SO_2 removal was only 55 to 60%.

For the 80 to 90+% SO_2 removal generally required, the usual practice is to feed an exess over stoichiometric (where stoichiometric is defined as the amount theoretically required to convert all the inlet SO_2 to calcium sulfite). The excess required depends on numerous factors including scrubber design, L/G, SO_2 concentration in the inlet gas, lime vs limestone, and the nature and concentration of the various species in solution. The technology is not advanced far enough to quantify the effect of these parameters so that the least-cost combination can be determined. The excess fed in most installations ranges from 10 to 50% over stoichiometric. Typical absorbent utilization, assuming 85% SO_2 removal, thus is 57 to 77%.

In some cases, basic constituents in the ash constribute absorbing capacity. At Four Corners, for example, the particulate scrubbers remove 20 to 30% of the SO_2 even without lime addition to control pH. There are some limits in using excess absorbent to achieve high SO_2 removal or to offset adverse effects of other parameters. With lime, for example, the excess raises the return pH and thus can cause carbonate or sulfate scaling. Moreover, the high pH can precipitate Ca^{++} (as $CaCO_3$) and Mg^{++} in the upper part of the scrubber, both of which can reduce absorption efficiency. With limestone, a large excess can raise pH at the scrubber outlet enough to produce sulfite scaling.

Slurry Circulation Rate: Use of excess absorbent not only increases absorbent cost but also produces more sludge. There is growing evidence that providing more retention time in the scrubber loop, particularly in the scrubber, is a better approach. In addition to the scrubber holdup obtained in overflowing tray types such as the marble bed, increase in L/G is also a good way of increasing slurry retention time in the scrubber. Most designers have increased the L/G considerably over the early levels thought to be adequate and some are now up to the 75 to 100 range. Pumping cost is higher but the limited data available indicate that high L/G is the best of the tradeoffs.

In this respect low inlet SO_2 (as in burning western coal) seems to have an advantage since the same recirculation rate as for high inlet SO_2 normally is used (for other reasons) and thus the slurry rate per unit of SO_2 absorbed is relatively high. This has been borne out in the Southwest, where at the Cholla plant, for example, good removal (90%) has been obtained with only 80% of the stoichiometric amount of limestone (with some help from basic species in the ash).

Reactor Tank Design: The size and type of reactor tank can affect scrubbing efficiency as well as scaling in the scrubber. Bisulfite species formed in the scrubber must have time in the reactor for adequate conversion to sulfite and a plug-flow reactor will assist in this. EPA tests with limestone have shown that under otherwise comparable conditions plug-flow resulted in 76% SO_2 removal as compared with 65.5% for a backmixed reactor. The optimum reactor tank size is a tradeoff among many considerations, none of them well quantified. From 10 to 15 minutes' retention time seems to be as good a choice as any for L/G's in the 50 to 60 range.

Concentration of Magnesium Species: The presence of dissolved magnesium species is highly beneficial to scrubbing efficiency. Since such species do not normally precipitate in the scrubber loop, they build up in concentration and, if not already in the sulfate form, become converted to magnesium sulfate by oxidation. Thus high levels of both magnesium and sulfate can be developed in the solution unless the amount of liquor leaving the system with the waste solids is so high as to keep the concentration at a low steady-state level.

As noted earlier, the magnesium and sulfate shift the solution equilibria in such a way that the concentrations of basic species in solution, especially $SO_3^{=}$ and $MgSO_3^{\circ}$ (which absorb SO_2 by conversion to HSO_3^-), are greatly increased. As a result the solution returned to the scrubber has a greater absorption capacity. It can be shown that this effect can be great enough to avoid the need for dissolving any $CaCO_3$ or $CaSO_3$ in the scrubber, which then eliminates the kinetics of solids dissolution in the scrubber as a controlling mechanism. With very high concentrations of $MgSO_3^{\circ}$, for example, the system approaches magnesia scrubbing.

The magnesium effect is just coming into the picture as a tool for improving lime/limestone processes and there are many questions still unresolved. The benefits appear to be greater for limestone scrubbing because Ca^{++} is more difficult to get into solution than in lime systems. However, supplying magnesium in the form of dolomite ($MgCO_3 \cdot CaCO_3$) or dolomitic limestone, which is the logical course, is questionable because of the low dolomite reactivity in a limestone system. Some of the magnesium dissolves, however, and this would accumulate to the desired high level unless the purge in the liquor leaving with the solids kept the concentration low. Further tests are needed to evaluate this situation.

A soluble magnesium salt can be added to limestone systems as an alternate to using dolomitic limestone. Magnesium sulfate is the logical choice because of its relatively low cost and good availability. In such case it would be desirable to separate the waste solids and wash them to minimize additive loss with the cake.

For lime systems dolomitic lime is probably the most economical source because the MgO would be relatively inexpensive and would be effective as an initial absorbent (in contrast to $MgSO_4$). The benefit would be less than in limestone scrubbing but should still be substantial, especially in allowing lower L/G or for making a single venturi adequate where two in series are needed for ordinary lime. If the magnesium is supplied as dolomitic lime or limestone there is the question of water pollution. Since magnesium is continually supplied, as compared with a single charge of $MgSO_4$, magnesium salts must be discharged with the waste solids. Since the magnesium compounds are much more soluble than their calcium counterparts, pond leaching or overflow would introduce more dissolved solids into ground water or watercourses. Various means for coping with this problem are available.

1. If the waste solids are stabilized as may be required (see later) the Mg should be tied up in a nonleachable complex.
2. If an impermeable pond lining is used to prevent leaching of calcium salts, as also may be required, it should also prevent magnesium leaching.
3. Precipitation of insoluble $Mg(OH)_2$ could be induced by raising the pH. If excess absorbent is used in lime systems, the slow increase in pH generally noted in sludge storage might eventually cause $Mg(OH)_2$ precipitation. More data are needed to evaluate this possibility.

Chloride Concentration: Since coal usually contains 0.1 to 0.3% chloride the concentration of chloride in the circulating slurry can also build up to a significant level. At Shawnee the chloride steady-state concentration reached levels as high as 10,000 ppm in closed-loop operation. Additional chloride can be introduced if cooling tower blowdown or sea water is used as makeup water.

Chloride has an opposite effect to magnesium in regard to basic species concentration and consequent absorption efficiency. The only methods for reducing the effect are to increase the purge rate or to add more magnesium to the system.

Mist Separation

In any wet scrubbing system the gas picks up mist droplets and carries them out of the scrubber unless a mist eliminating device is interposed. Such separation is essential because the mist would otherwise corrode and scale downstream equipment in addition to requiring additional energy in the gas reheater to evaporate the droplets.

Fortunately the mist drops are fairly large, on the order of 40μ or so, and thus can be removed adequately by impact on surfaces that change the direction of gas flow. The most common device is a set of baffles or slats set in such a way as to impart a zigzag flow to the gas over a distance from a few inches to a foot or so. The assembly is called a chevron mist eliminator. Imparting a cyclonic motion to the gas is also effective since the centrifugal force throws the mist out onto the collector wall; both vane-type and cyclonic chambers are used.

Since chevrons are relatively inexpensive, have low pressure drop, and take up little room they are used in most of the lime/limestone scrubbing installations. The usual arrangement is to install the chevron in the top of the scrubber in a horizontal position, perpendicular to the gas flow. At least three passes (individual baffles, each turning the gas to a new direction) are normally used, and there may be two or more banks (sets of baffles) with some distance between. The main drawback to the horizontal position is that the upflowing gas tends to hold separated mist in the chevron rather than allowing it to drain back into the scrubber. As a result mist can both stick and be re-entrained into the gas. Tests at TVA and elsewhere have shown that much better performance is obtained if the slats are slanted on the longitudinal dimension or set vertically so that the mist can drain off. This can be accomplished either by placing the unit in a horizontal duct after the scrubber or using a slanted arrangement in the top of the scrubber something like the rafters of a house.

If a horizontal chevron is used the gas velocity is limited to about 8 feet per second to avoid excessive hangup and re-entrainment, whereas tests have shown that 10 fps is necessary to get adequate impact on the baffles and that up to 20 fps is feasible if the unit is in a vertical position. The vertical type will be used in some of the installations under construction including Widows Creek (TVA) and Duck Creek (Central Illinois Light and Power), and Ontario Hydro has made pilot-plant studies. Combustion Engineering has done development work on the slanted type and has used it in all installations.

The cross-flow scrubber is especially suited to vertical chevron installation because the unit can be mounted in the end of the scrubber without any duct bend. The bend required for a vertical scrubber gives trouble because soft solids sticking in it may change the gas flow pattern enough to cause nonuniform gas flow through the chevron.

Even with the best operation, solids accumulate slowly in the duct after the demister. Another advantage of vertical positioning is that these deposits cannot fall back down into the demister and block it. It is reported that a fixed-vane demister will be used at St. Clair (Detroit Edison). A similar type has been used by STEAG in Germany with good results. Cyclonic chamber collectors are quite effective, even at high gas velocities, but do not appear to have been used in SO_2 scrubbing. They are relatively large and probably more expensive than the other types.

Mist eliminators must be washed because the solids-laden drops otherwise would stick and accumulate on the impact surfaces. Adequate washing is not easy to accomplish. Getting the wash liquid to all areas with sufficient intensity is difficult, not only because of structural considerations but also because not enough of the proper type of wash liquid may be available. There are several possibilities.

Slurry Wash: Use of slurry is attractive because there is plenty of it. Slurry wash is used at Mohave and has been tested by Ontario Hydro with good results; both demisters are vertical chevrons. However, the slurry wash is not the only one; in both cases the areas near the end of the zigzag path, where the wet-dry interface is located, are washed with fresh water to prevent soft solids build up.

Supernatant Liquor Wash: Although some fresh water can be added to the scrubber system (see later discussion of the water balance), in some installations this is used for other purposes such as in pump seals and for humidification. The only other clear liquor available, supernatant from the clarifier or settling pond, is used for demister wash. This has generally resulted in sulfate scaling on the chevron baffles, although the situation is affected by many factors. Supernatant liquor washing may be feasible if the unsaturated-sulfate mode of operation is used. However, if the liquor hangs very long in the chevron it could absorb enough SO_2 and develop enough sulfate to become saturated. A slanted or vertical chevron would be preferable. Some have used a mixture of fresh water and supernatant to reduce scaling.

Recycled Fresh Water Wash: To conserve the limited amount of fresh water available some have installed a liquor-collecting device under the chevron to catch the wash liquid so that it can be recycled, thus increasing the amount of wash liquor available per square foot of demister area per minute. Fresh water is added to the loop and blowdown removed.

The wash tray principle will be used by UOP in installations under construction and is being tested at Shawnee. Results have not been as good as expected. Deposition still occurred in the demister (but to a lesser extent), apparently because of the accumulation of solids, both suspended and dissolved, in the wash loop. There was also heavy accumulation of soft solids on the underside of the wash tray, requiring soot blowing for removal. It appears that wash recycling will delay demister plugging but may not prevent it completely.

Intermittent Fresh Water Wash at High Intensity: The above systems generally involve continuous washing at ordinary spray nozzle pressure. Another method is to wash intermittently but with an intense, jet-type spray—similar to soot blowing. Such a system has been used successfully by Combustion Engineering at Paddy's Run and in the Mohave crossflow system by Southern California Edison. Combustion uses a standard soot blower arrangement and SCE a set of sprays that are turned on and off in rotation. In both cases a single short wash period (30 seconds to 10 minutes) every 8 hours is adequate.

This appears to be the best of the several systems. Apparently soft solids can be allowed to accumulate for some time and still be removable by a jet. The use of once-through fresh water prevents scaling, and the amount of water required, which is small because of the limited wash duration, is well within the amount available in a closed-loop system.

Stack Gas Reheat

One of the major problems in scrubbing stack gas to remove pollutants is that the gas is cooled and saturated with water vapor in the scrubber operation, which brings on several difficulties. From the environmental standpoint the main drawback is the adverse effect on plume dispersion. Cooling the gas below its usual temperature reduces both the momentum and buoyancy which cause the plume to rise above the stack after it is emitted.

The effective stack height is thus reduced and ambient concentrations are higher than they otherwise would be. This should not be serious if the objectionable components are removed essentially completely, but such efficiency is unlikely. For a typical degree of removal, say 80%, the ambient concentration is not reduced by 80% but by some lesser percentage because of the offsetting effect of gas cooling.

Other adverse effects of cooling are a higher incidence of steam plume formation in cool weather and the possibility of acid rain—condensation of droplets near the stack that absorb residual SO_2 and bring it to the ground. Because of these problems, plus the need to protect downstream equipment from acidic condensate, reheating the gas has become standard practice in modern scrubber installations.

The reheat problem is confused by the fact that no rules have been established by regulatory authorities as to how much reheat is needed. Therefore the various scrubber installations now operating or under construction have a wide range of reheat temperature. Various factors have entered into selection of the reheat level.

Protection of Downstream Equipment: From the practical design standpoint, reheat may be indicated to protect the fan and stack from condensation. The gas leaving the scrubber is saturated and at the wet-bulb temperature. Incidental cooling after the scrubber causes condensation, and the condensate becomes acidic by absorbing some of the residual SO_2 from the gas. Moreover, if there is no reheat then mist droplets not caught in the scrubber mist eliminators will collect in ducts, fan, and stack with resulting corrosion.

The degree of reheat necessary to avoid condensation and to evaporate mist droplets varies with the situation (amount of heat lost through equipment walls, mist eliminator efficiency, gas temperature at scrubber outlet) and can be calculated if adequate data are available. Although no general rule has been established, several companies have chosen 50°F as the reheat increment. A major consideration is how much heat is required to evaporate mist droplets. If the fan is adequately protected from mist it will add to the reheat because of the energy imparted to the gas. A temperature increase on the order of 10°F is usually obtained.

Inadequate reheat has been a major problem in some installations. At La Cygne (Kansas City Power and Light), for example, the reheaters were designed to increase the gas temperature about 25°F (to 150°F). This has not been sufficient to evaporate the mist, with the result that the reheaters have corroded severely and unevaporated mist has damaged the fans.

Even with reheat, some designers fear to put the fan downstream of the scrubbers because of possible damage. Placing them ahead of the scrubbers has drawbacks, of course, including larger fan size and greater wear from dust if the scrubbers are being depended on for any significant degree of dust removal.

Wet Stack Operation: Because of the uncertainty in protecting downstream equipment by reheat, plus the high cost of installing and operating a reheat system, some designers have gone to a wet system with no reheat at all. This requires the use of corrosion-resistant materials and a wash system on the fan to prevent solids depostion.

The first installations of this type were in connection with scrubber systems for removal of particulate only. Most of these went through a period of severe problems before an adequate construction material could be identified and developed. At one plant the fans are now made of Inconel.

The ductwork and stack are also subject to corrosion. Metal expansion joints have had to be replaced with a nonmetallic type in some installations. Acid-proof brick in some stacks has also failed, apparently because the acidic condensate corroded out the mortar joints. Notwithstanding these problems, it is reported that at least two utilities will use wet stacks in major SO_2 scrubber installations now in the planning stage.

Acid Rain: If the gas is not reheated or is reheated to an inadequate level, mist and condensate in the gas can cause an acid rain condition. The leading examples of this probably are the Battersea and Bankside SO_2 scrubbing installations in England, where large quantities of water are used to scrub the gas and there is no reheat. The systems have been criticized on the basis that, among other drawbacks, the plume brings an acidic mist to ground level.

Steam Plume: Since the gas contains more water vapor than in normal plant operation formation of a steam plume will occur more often with scrubbing than without, even if the gas is reheated. This is a major consideration in some situations, especially if the plant is in an area of high population density and the neighbors are sensitive to any kind of visible emission. In such areas the gas is reheated in some cases to a temperature about as high as at the scrubber inlet, which for power plants is usually in the range of 250° to 300°F.

This practice is especially prevalent in Japan, where in the four major scrubber installations operated by utilities the reheat level at one is 250°F and at the other three 285°F. The cost is quite high, about 5% of the total fuel consumption, but is considered advisable for promoting good community relations. In the United States no one seems to have been concerned this much with steam plumes. The plumes produced by cooling towers seem generally acceptable.

Effect on Ambient Concentration: The considerations covered up to this point are those with which the designer is most directly concerned–protection of downstream equipment, acid rain that could have an adverse effect on plant personnel and equipment, and a steam plume that might not be acceptable locally. Beyond these is the problem of the effect of reheat (or lack of it) on ambient concentration where the plume finally reaches the ground–which may be several miles from the plant. This is mainly a concern of regulatory authorities who set maximum ambient levels to protect both people and vegetation.

The Federal SO_2 regulation for new coal-fired plants is a maximum emission of 1.2 lb per million Btu. Nothing is said in the regulation as to degree of reheat (or, for that matter, height of stack). Thus as far as the emission regulation is concerned the plant designer can get the economy of no reheat if he is willing to accept the other trade-offs.

There is also the Federal ambient standard, however, which applies to all sources. In the United States it is the responsibility of the states to set such emission regulations as to ensure, for the particular area and situation involved, compliance with the ambient standard. Thus if scrubbers are to be installed in a given plant, it becomes a matter–theoretically at least–of predicting whether or not, for the state emission regulation in effect, the ambient standard will be met at the proposed degree of reheat. If the prediction (based on plume dispersion models) indicates that the ambient standard will be exceeded then presumably the emission standard would have to be reduced or the degree of reheat increased.

TVA has carried out comprehensive studies of plume dispersion from coal-fired power plants and has developed mathematical models, refined by actual data, that predict maximum ambient concentrations under any given set of conditions. Because of the gas cooling during scrubbing, calculations based on these models show that reduction in ground level concentration is always less than the reduction accomplished in the scrubber.

It should be noted that the higher water vapor content in the gas offsets to some extent the adverse effects of gas cooling. Since the water vapor has a lower density than other constituents of the gas it makes the plume more buoyant. This effect was taken into account in calculations. The SO_2 removal increases, degree of reheat has less and less effect on reduction in ground-level concentration (which is a logical conclusion even without the calculated values). At 95% removal, for example, even at reheat to only 145°F the ambient concentration is reduced by 91 to 93%. At 65% removal, however, the reduction in the ambient at 145°F is only 37 to 50%.

It is concluded that for the high degrees of SO_2 removal attained in the best-developed scrubber systems (85 to 95%), reheating is not likely to be economically justified except in marginal situations where the inlet SO_2 to the scrubber is so high that even with high SO_2 removal the ambient concentration is still close to exceeding the standard. This could be true for very large power plants or for operations like ore smelting where the inlet SO_2 concentration is quite high.

There are some considerations, however, which may make the situation worse than it appears. If there is no reheat then the gas leaving the stack can already have a load of mist, in which case evaporation of the droplets as the plume becomes mixed with air can cool the plume and further reduce buoyancy. This was considered also in the TVA study, with the conclusion that not enough data are avilable to quantify the effect.

Moreover, very little NO_x (probably less than 15%) is removed in SO_2 scrubbing. Thus NO_x ambient concentration would be greatly increased unless the gas were reheated to something like 250°F. Calculations show that at 20% NO_x removal the ambient concentration is increased by almost 50% if the gas is not reheated.

Reheater Design and Operation: Four types of reheating have been used in conjunction with scrubbers and several others have been proposed. Each has good and bad points and it is not clear yet which is best.

Direct Firing – The simplest system, with the lowest investment, is burning gas or oil and mixing the hot combustion products with the scrubbing gas. The Paddy's Run system (Louisville Gas and Electric) uses natural gas and light oil will be burned in the Bruce Mansfield plant (Ohio Edison) when it is completed. In Japan practically all the scrubbers have oil reheat.

The main problem, of course, is availability and cost of gas and oil. In Japan, where most of the boilers are fired with oil, reheating with oil is logical because the overall oil requirement is increased only by 2 to 5%. In the United States, however, coal is the principal fuel and oil or gas would not only be difficult to obtain (in most situations) but would also be costly and would usually require a new storage or pipeline installation. It seems unlikely that gas or oil will be used in very many U.S. plants.

Direct firing requires some care in mixing the hot combustion gas with the cool scrubbed gas. At Mitsui Aluminum in Japan, for example, there is a special mixing chamber for this; the hot gas enters at the top and the scrubber gas at the sides.

In-Line Steam-Heated Exchangers – The most popular method in the United States is passing steam (extracted from the boiler) into a tubed exchanger mounted in the gas stream after the mist eliminator. The point of extraction varies with plant situation, depending on the trade-offs involved. The tubes are usually arranged in banks with room for sootblowers between the banks. At Will County, for example, there are three banks each three tubes deep.

The main problems have been corrosion and formation of deposits that reduce heat transfer. Corrosion has been such a problem that even high-alloy materials have been unsatisfactory under some conditions. The results obtained in TVA's test program on reheater materials are probably the most thorough and definitive available. A large number of materials have been tested in TVA's pilot plant, including E-Brite 26-1, Cor-Ten B, Carpenter 20 types, Incoloy 825, Inconel 625, various 300 series stainless steels, and Hastelloy C-276. The Inconel and Hastelloy gave the best service but even they were subject to pitting and are considered to be unsatisfactory.

Generally similar results have been obtained in the less thorough test experience from the operating plants. At Will County 304 SS failed after a few months operation (not many actual hours because of downtime for various reasons). The Cor-Ten used in the top two banks (which presumably are less subject to corrosion because they do not evaporate mist) was then moved down to the first bank position, where it failed in 2 weeks. At other plants 316 SS has failed. The situation has been so bad at La Cygne that some of the reheaters have been removed and hot air from the air heater routed to the scrubber exit ducts to protect the downstream fans.

In contrast, there is the major anomaly of the Lawrence and Hawthorne experience (Kansas Power and Light), plus that at the three plants of Public Service of Colorado (Valmont, Cherokee, and Arapahoe), where carbon steel tubes have given acceptable service for varying periods including over 5 years at Lawrence.

No good explanation of this appears to be available. Obviously some difference in the operating conditions is responsible but it is not clear what the significant differences are. Factors that have been postulated to be significant include amount of mist impinging on the tubes, distance between the mist eliminator and reheater, temperature inside the tubes, and adequacy of sootblowing. There is some indication that the last of these may be the controlling factor.

At Lawrence, for example, a 500 hp air compressor is required for the air used in blowing the reheaters; the hot compressed air blown over all surfaces may keep the reheater more free of deposits than at those plants where corrosion has been a problem. At Will County, for example, much of the corrosion is of the pitting type and occurs under deposits on the tubes. Another difference at Lawrence is that the heating medium is hot water rather than steam.

Hot-Air Reheat: The problems with in-line exchangers have led at least two utilities to shift to steam-heated exchangers (carbon steel) over which air is blown. The heated air is then introduced into the scrubber gas much like the combustion gases in direct-fired reheaters. Southern California Edison uses such a system on the horizontal scrubber at Mohave and TVA has recently adopted the indirect method for Widows Creek.

Hot-air reheat obviously requires more energy because the air must be heated as well as the stack gas and a separate fan is required to move the air. The amount of extra energy required depends mainly on the temperature of the available steam. There is an intangible advantage, however, that offsets the higher energy requirement to some extent. The dilution resulting from the air addition reduces the incidence of steam plume formation and also presumably gives better plume dispersion.

No cost comparison can be made between the two systems at the present time. If it is assumed that the Lawrence-type system can be installed in other plants with assurance that it will work as well as at Lawrence, then it is likely to be less expensive even with the heavy soot-blowing requirement. Because of the uncertainty and the failures at Will County, La Cygne, and TVA, however, the indirect system seems preferable since there is no chance for corrosion and no necessity for sootblowing to keep the unit from plugging and the tubes from losing heat transfer efficiency.

Sludge Disposal

Disposition of the waste solids generates problems both in handling and conveying and in achieving an environmentally acceptable final disposal procedure. Various approaches are being pursued.

Ponding: The simplest procedure is to pump the bleed slurry stream to a waste pond, allow the solids to settle, and discharge the supernatant to a watercourse, similar to the usual fly ash disposal method. This is allowable in many parts of the United States and in other parts of the world. Many of the operating U.S. installations are of this type, including Hawthorne, Lawrence, La Cygne, and Cholla.

State and local water pollution regulations differ so that it is difficult to generalize on open-loop operation. The actual dissolved solids content of the discharged liquor will vary widely depending on conditions; from TVA pilot-plant data it has been estimated that open-loop operation under Widows Creek conditions would give an overflow containing about 4,500 ppm dissolved solids, of which about half would be calcium sulfate. For the 550 Mw system involved and the flow of the Tennessee River at that point, the dissolved solids content of the river would be increased only slightly—well within Alabama standards.

TVA plans to operate closed loop, however, as do several other utilities building scrubbers, because of the tighter water pollution restrictions likely to be in force in the future and the consequent need for learning how to operate closed loop.

The main advantage of open-loop ponding, in addition to any saving from not having to pump the supernatant back to the scrubbers, is in the scrubber operation as discussed earlier. The main drawback, in addition to environmental considerations, is that in arid areas the water supply may not allow the discharge.

For both open- and closed-loop ponding, the first consideration is how to get the solids content of the slurry high enough to avoid excessive pumping cost. TVA plans to run the scrubber at 15% solids, which is high enough to give economical pumping for the plant-to-pond distance involved. Others will operate in similar fashion. If, for some reason, the scrubber is operated at a low slurry-solids content, or if the distance to the pond is long, it may be more economical to concentrate the bleed stream before pumping to the pond. This is being done at Mohave (for the horizontal module) and will be the practice at Mansfield (Ohio Edison). The concentration is done in thickeners, which give a 35 to 40% solids underflow at Mansfield and up to 60% or more at Mohave.

The difference in the solids content of the underflow at Mansfield and Mohave illustrates one of the problems in ponding. For the relatively high inlet SO_2 at Mansfield, the crystals are mainly calcium sulfite which crystallizes in the form of thin platlets that do not settle very compactly. The 35 to 40% solids is the best the thickener can do even with long retention time. At TVA, where the pond will serve as the thickener, about 40% solids is the maximum expected in the pond after extended settling; in pilot-plant tests there was no improvement between 42 days storage and 96 days.

The main drawback to this is the large pond volume required. This will vary depending on the several factors that affect size and nature of the crystals but is on the order of 50 ft^3 per ton of dry solids at the 40% water level. In comparison most fly ashes will require only about 20 ft^3 per dry ton. For Widows Creek (550 Mw) it is estimated that for the remaining life of the plant (about 25 years) 9.3 million cubic yards of sludge storage must be provided, which is equivalent to a 240-acre area filled to a depth of 25 feet.

There has been much research aimed at reducing the required pond volume by getting better compaction. In TVA tests occasional stirring of the settled layer gave significant improvement but this may not be a practical method. Dewatering by filtration or centrifuging reduces volume, of course, as will be discussed later. The most promising method for simple ponding, however, may be oxidation of the calcium sulfite to sulfate since the sulfate crystallizes as relatively large, blocky crystals that settle compactly.

TVA has done extensive work on sulfite oxidation and in Japan, where gypsum is produced because it can be sold, oxidation is practiced commercially. One of the main problems is that the pH must be reduced to get the sulfite into solution so that it can be oxidized. The Japanese normally add sulfuric acid to adjust the pH, which is expensive. Several other approaches have been tested but all of them add cost. One of the more promising involves use of magnesium in the system; since $MgSO_3$ is relatively soluble the oxidation can be carried out at normal scrubber pH.

At this time there is no proven oxidation process that appears to be economically acceptable but work is continuing. In some low-sulfur systems such as the horizontal scrubber at Mohave, a high degree of oxidation is normal because of the high O_2:SO_2 ratio in the gas. The benefit of this is evident because the bleed slurry can be thickened to as much as 65% solids.

If the slurry is thickened at the plant to reduce pumping cost to the pond there is some limit to the solids concentration that can be pumped. At 35 to 40% solids, pumping appears to be satisfactory with any type pump. At 40 to 60%, such as pumped at Mohave, a positive displacement type such as the Monyo is indicated. One problem is the thixotropic nature of the sludge; if it is kept stirred the viscosity stays fairly low but if allowed to become quiescent it thickens quickly. Thus a heavy slurry allowed to stand for a period in the pipeline could become so thick that a positive displacement pump would likely be needed to get it going again. Provisions for slushing with clear water will also be provided at Mansfield.

Some minimum flow velocity is required in the pipeline to prevent sedimentation and sticking. Little information is available on the critical limits. At Mansfield a flow rate of 7 ft/sec or higher is planned.

For settling either in a thickener or pond, the initial settling rate is important since it must be fast enough to produce clear supernatant at a rate adequate for recycling. Fortunately the slurry settles fairly rapidly in the early phases although it slows down later. In a few instances a flocculent has been used, either because of inadequate thickener size or because a very clear supernatant liquor was required for use in pump seals.

Even with simple ponding there are several options available in regard to pond management. Some divide the pond into sections, some start with it either partly or completely filled with water, and the overflow can be either through weirs or downcomers. There is not sufficient information yet to optimize the system. If the solids contain a high proportion of sulfite they will be so thixotropic and structurally unstable that retaining walls will be necessary; normally this type of pond is constructed by excavating and forming a wall around the excavation by piling the dirt around the periphery.

If the solids were mainly sulfate, however, there would be the interesting possibility that the pond could be managed in a manner similar to the phosphoric acid industry, where the by-product gypsum is so stable that after settling and air drying it is used to build up the pond walls. The practice seems to be environmentally acceptable and might be used in lime/limestone scrubbing to give much lower ponding cost.

In areas where water pollution prevention is emphasized, the possibility of leaching from the pond may come into question. Actually, test data from STEAG (Germany) and Phillips (Duquesne Light), although limited, indicate insignificant leaching even without any special pond sealing procedures. Nevertheless, TVA plans to seal with clay at Widows Creek and there is pressure in other areas to make the pond bottom completely impermeable by using a plastic liner. The situation is in a state of flux and a considerable period probably will elapse before it is clarified.

Landfill Disposal: In addition to simple pond disposal, several other approaches are being considered. They are somewhat difficult to categorize but can be fitted into the general class of landfill disposal where the term landfill includes (1) dumping into suitable excavated areas (mines, quarries) or natural depressions, (2) piling the solids at some point near the main plant, and (3) even ponding when the sludge is stabilized (see later description) so that the area can be used for other purposes after the pond is filled (as opposed to the likely permanent dedication to waste disposal when ordinary sulfite sludge is ponded). The discarded sludge may or may not be covered with earth later, depending on the type of disposal area and on local regulations.

Landfill disposal in the mine is especially applicable to plants in arid areas such as Four Corners, where water is at a premium and ash disposal in the mine is already practiced. Ponding to allow time for the sludge to stabilize is practiced (or will be) at Mohave, Phillips, and Mansfield. Simple piling has not been practiced yet but may be.

The main characteristic of landfill disposal is that the sludge must be stabilized, that is, converted to a form that is structurally stable and relatively nonleachable. For this, dewatering to a higher degree than that normal in simple ponding is desirable, since the presence of water makes stabilization more difficult. The dewatering can be carried out in several ways.

Filtration – In addition to small-scale test work by several organizations, Louisville Gas and Electric operates filters at Paddy's Run. Depending on conditions, sulfite-type sludges can be dewatered by filtration to roughly 50 to 70% solids.

Centrifuging – Tests indicate that centrifuging will give a higher degree of dewatering but erosion may be a problem. Centrifuging will be tested in the vertical module installation at Mohave. Centrifuges are widely used in Japan for sulfate sludge.

Thickening – In the ponding-type stabilization systems at Mohave, Phillips, and Mansfield, dewatering occurs in the thickener at the plant and to some further extent in the pond. Thus the amount of water is relatively high and stabilization is correspondingly difficult.

Mixing with Dry Material – A low water:solids ratio can be obtained merely by mixing dry material such as dry-collected fly ash with the sludge, in which case thickening alone will give a low water concentration in the final mix. This type of operation is planned for several plants.

A major consideration in dewatering is the relative cost of transporting the material as a wet solid versus pumping a slurry. One of the companies that offers a stabilization system considers pumping to be so much more economical that the adverse effect of the water on stabilization can be tolerated. An alternative would be to pump the slurry to the disposal area and do the dewatering there. No such system has been announced but it may have merit. Various stabilization processes have been proposed but only two appear to be receiving serious consideration.

Lime-Fly Ash-Sludge – IU Conversion Systems mixes fly ash and lime with the sludge to give a mix in which chemical reactions of the cementitious type take place. Normally all the fly ash available would be used since it supplies much of the reactant. The amount of lime added would then depend on several factors such as (1) degree of stabilization needed, (2) amount of water present, and (3) amount of lime in the fly ash. The proportions must be tailored to each situation.

The IUCS system is basically one of producing a transportable solid. The usually better economics of slurry pumping might be obtained, as noted earlier, by pumping to the pond, dewatering, and mixing lime, dewatered sludge, and dewatered ash. The main problem with this would be in a system such as at Mansfield where both ash and SO_2 are collected in the scrubbers. The mix would not dewater to a water content as low as the dewatered sludge-dewatered ash and dewatered sludge-dry ash combinations. Mixing procedures and time of setting for the IUCS method are not very well known. The first major test of the process will be at the Mohave vertical scrubber installation. Tests are also planned at Shawnee.

Sludge-Additive – The Dravo Corporation offers a process in which a proprietary additive (Calcilox) is mixed with the sludge. Very little is known about the nature of the additive, how much is required, how long the mix requires to set up, and what the final product characteristics are. Dravo favors pipeline transport and has used it in their three installations (Mohave, Phillips, and Mansfield). Fly ash is not required as in the IUCS process but will be present incidentally at Mansfield.

It is known that excess water is objectionable, however, so there is a tradeoff between the cost of dewatering and solids transport in the IUCS process and the presumably higher raw materials cost for Dravo plus the cost of the pond. The additive also seems to have some other drawbacks on which little information is available. The Dravo system will also be tested in the Shawnee program, which hopefully will provide some badly needed comparative data.

Utilization: Since discarding is expensive and troublesome, a considerable amount of research has been done on developing uses for the sludge. In Japan the general practice is to oxidize to gypsum and use it in wallboard and cement manufacture, a profitable operation since natural gypsum is in short supply in Japan. In the United States, where there are ample natural deposits, this is less attractive but there is some current consideration of the possibility. Even if this develops, however, the tonnage of sludge consumed would likely be a very small percentage of the total.

IUCS has also done work on converting the sludge to aggregate and other building products. The procedure is similar to that for stabilizing to acceptable landfill properties but would be more expensive because of the more rigorous strength requirements.

Some consideration has also been given to recovering sulfur from the sludge by reducing at high temperature to give sulfuric acid or sulfur as the end product. Other recovery processes now being developed are likely to be more economical.

Water Balance

The shortage of water in closed-loop systems has been mentioned in several of the preceding sections. Since it is a general problem, the water balance will be discussed separately.

A lime/limestone scrubbing system consists of two loops (scrubber and waste disposal) into which water is being introduced at some points and removed at others. Water inputs to the system are (1) free moisture in the lime or limestone, (2) water of constitution when hydrated lime is used (unless hydrated at the power plant with process liquor), (3) rainfall if the disposal system involves ponding, and (4) fresh water makeup.

Water losses are (1) humidification of the stack gas as it passes through the scrubber, (2) mist carried through the demister, (3) water of hydration in the product solids, (4) evaporation from the pond (if one is involved), (5) seepage from the pond, (6) intentional pond overflow in open-loop systems, and (7) unaccounted-for losses (usually leakage from equipment).

The term closed-loop is a misnomer if applied to the scrubber loop only (as many do), because from the practical standpoint no scrubber loop can be fully closed. Some liquor always stays with the discarded solids and leaves the scrubber area. The only way to achieve a fully closed scrubber loop would be to wash all the liquor out of the solids with fresh water and return it to the scrubber. This is not feasible because of several operating reasons, one of which is that there must be some scrubber blowdown to rid the system of chlorides and other soluble constituents brought in with the fuel and limestone.

If the solids disposal loop is included in the consideration then the term closed-loop is applicable, assuming that the liquor stays with the solids and does not escape to watercourses or into ground water.

Even with closed-loop operation there can be a difference in the amount of fresh water addition allowable in the scrubber, depending on how the waste solids disposal system is designed and operated. The smallest allowable addition is in systems that involve filtration or centrifuging to separate the solids, thereby dewatering to a greater extent than in pond settling.

In practice, the tightest water balances are in Japan where the calcium sulfite in most plants is oxidized to calcium sulfate. Since the gypsum dewaters easily to 10 to 15% water content, the amount of liquor leaving with the product solids (replaceable with fresh water) is relatively low. In most Japanese plants some water discharge is required, however, because the small amount of liquor left in the gypsum is not an adequate purge.

Simple ponding allows a relatively large amount of fresh water addition, both because of the larger amount of water left in the solids and evaporation from the pond. Various considerations affect the water balance.

1. Scrubber evaporation depends on the volume, water content, and temperature of the inlet gas, which can vary widely with fuel, amount of excess air used in the boiler, and air leakage in the air heater. For Widows Creek, assuming 30% excess air (including air heater leakage) evaporation is estimated to be about 360 lb water per hour per Mw.

2. Residual liquor in the solids varies widely with the characteristics of the crystals and the resulting water content of the settled layer. The total weight of sludge is also an obviously important consideration. For Widows Creek the estimate is 475 lb water per hour per Mw (somewhat high because of extra tonnage due to fly ash). The net effect of this is complicated by whether there was any water in the pond when the system was started up.

If so, part of the liquor remaining with the solids presumably would be from the initial fresh water charge and thus less liquor would leave the scrubber system. In effect, fresh water is added but not in a way that helps scrubber operation. When the slurry is dewatered by filtering or centrifuging the residual water can range from 12 lb per hour per Mw in Japanese practice to about 67 lb at Paddy's Run. In both cases the absence of fly ash reduces the sludge tonnage and therefore the quantity of water.

3. Pond evaporation varies widely with climate and surface area:volume ratio. For Widow's Creek evaporation is estimated at 283 lb per hour per Mw.

4. Pond seepage obviously depends on several factors. The TVA design estimate is 60 lb per hour per Mw.

5. Rainfall (249 lb per hour per Mw at Widow's Creek) is a consideration because it collects in the pond and partially replaces the evaporated water. The net effect is uncertain. To the extent the rainfall is evaporated before it goes to the scrubber it offsets the desirable evaporation from saturated liquor in the pond; to the extent it mixes with saturated liquor in the pond it is less useful for mist eliminator washing in the scrubber (but still usable for pump seals and to reduce scrubber scaling); and to the extent it dissolves crystals in the pond it loses effectiveness as fresh water addition.

Ideally the rainfall should stratify in the pond in order to avoid mixing and crystal dissolution and pass quickly to the scrubber before any of it evaporates. This cannot occur completely, of course, but may be approached.

Thus fresh water addition in closed-loop systems can vary over a wide range from about 370 to 1,460 lb per hour per Mw based on the above figures. The main question is whether all the 1,460 lb is effective as fresh water.

Materials of Construction

Equipment surfaces are subjected to a wide range of conditions in lime/limestone systems and thus corrosion and erosion vary from point to point. There are so many factors and so many contradictions in full-scale operation that generalization is difficult. The following are presented with some reservations as to adequacy of supporting data.

1. Stainless steel is suspect in systems attached to coal-fired plants because of the chloride that accumulates in the scrubber slurry in closed-loop operation (as high as 10,000 ppm).
2. Plastic linings tend to crack.
3. Rubber linings give superior service.
4. Mild steel is adequately resistant in several parts of the system.
5. On metal parts exposed to gas, there is a tendency to pitting under any solid deposits on the surface.
6. Ash in the slurry aggravates erosion. (However, TVA results differ; ash characteristics are no doubt important.)

The general consensus on selection of materials for different pieces of equipment appears to be as follows. However, it should be noted that the experience is from only a few installations, some of which were probably not operated properly. The body of experience currently being obtained may change some of the conclusions.

1. Prehumidifier – A difficult area, since the high temperature rules out some materials. Stainless steel seems satisfactory if water is used but has pitted at Mohave where recycled liquor is sprayed in to cool the gas. TVA tests show Hastelloy C-276, Inconel 625, and Incoloy 825 to be superior.
2. Scrubber Body – A few scrubbers have been built of mild steel but this appears inadvisable over the long term. Most are lined with rubber or fiber-reinforced plastic (FRP). Both have been generally satisfactory and FRP is cheaper and easier to repair. There have been some reports of FRP cracking but it does not appear to be a serious problem.
3. Scrubber Internals – These are a problem because stainless steel pits and FRP may not be suitable or strong enough. The high-Ni alloys should be suitable but would be expensive.
4. Nozzles – Silicon carbide appears best.
5. Mist Eliminator – FRP resists corrosion but is somewhat fragile.
6. Reheater – As noted earlier, the situation is very confused; even high-Ni alloys were unsatisfactory in TVA tests and at Will County and La Cygne, whereas carbon steel does well at Lawrence, Hawthorne, and Valmont. Until the discrepancy is cleared up hot air reheat appears best.
7. Ducts – In wet areas, as in wet stack systems and between the demister and reheater, ductwork is subject to corrosion–especially in the expansion joints. Various materials have been used with varying results. Metal expansion joints have been generally replaced with nonmetallic materials.
8. Fans – Those which are downstream from the scrubber are vulnerable, especially in wet systems. Coatings have generally come off. Inconel is working well at Four Corners.
9. Stacks – Those which are subjected to wet gas have been subject to mortar joint damage. The best solution to wet stack protection is not yet apparent.
10. Pumps – Rubber lining seems to be the best approach. Most alloys have failed but a few of the high-Ni, hardened type may be acceptable.
11. Lines – Rubber lined seem best.
12. Tanks – Mild steel is acceptable.

NEW DEVELOPMENTS

The techniques discussed so far in this chapter can be regarded as the established technology, although much further improvement is needed. The three methods described, boiler injection, limestone calcination in a separate facility, and limestone addition to the scrubber, have all been used in operating plants. The problems encountered, including limited absorption, scaling, and corrosion, have led to proposals for new departures, some radically different, for improving operation. The use of added magnesium salts to improve operation has been described. Other departures are as follows.

Use of Weak Acids

Since the rate of limestone dissolution seems to be of major importance in use of limestone slurry, addition of some reagent to dissolve the limestone more fully and rapidly seems logical. In work at TVA, weak acids intermediate in strength between carbonic (H_2CO_3) and sulfurous (H_2SO_3) have been tested. The objective is to get high Ca solubility by forming a soluble calcium salt with the added acid, with the dissolved salt then giving up the calcium to H_2SO_3 in the scrubber. The reactions presumably are as follows:

$$2HR(aq) + CaCO_3(s) \longrightarrow CaR_2(aq) + CO_2(g) + H_2O$$

R is the anion of the added acid, presumed to be monobasic.

$$CaR_2 \longrightarrow Ca^{++} + 2R^-$$

$$SO_{2(g)} + H_2O \longrightarrow H_2SO_{3(aq)}$$

$$H_2SO_{3(aq)} \longrightarrow H^+ + HSO_3^-$$

$$HSO_3^- \longrightarrow H^+ + SO_3^=$$

$$Ca^{++} + SO_3^= \longrightarrow CaSO_{3(s)}$$

$$2H^+ + 2R^- \longrightarrow 2HR$$

Thus the added acid, after solubilizing Ca, is regenerated in the scrubber. The waste $CaSO_3$ (plus any $CaSO_4$) must be washed to avoid loss of the acid. There are many acids intermediate in strength between H_2CO_3 and H_2SO_3, both organic. Few of them, however, have the desired combination of properties:

(1) Low cost.

(2) Low vapor pressure.

(3) Adequate solubility of calcium salt.

(4) Low molecular weight, to reduce the weight required.

(5) Intermediate acid strength (degree of ionization). If too close to that of H_2CO_3, reaction with $CaCO_3$ is inhibited; if too close to that of H_2SO_3, reaction of the dissolved calcium salt with SO_3 is restricted.

(6) Good stability over a long period of operation.

Of the several acids tested, benzoic acid (C_6H_5COOH) appears to have the best combination of the properties listed. Addition of the acid to the scrubber circuit, in amounts up to stoichiometric, improved scrubber operation significantly. There was no evidence of volatilization or decomposition, and loss of acid with the discarded solids was prevented by dewatering the solids, washing, and feeding the wash water back to the scrubber.

Calcium Chloride Addition

The Hölter process in Germany and the Kobe Steel (CAL process) in Japan involve use of $CaCl_2$ solution (up to 30% concentration) to carry the lime or limestone in the scrubber loop. Hölter uses limestone and Kobe lime. It is claimed that the solubility of the lime or limestone is increased by the $CaCl_2$. Kobe, for example, has data showing a six-fold increase in lime solubility as the $CaCl_2$ chloride concentration is increased from zero to 30%. At higher concentrations the solubility falls off sharply. With the increased solubility, the system can be operated on a soluble lime basis even at relatively low L/G, which is claimed to be advantageous. The high viscosity is said to reduce mist formation and at the reduced vapor pressure less water is evaporated into the gas.

The $CaCl_2$ process has some obvious disadvantages and the chemistry is still not clear. Nevertheless, both companies are proceeding with installation of full scale plants.

Rotary Drum Scrubber

Most of the scrubbers used in lime/limestone scrubbing are of more-or-less conventional design. An exception is the rotary-drum scrubber, in which the gas flows horizontally through the turning drum and lime slurry is lifted by the rotation and dropped through the gas stream.

The rotary drum is fitted with chains inside, the two ends of each chain fastened in such a way that it traverses the open space during the rotational period and also rubs against the walls. Since a large number of chains is used the links in effect serve as a mobile packing. The scrubber slurry is carried as a pool in the drum, contained by rings at each end. The slurry is pumped in at one end, flows by gravity through the drum countercurrent to the gas flow, and overflows at the other end. The chains, assisted by lifting flights, carry the slurry to the top and cascade it down through the gas. The following advantages have been claimed for the rotary-drum scrubber.

1. Absolute reliability because the unit cannot plug. Even when the rotation is stopped (as for external maintenance) and the flow of hot gas continued, it is claimed that the dried solids caked on chains and walls break up and drop off when rotation is resumed.
2. Ability to remove dust as well as SO_2 without any danger of plugging.
3. Lower pressure drop than for vertical mobile-bed scrubbers such as the TCA and Hydrofilter (marble bed).
4. Low pumping power since the head is low and no circulation through the scrubber is necessary. (Only make-up lime slurry is fed and the bleed stream removed.)
5. No mist eliminator to plug. The lifting flights are omitted from a section at the end of the scrubber, in which the chains themselves serve as a mist eliminator.
6. No nozzle erosion such as usually encountered in spray scrubbers.
7. No need to slake lime since the drum acts as a slaker.
8. No need for a delay tank.
9. Suitability of carbon steel as the construction material.

10. No need for prehumidification.

Of these claimed advantages, reliable operation seems justified because of the nonplugging performance in the various pilot-plant tests. However, it should be noted that scaling in the scrubber proper has been brought under control in most of the recent scrubber installations, especially the mobile bed and spray types. Mist eliminator plugging is a different matter; if the chains can serve the mist eliminator function a major advantage is gained.

The dust removal ability of the scrubber does not appear to have been well established and SO_2 removal efficiency is also somewhat questionable; the chains may not be as good for mass transfer as other types of packing. Pressure drop is low but not as low as for the Southern California Edison horizontal scrubber. Gas velocity is also not as high as for the horizontal type.

The other advantages cited all seem valid. Use of unlined carbon steel appears to be feasible because the rotary drum must have a relatively thick wall anyway for structural reasons and therefore can survive a medium corrosion rate for several years if it is general corrosion rather than pitting. This seems likely since the chains prevent formation of scale under which pits usually develop. Moreover, the pH reported from pilot-plant tests is somewhat higher than for most lime scrubbers and therefore the solution is less corrosive. The scrubber is to be installed on an industrial boiler, which should supply more complete data on feasibility.

ECONOMICS

The cost of lime/limestone scrubbing is especially important as the method is considered to be the basic throwaway process, against which recovery processes must compete. The most complete cost analysis available is a conceptual design and cost study made by TVA for EPA in 1974.

Investment

The cost of a lime/limestone scrubbing installation varies widely with many factors, including type of fuel, type of absorbent, size of the installation, new vs existing, sulfur content of fuel, degree of SO_2 removal, method of sludge disposal, and various site-related considerations. Generalizations on cost are therefore of little value. In the past two or three years the rapid inflation in costs has been an added factor that makes it difficult to tie down the investment required for a scrubber installation.

The TVA estimates, based on early 1974 prices, ranged from $23 per kw for large new boilers burning oil of relatively low-sulfur content to $65 per kw for smaller boilers (200 Mw; existing) burning high-sulfur coal.

The estimates were based on a somewhat idealized situation, assuming a fully proven technology, no overtime in construction, an average sludge disposal system, and no allowance for capacity replacement to offset the power requirement. Allowance for these factors, plus some others, ran the cost of a 500-Mw system from $50 per kw to over $100.

Actual experience in recent plant installations indicates a general cost range from $75 to $125 per kw. This is a changing situation, however; costs are likely to go up somewhat with inflation and to come down as the technology matures.

Operating Cost

The TVA estimated operating costs ranged from 1.3 to 3.4 mills per kw hr, which is equivalent to $3.50 to $9.10 per ton for a typical high-sulfur coal. This assumed an average price for lime/limestone and for sludge disposal. In nonaverage situations the cost could be much higher.

THROWAWAY OPERATION
INDIRECT LIME/LIMESTONE SCRUBBING

In direct lime/limestone scrubbing, the problems of sulfite-sulfate scaling and sticking of slurry solids in the scrubber have led process developers to other routes that avoid these difficulties. One such method (see Figure 3.1) is to absorb the SO_2 in some soluble medium and then treat the scrubber effluent solution with lime or limestone in a reactor outside the scrubber loop. The usual absorbent is a solution of an alkali such as a sodium or ammonium salt, from where comes the common term, double alkali. However, other absorbents such as water, ammonium sulfate solution, and aluminum sulfate solution are used, all with regeneration by lime or limestone in a separate reactor vessel.

The indirect method is a relatively recent development, coming to the fore mainly after the early troubles with direct scrubbing were encountered. The recent progress in making direct lime/limestone work more reliable and efficient has introduced some question as to justification for the indirect route, but intensive test work continues and there have been some commercial applications on industrial boilers plus two or three utility boilers in Japan. The technology can be classified as follows:

Sodium Salts: In the U.S. there are several processes under development that involve scrubbing with NaOH or Na_2SO_3 solution and regenerating with lime or limestone. The processes include A.D. Little (with Combustion Equipment Associates), FMC, Envirotech, General Motors, and Zurn. In Japan, Kureha-Kawasaki and Showa Denko have built full-scale plants.

Ammonium Salts: Test units with ammonium sulfite or hydroxide as the absorbent have been operated by Kuhlmann (France), the Smelter Control Research Association (U.S.), and Showa Denko and Nippon Kokan KK in Japan. A full-scale system is under construction in Japan.

Water: Chiyoda Engineering (Japan) uses water (actually very weak sulfuric acid containing a catalyst) as the absorbent and limestone as the regenerant. Several commercial plants are operating in Japan.

TABLE 4.1: INDIRECT LIME/LIMESTONE SCRUBBING—FULL SCALE AND TEST INSTALLATIONS[a]

Developer	User and site	Plant type	Size, Mw equivalent	Absorbent	Regenerant	Status[b]
Full-scale						
FMC	FMC, Wyoming	Industrial power boiler, coal	150	NaOH	$Ca(OH)_2$	C
		Same, coal	57	NaOH	$Ca(OH)_2$	C
General Motors	GM, Ohio	Same, coal	32	NaOH	$Ca(OH)_2$	O
Kureha-Kawasaki	Tohoku Electric, Japan	Utility boiler, oil	150	Na_2SO_3	$CaCO_3$	O
	"	"	350	"	"	O
	Kyushu Electric, Japan	"	500	"	"	C
	Shikoku Electric, Japan	"	900	"	"	C
Showa Denko	Showa Denko, Japan	Industrial boiler, oil	150	"	"	O
	Showa Pet. Chem., Japan	"	62	"	"	O
	Kan. Chem., Japan	"	93	"	"	O
	Poly Plastic, Japan	"	65	"	"	O
	Kyowa Pet. Chem., Japan	"	46	"	"	O
	Toho Zinc, Japan	H_2SO_4 plant	43	"	"	O
	Nippon Mining, Japan	"	37	"	"	O

(continued)

TABLE 4.1: (continued)

Developer	User and site	Plant type	Size, Mw equivalent	Absorbent	Regenerant	Status
Chiyoda	Fuji Kosan, Japan	Industrial boiler, oil	50	Dil. H_2SO_4	$CaCO_3$	O
	Daicel, Japan	"	31	"	"	O
	Mitsuibish Chem., Japan	"	130	"	"	O
	Mitsubishi Pet. Chem., Japan	"	230	"	"	O
	Denki Kogaku	"	37	"	"	C
	Hokuriku Electric, Japan	Utility boiler, oil	250	"	"	O
	"	"	350	"	"	C
Nippon Kokan	Nippon Kokan, Japan	Sintering plant	335	NH_4OH	$Ca(OH)_2$	P
Kurabo Engineering	Kurarey, Japan	Industrial boiler, oil	31	$(NH_4)_2SO_4$	$Ca(OH)_2$	O
	Daicel, Japan	"	53	"	"	C
Dowa Mining	Dowa Mining, Japan	H_2SO_4 plant	82	$Al_2(SO_4)_3$	$CaCO_3$	O
Hitachi	Tokyo Electric, Japan	Utility boiler	150	Carbon	$CaCO_3$	O
Test units						
CEA-ADL	Gulf Power, Scholz station	Utility boiler, coal	20	NaOH	$Ca(OH)_2$	O
Chiyoda	"	"	20	Dil. H_2SO_4	$CaCO_3$	O
SCRA	Magma Copper, Arizona	Smelter	2	$(NH_4)_2SO_3$	$CaCO_3$	C
Envirotech	Utah Power, Utah	Utility boiler, coal	1	NaOH	$Ca(OH)_2$	O
Kuhlmann	EDF, France	Same, oil	30	NH_4OH	$Ca(OH)_2$	c
Nippon Kokan	Nippon Kohan, Japan	Sintering plant	46	NH_4OH	$Ca(OH)_2$	O
Monsanto	Ind. P and L, Indiana	Utility boiler, coal	1.5	Amine	$Ca(OH)_2$	C

[a] In early 1975.

[b] O = operating; C = under construction; P = planned.

[c] Project completed.

Ammonium Sulfate Solution: Absorbent in the Kurabo process (Japan), with lime required for regeneration. A full-scale unit is under construction.

Aluminum Sulfate: Used by Dowa Mining in Japan; limestone is the regenerant. A full-scale system is operating.

Carbon: The Hitachi (Japan) and Lurgi (Germany) processes involve adsorption in carbon (which forms H_2SO_4 in the pores) followed by washing to remove the acid. Limestone is used for neutralization.

CURRENT STATUS

Table 4.1 lists the various installations, both full-scale and test units. The full-scale list is limited to those larger than 30 Mw equivalent capacity; there are many more smaller than this, particularly in Japan.

It is obvious that the bulk of the effort, especially in regard to commercial use, is in Japan. In the U.S., there are a few installations on industrial boilers but none are known to be planned for use in the power industry. As noted earlier, the increasing reliability and effectiveness of direct scrubbing has reduced the attractiveness of the more complicated indirect processes. Nevertheless, the indirect type may gain in the future because of the high SO_2 removal efficiency and the ability of some processes to make gypsum as the product.

Although definitive cost estimates are not available, the indirect type of operation seems likely to cost more than the direct method because (1) more equipment is required (in most processes), (2) make-up alkali (or other absorbent) is required, and (3) residual absorbent left in the product solids will make the disposal problem more difficult.

SODIUM-BASED PROCESSES

There are more processes based on use of a sodium compound as the absorbent than any other, mainly because materials such as NaOH, Na_2CO_3, and Na_2SO_3 are very effective absorbents, have relatively low back-pressure in the scrubber, and are readily available. (See Chapter 7 for further discussion of alkali scrubbing.)

The main drawback is that the sulfate formed incidentally by oxidation in the scrubber and in other parts of the system is more difficult to regenerate than when other absorbents are used. Much of the research in the area is concerned with this problem.

ADL-CEA

Arthur D. Little, Inc., has worked wtth Combustion Equipment Associates in developing and commercializing systems based on NaOH-Na_2SO_3 scrubbing and lime/limestone regeneration. Much of the development work was done under contract with EPA.

Process Chemistry: In the absorption section, asborption of SO_2 in sodium sulfite solution produces a bisulfite scrubber effluent solution according to the overall reaction:

$$Na_2SO_3 + SO_2 + H_2O \longrightarrow 2NaHSO_3$$

Depending on the type of operation, the feed to the absorber may also contain some sodium hydroxide (formed in the regeneration section or used as sodium makeup) and/or some sodium carbonate (used as sodium makeup to the system). Both sodium carbonate and hydroxide form sodium sulfite in the scrubber.

$$Na_2CO_3 + SO_2 \longrightarrow Na_2SO_3 + CO_2$$
$$2NaOH + SO_2 \longrightarrow Na_2SO_3 + H_2O$$

The absorber feed solution will also contain sodium sulfate in solution and may contain some sodium bisulfite if neutralization is not completed in the regeneration section. The sulfate is formed in the scrubber by reaction of sulfite with oxygen in the flue gas:

$$2Na_2SO_3 + O_2 \longrightarrow 2Na_2SO_4$$

The rate of oxidation is a function of the absorber design, oxygen concentration in the flue gas, flue gas temperature, and the nature and concentration of the species in the scrubbing solution. As an example, for flue gas containing about 4 to 5% O_2 and 2,500 ppm SO_2, approximately 10% of the SO_2 removed from the flue gas will normally be oxidized to sulfate.

Under similar conditions of absorber design, oxygen concentration, and solution characteristics, but at a much lower SO_2 inlet concentration (as in burning low-sulfur coal), a similar rate of oxygen absorption-reaction will produce a much higher equivalent percentage of SO_2 oxidized, perhaps as high as 50% of the SO_2 removed. At steady state, this sulfate must leave the system, either as calcium sulfate or as a purge of sodium sulfate, at the rate at which it is being formed.

Regeneration of sodium sulfite-sulfate scrubber effluent solutions can be considered as a sequential reaction first involving neutralization of the bisulfite:

$$2NaHSO_3 + Ca(OH)_2 \longrightarrow Na_2SO_3 + CaSO_3 + 2H_2O$$
$$2NaHSO_3 + CaCO_3 \longrightarrow Na_2SO_3 + CaSO_3 + CO_2 + H_2O$$

In theory, the neutralization should go to completion with lime but not with limestone. The lime reaction can go even further, to an equilibrium OH^- concentration.

$$Na_2SO_3 + Ca(OH)_2 \longrightarrow 2NaOH + CaSO_3$$

The usual form of calcium sulfite produced is the hemihydrate, $CaSO_3 \cdot \frac{1}{2}H_2O$. Some sulfate is also precipitated, the amount depending on sulfite and sulfate concentration and on pH.

$$Ca^{++} + SO_4^{=} \longrightarrow CaSO_4$$

The rate of precipitation must equal the rate of oxidation if purging of soluble sodium sulfate is to be avoided. In practice, even if such a balance is established, the washed calcium sulfite-sulfate will contain a small amount of soluble sodium salts and some sodium makeup to the system will therefore be required.

The system can be designed to take care of any level of oxidation up to 100%. For example, 0.5 molar sodium sulfate solution can be reacted with lime as follows:

$$Na_2SO_4 + Ca(OH)_2 + 2H_2O \longrightarrow 2NaOH + CaSO_4{\cdot}2H_2O$$

The product solution then will contain approximately 0.1 to 0.15 molar sodium hydroxide and 0.45 molar sodium sulfate. If this solution is recycled to the absorber, where 100% oxidation recurs, then a 0.5 molar sodium sulfate solution is returned to the regeneration system and the cycle is repeated. However, the 0.1 to 0.15 molar solution is a very dilute absorbent, requiring large tanks, pumps, and lines.

Operating in such a way that all of the sulfate is precipitated as $CaSO_4{\cdot}2H_2O$ is called a dilute operating mode; an active sodium concentration less than or equal to 0.15 molar active Na^+ is required, where active sodium is considered to be sodium sulfite-bisulfite, carbonate-bicarbonate, or hydroxide.

Concentrated modes are those involving active sodium concentration greater than 0.15 molar active Na^+. In this mode, the sulfate precipitates with the sulfite in a manner similar to the unsaturated sulfate operating mode in direct lime/limestone scrubbing (see Chapter 3). Because of this, there must be some limitation on the degree of sulfite oxidation, depending on the Mg^{++} and Cl^- concentrations.

The sulfite and sulfate concentrations, besides affecting sulfate precipitation, also have a major effect on the physical properties of the calcium-sulfur salts and, as a result, on the settling and filtration characteristics of the solids produced. A high degree of oxidation can also be accommodated by using sulfuric acid.

$$Na_2SO_4 + H_2SO_4 + 2CaSO_3{\cdot}\tfrac{1}{2}H_2O + 3H_2O \longrightarrow 2NaHSO_3 + 2CaSO_4{\cdot}2H_2O$$

The H_2SO_4 drops the pH (to 2-3), thereby converting sulfite to bisulfite and bringing it into solution. The net effect is that the solubility product for calcium sulfate is exceeded with resulting precipitation. This is the only method that allows both high oxidation and use of a concentrated operating mode. However, the acid is expensive and additional lime or limestone is required to neutralize it.

Dilute Mode Operation: One of the problems in Na-based double alkali processes is that the calcium going into solution in the regenerator-reactor may not all reprecipitate. When the dissolved calcium is introduced into the scrubber in the recycle stream, scaling may occur, in which case one of the major advantages expected for indirect lime/limestone over direct scrubbing is lost. Because of this, a softening step is usually included in dilute processes to keep the calcium out of the scrubber.

Two processes have evolved, one using makeup sodium in such a way as to achieve the desired degree of softening and the other a lime/limestone double loop-sulfite softening. The sodium carbonate method is shown in Figure 4.1. Addition of Na_2CO_3 to the thickener precipitates enough calcium to keep the calcium content of the liquor to the scrubber well on the safe side of saturation (about 100 ppm below saturation). It is expected that the Na_2CO_3 makeup requirement will be at least this high because of losses in the filter cake.

The double loop system involves use of $CaCO_3$ in a first reactor to precipitate as much sulfite and bisulfite as possible (about 70%); very little sulfate is precipitated. Part of the filtrate is then treated with lime in a separate loop to precipitate the sulfate and the rest of the sulfite-bisulfite. After separating the solids, the liquor from this step is recirculated to the limestone reactor where precipitation of calcium sulfite lowers the dissolved calcium concentration to a safe level.

The system has the advantage that about 60% of the regeneration is done with limestone, which is cheaper than lime. However, the additional equipment required and the necessity for storing and handling two reactants are serious drawbacks. A cost estimate has indicated the carbonate method to be the most economical of the two methods.

Concentrated Mode Operation: The concentrated mode (Figure 4.2) is a little simpler than the dilute operation because the thickener is replaced with a small mix tank. Moreover, Na_2SO_4 can be used as makeup whereas it cannot in the dilute system (carbonate softening). The flowsheet shows a venturi scrubber for quenching the gas, which may or may not be necessary.

In tests with lime as the regenerant, the system has been operated successfully (pilot-plant tests) with a sulfite-bisulfite concentration in the 0.4 to 0.45 molar range, some four times as concentrated as the dilute system. The sulfate concentration in the scrubber effluent ranged from 0.5 to 0.8 molar and overall oxidation rate, under optimum conditions, was about 11%.

The sulfate precipitated with the sulfite was almost enough to reach a balance; it is expected that at higher sulfate concentrations a balance would be reached but this remains to be demonstrated. One problem is that high sulfate concentration adversely affects the settling properties of the product solids. The tentative conclusion from the tests was that only about 20 to 25% oxidation can be tolerated.

Under these conditions the calcium in the liquor returned to the scrubber is quite low (30 to 60 ppm) and scaling would not be expected. None was encountered in the pilot-plant tests.

Tests with limestone as the regenerant have not been very conclusive. The use of limestone introduces other considerations such as wide variations in regenerant reactivity. The tentative conclusion is that a limestone system may be feasible but there are several unresolved questions.

Further Evaluation: The CEA-ADL-EPA unit (20 Mw) at Gulf Power's Scholz station in Florida was started up in early 1975. The test program planned by Southern Services (in direct charge of the program) and other participants is designed to resolve the several unanswered questions from the pilot-plant tests.

FIGURE 4.1: SODIUM-BASED DOUBLE ALKALI PROCESS—DILUTE MODE, Na_2CO_3 SOFTENING

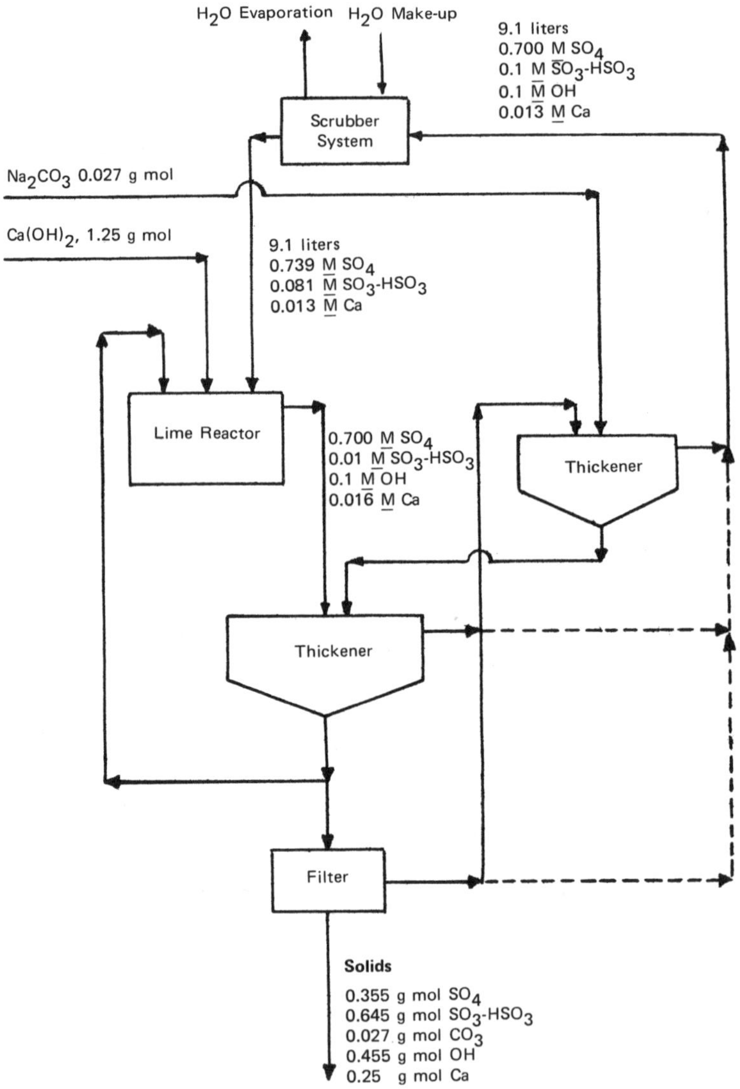

Basis: 1 g mol of SO_2 absorbed
35% oxidation rate (all assumed in scrubber for convenience)

Source: C.R. LaMantia et al (ADL) and N. Kaplan (EPA) at the EPA Symposium, Atlanta, 1974

FIGURE 4.2: SODIUM-BASED DOUBLE ALKALI PROCESS—CONCENTRATED MODE

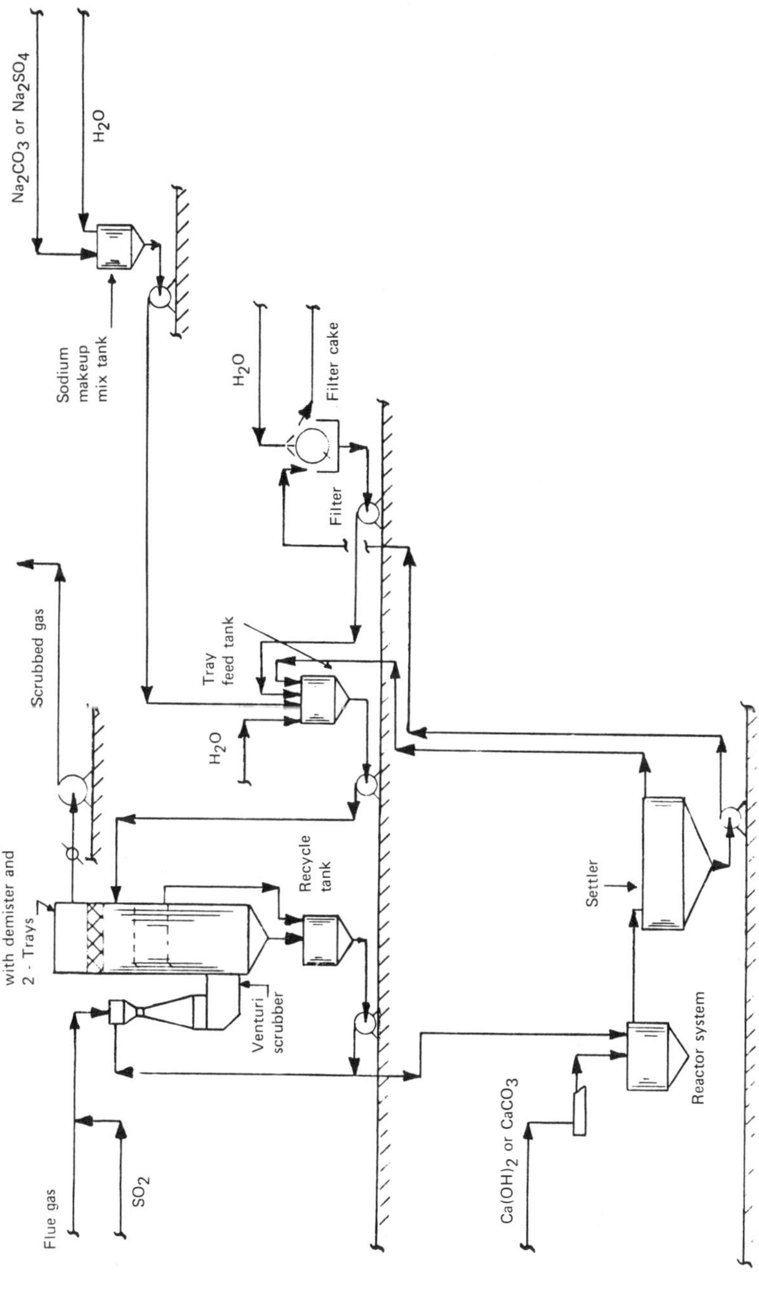

Source: C.R. LaMantia et al (ADL) and N. Kaplan (EPA) at the EPA Symposium, Atlanta, 1974

General Motors

The GM system is one of the first to be used commercially in the U.S. An installation on a 32 Mw equivalent boiler was started up at GM's Parma, Ohio, plant in March 1974.

The process is a dilute mode type. A simplified flowsheet of the Parma system is shown in Figure 4.3. The regenerated solution has an OH^- concentration of about 0.1 M, about the same as in the ADL pilot-plant tests on dilute mode operation. The Parma system has had several operating problems. As of November 1974, the status was still regarded as in a shakedown stage.

FIGURE 4.3: GENERAL MOTORS DOUBLE ALKALI PROCESS

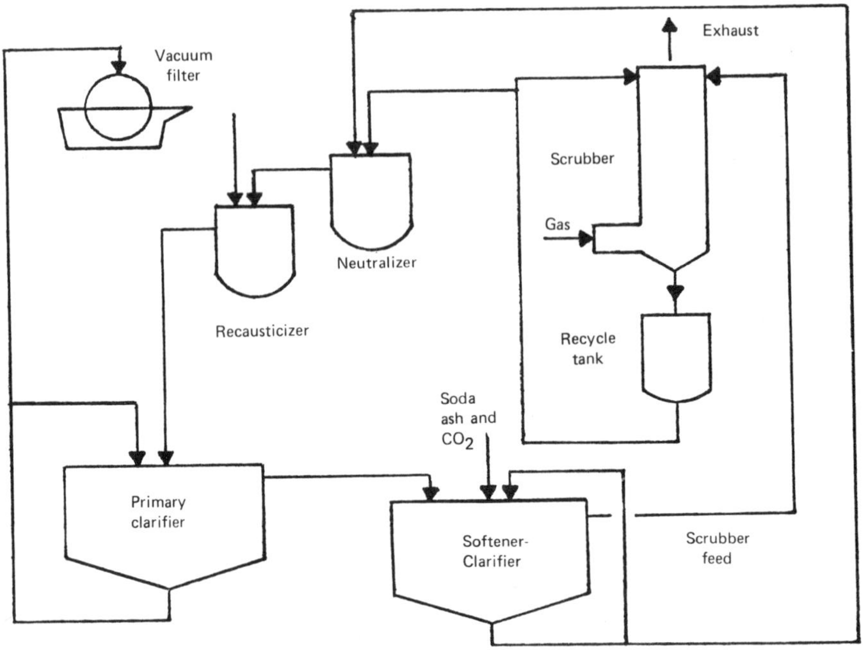

Source: C.R. LaMantia et al (ADL) and N. Kaplan (EPA) at the EPA Symposium, Atlanta, 1974

FMC

The FMC double alkali system is based on sodium sulfite scrubbing followed by regeneration with lime. In comparison with other double alkali processes, the active sodium concentration in the scrubber is relatively high. As a result, there is little or no sulfate regeneration; the sodium sulfate formed by oxidation in the scrubber is purged with the calcium sulfite-sulfate sludge. This may or may not cause water pollution, depending on the situation. It also increases running cost because of the make-up sodium carbonate required. On the other hand, the higher solution concentration and other features reduce investment to the point that overall operating cost, including capital charges, is probably lower than for any other double alkali process.

FMC developed the process initially for its plant in Modesto, California, where barium and strontium sulfate ores are roasted in high-temperature kilns. The high SO_2 content of the roaster gas caused scaling when lime/limestone scrubbing was attempted, so the research effort was shifted to double alkali.

In December 1971 a full-scale installation (gas flow equivalent to about 6 Mw) was started up. The operation is reported to be both efficient and reliable; the operating time logged is over 22,000 hours, without any occasion to shut down the roasting system because of scrubber malfunction. The load factor has been over 95%. The chemistry involved is relatively simple. In the regeneration tank, only enough lime is fed to convert the bisulfite to sulfite.

$$2NaHSO_3 + Ca(OH)_2 \longrightarrow CaSO_3 + Na_2SO_3 + 2H_2O$$

The $CaSO_3$ is separated and the Na_2SO_3 recycled, plus that not converted to bisulfite in the scrubber.

If particulate removal is required, FMC uses a venturi scrubber for both particulate and SO_2 (Figure 4.4). If particulate is not involved, a simple baffle-type scrubber or spray cyclone is employed.

A typical sulfite-bisulfite concentration is about 1.0 molar as compared to 0.1 to 0.15 in dilute systems. Since the absorbent is Na_2SO_3 rather than NaOH, the pH of the solution to the scrubber is relatively low, from 6.2 to 6.8. The exit pH is usually no more than 0.1 pH unit below the inlet, as compared to a wide swing in pH through the scrubber in dilute systems.

The scrubber effluent bleed stream is treated with lime and the resulting calcium sulfite separated by filtration. Sodium sulfate must be purged by allowing it to leave the system in the filter cake.

The relatively high Na_2SO_3 concentration keeps the dissolved calcium low, about 50 ppm, thus avoiding the need for a softening step to avoid calcium sulfate scaling in the scrubber.

Operation with a relatively high dissolved salt concentration inhibits oxidation of sulfite to sulfate, to the extent that total production of sulfate is said to be no higher than in dilute systems incorporating sulfate regeneration. Dilute systems, as noted earlier, may oxidize 50% or more of the sulfite during passage of the solution through the scrubber.

The main drawback cited by FMC is that the high dissolved salt content magnifies the reagent loss from leaks or in filter cake. For a 500 Mw boiler (3.5% S in coal; 80% capacity factor), the soda ash consumption is said to be about 24,000 tons per year, or a loss of about 15%.

FIGURE 4.4: FMC DOUBLE ALKALI PROCESS — CONCENTRATED MODE

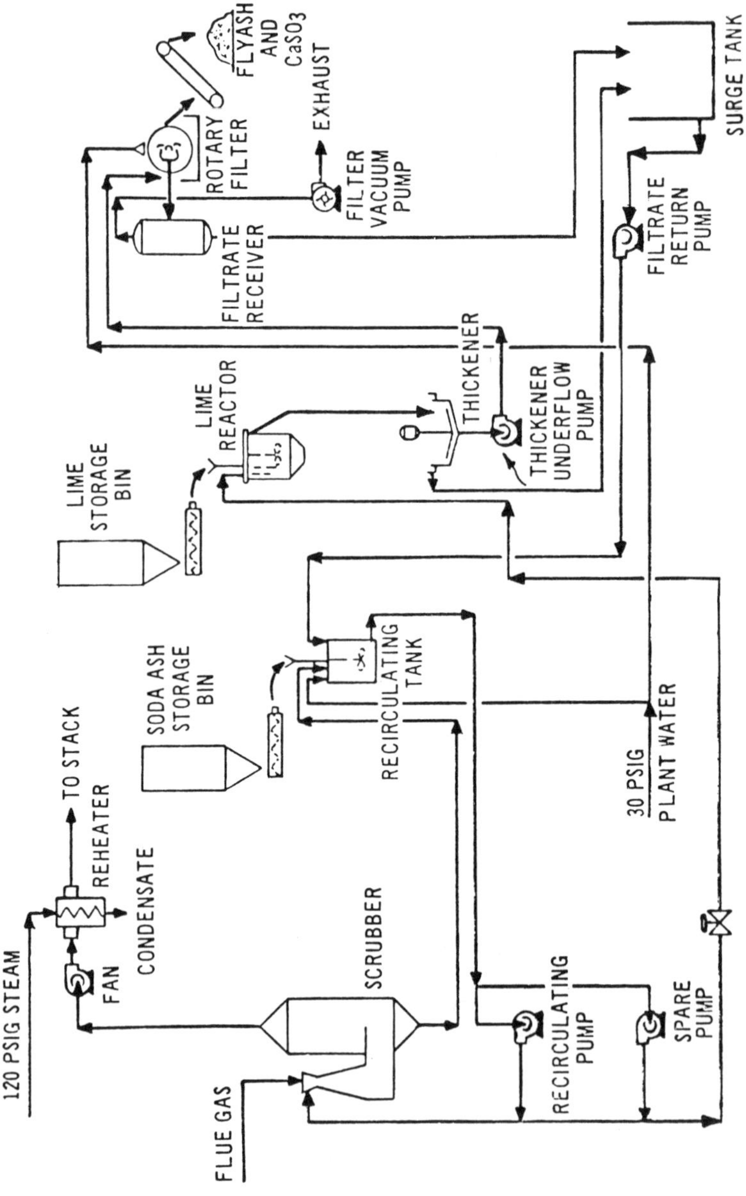

Source: C.R. LaMantia et al (ADL) and N. Kaplan (EPA) at the EPA Symposium, Atlanta, 1974

FIGURE 4.5: ENVIROTECH DOUBLE ALKALI PROCESS—DILUTE MODE WITH SULFITE OXIDATION

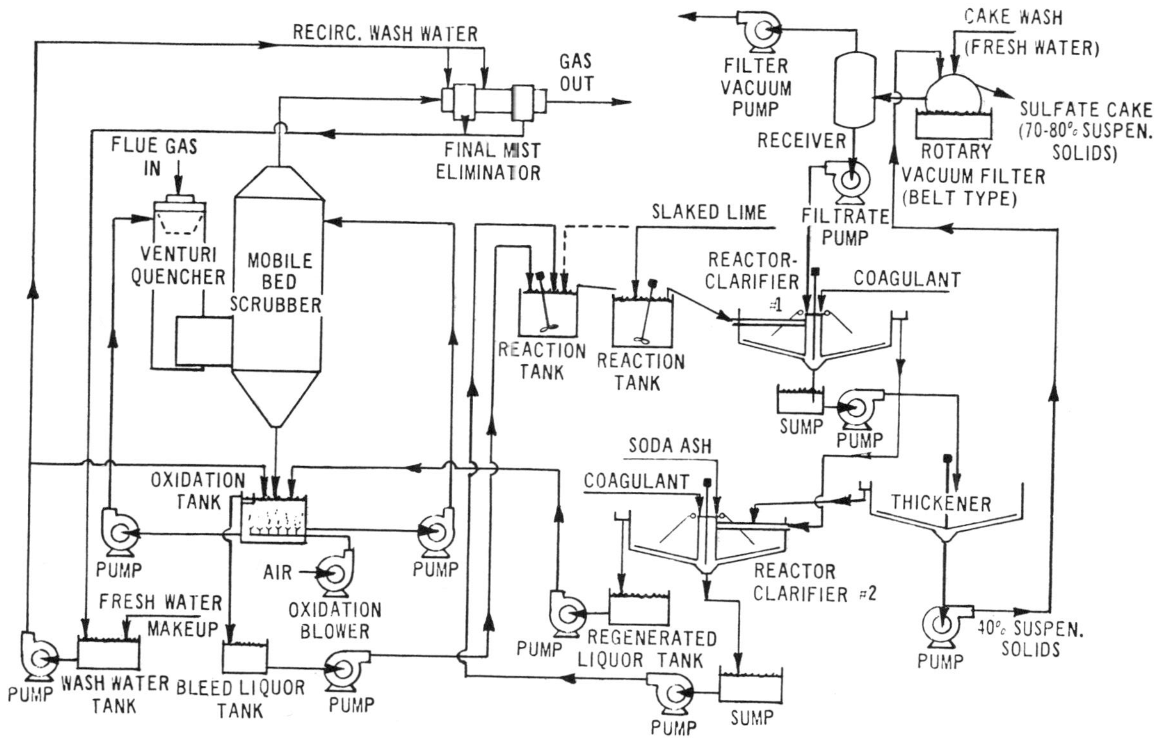

Source: C.R. LaMantia et al (ADL) and N. Kaplan (EPA) at the EPA Symposium, Atlanta, 1974

Envirotech

The Envirotech process (Figure 4.5) is of the dilute type. The unique feature is oxidation of the scrubber effluent solution before reaction with lime. As a result the sludge has superior properties in regard to settling rate and ease of dewatering.

The process has been tested mainly on low-sulfur Western coal, for which a high degree of oxidation in the scrubber is to be expected because of the relatively high O_2:SO_2 ratio in the gas. Thus the system cannot be operated in a concentrated mode anyway, except with a Na_2SO_4 purge as in the FMC process, so the completion of oxidation outside the scrubber to get better sludge properties is a logical step.

Kureha-Kawasaki

The double alkali process developed by the Kureha and Kawasaki companies in Japan is one of the more important ones in the world because of the degree to which it has been adopted commercially. A full-scale system is operating at Tohoku Electric's Shinsendai station and several others are planned (see Table 4.1).

The process (Figure 4.6) is somewhat complicated, mainly because of the oxidation required to give gypsum as the product and because the sodium sulfate is decomposed in a separate, special unit. It is a leading example of a concentrated system using H_2SO_4 for sulfate regeneration.

The gas passes from an electrostatic precipitator through an FD fan and then into a precooler. There is no bleed from the precooler loop; instead, residual liquor flows on into the main scrubber. The scrubber is a special grid-packed type, operated at an L/G of 1 kg/kg (about 12 gal/mcf). The pH at the scrubber inlet is 7.0 and the effluent 6.5; the dissolved solids concentration is 20%. Oxidation is about 5% (3% O_2 in gas).

The scrubber bleed stream is first filtered to remove carbon and then passed through five reactors in series where limestone reacts with the $NaHSO_3$ to precipitate $CaSO_3 \cdot 0.5H_2O$.

$$2NaHSO_3 + CaCO_3 \longrightarrow CaSO_3 \cdot 0.5H_2O + Na_2SO_3 + CO_2 + 0.5H_2O$$

Residence time for adequate reaction appears to be high because of the several reactors required. The reactors are closed and steam heated (operating temperature not reported). The pH at the outlet is 7.3.

The calcium sulfite (5% slurry) is separated on a vacuum filter and washed to remove Na_2SO_3. The filter cake (about 60% water) is repulped to 10% solids, H_2SO_4 added to reduce the pH, and the slurry oxidized with air in a special oxidizer. The slurry is dewatered to less than 10% water by centrifuging and the separated liquor is recycled to repulp the $CaSO_3$. A sidestream of liquor from the scrubber is treated to decompose Na_2SO_4 by reaction with H_2SO_4 and $CaSO_3$ in a special reactor.

$$Na_2SO_4 + 2CaSO_3 \cdot 0.5H_2O + H_2SO_4 + 3H_2O \longrightarrow 2CaSO_4 \cdot 2H_2O + 2NaHSO_3$$

FIGURE 4.6: KUREHA-KAWASAKI DOUBLE ALKALI PROCESS AT TOHOKU ELECTRIC'S SHINSENDAI STATION

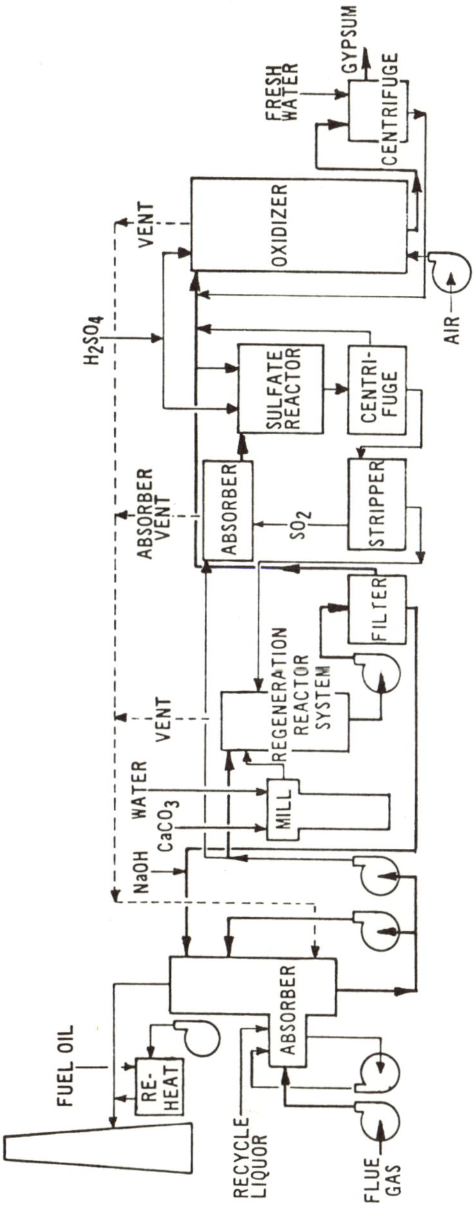

Source: C.R. LaMantia et al (ADL) and N. Kaplan (EPA) at the EPA Symposium, Atlanta, 1974

The slurry from the desulfation unit is filtered and the solution then passed through a steam-heated stripper to reduce the vapor pressure of SO_2. The gypsum is washed to reduce the Na^+ content to less than 10 kg NaOH per metric ton of gypsum, which is considered to be adequate for wallboard use.

There have been no scaling or plugging problems. The Ca^{++} content in the return liquor is <10 ppm. About 10% $SO_4^=$ in the scrubber loop is regarded as the maximum for good absorption. The product gypsum contains only about 5% water. Crystal size is on the order of 30 x 200μ. Kawasaki claims to grow the largest crystals of any gypsum producer because of the seeding induced by the gypsum crystals formed in the desulfation reactor.

The KK system is relatively complex and so many pieces of equipment are required that the installation covers a relatively large area. For U.S. conditions, where high-quality gypsum is not required as the product, a simpler system would appear more appropriate.

Showa Denko

The Showa Denko double alkali process in Japan (Figure 4.7) is quite similar to the Kureha-Kawasaki. Commercial adoption is also similar; several installations are operating or under construction (Table 4.1). The main difference as compared to Kureha-Kawasaki is that a special vertical cone scrubber is used; the liquor is entrained in the entering gas and carried upward with the gas through a draft tube. There is no prescrubber or cooler as in other systems. Like the Kureha-Kawasaki method, the process is complicated and a large equipment assembly is required. There is no apparent advantage of the special scrubber over other types.

FIGURE 4.7: SHOWA DENKO SODIUM-LIMESTONE PROCESS

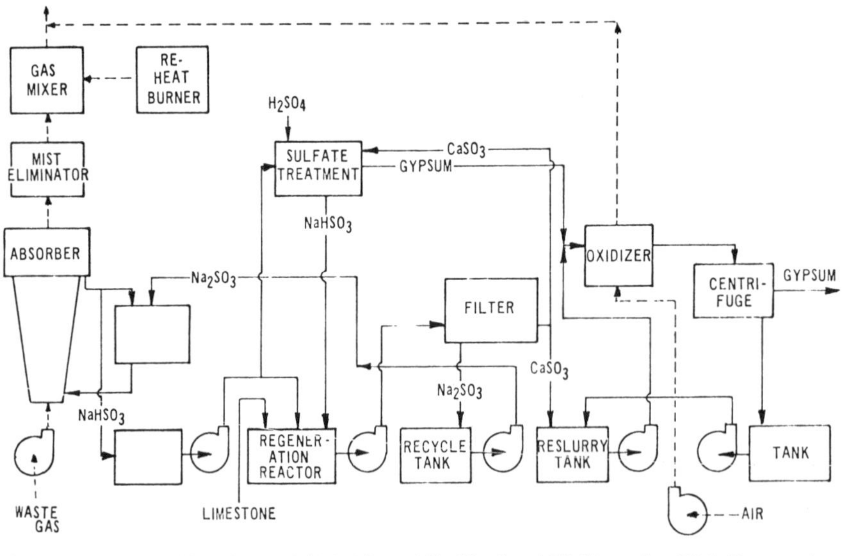

Source: C.R. LaMantia et al (ADL) and N. Kaplan (EPA) at the EPA Symposium, Atlanta, 1974.

AMMONIA-BASED PROCESSES

Ammonium salts have not been used as much as the sodium type in double alkali processes. Drawbacks are the volatility of ammonia and the tendency to form an ammonium sulfite fume in the scrubber. On the other hand, the system can be operated at higher concentration, ammonium sulfate can be separated as a useful product, and if sulfate regeneration is required it is easier to decompose than in the sodium system.

Nippon Kokan

The leading process is that developed by Nippon Kokan KK, a steel company, in Japan. A 46 Mw equivalent test unit operating on sinter plant gas is in operation and a 335 Mw equivalent commercial system is under construction.

In the test unit, the gas scrubbed contains 200 to 800 ppm SO_2 with the oxygen content varying from 12 to 15%. The gas is first washed in a cooler to increase SO_2 removal efficiency. The humidified gas is then scrubbed with ammoniacal solution of approximately 1.0 M $SO_3^{=}$-HSO_3^{-} concentration.

Fuming in the scrubber is a continuing problem. The fume (collected after the stack) consists primarily of $(NH_4)_2SO_4$ (0.25 g/m^3 concentration; approximately one-half of this is water vapor). NKK plans to install a mist eliminator (low pressure drop) to remove the fume.

In the regeneration section a 10% slurry of slaked lime is added to the decomposition tank. The tank is heated with steam to 98°C. The offgas ammonia is absorbed in a three-stage spray tower. The top two stages use low pH bisulfite liquor while water is recirculated on the bottom stage.

The calcium salts formed in the decomposition tank (pH = 11.2) are sent to the pH adjusting tank, where the pH is adjusted to approximately 4 by the addition of sulfuric acid. In this tank part of the calcium sulfite is converted into water-soluble bisulfite and is sent to the oxidation tank in which bisulfite is oxidized to gypsum by use of air. The gypsum slurry is fed to the filter through a slurry tank. The filtrate is neutralized with lime, clarified, and returned to the absorption system.

The NKK system is a concentrated-mode type yet sulfate regenerates easily because ammonia is evolved in the reaction between $(NH_4)_2SO_4$ and $Ca(OH)_2$. The system is relatively complicated, however, and the fuming is a problem. Work by TVA, EPA, and Air Products in the U.S. has shown promise of eliminating the fume problem by close control of scrubber liquor composition and scrubber operation.

Kuhlmann

Ugine Kuhlmann in France did pilot-plant work some years ago on an ammonia-based double alkali process very much like the NKK. A 25 Mw test facility was built at the St. Ouen station of Electricite de France (EDF). The station fuel was changed to low-sulfur oil soon after, however, so the test unit was moved to EDF's station at Champagne-sur-Oise. After about a year of testing, the unit was shut down and the regeneration section dismantled to make room for testing another process. The main trouble was plugging of the ammonia stripping column in an upper packed section.

The troubles with the Kuhlmann test unit do not appear to have been intrinsic to the process. The stripper plugging, for example, has not been a problem in the NKK installation.

Smelter Control Research Association (SCRA)

The SCRA is constructing a small pilot plant in Arizona to test an ammonia-based double alkali process in which the regenerant is limestone. A cost estimate indicated that this would be the most economical method for use on high-SO_2 waste gas from smelters, especially reverberatory furnaces in copper plants.

The process has the advantage that the limestone reacts mainly with bisulfite and the problems of stripping out ammonia and then recovering it are therefore avoided.

$$2NH_4HSO_3 + CaCO_3 \longrightarrow CaSO_3 + (NH_4)_2SO_3 + CO_2 + H_2O$$

OTHER ABSORBENTS

There have been various efforts to find an absorbent less volatile than ammonia but which gives a sulfate easier to decompose than sodium sulfate. Three such processes have reached a position of significance.

Monsanto

The Monsanto process uses an ethanolamine as the absorbent and lime as the regenerant (Figure 4.8). A 1 Mw pilot plant was operated for about 9 months in 1973 at the Stout station of Indianapolis Power and Light.

Although sulfate regenerates easily in the process, even at fairly high concentrations of active species, calcium dissolves in the lime-treated liquor to the extent of about 0.02 M. To avoid scaling in the scrubber, the thickener overflow is passed to a reactor where it is mixed with the scrubber effluent flow, thereby precipitating calcium and reducing its concentration to less than 50 ppm.

Kurabo

Kurabo Industries, Ltd., in Japan has developed a process in which a slightly acidic $(NH_4)_2SO_4$ solution is the absorbent and lime the regenerant. The scrubber effluent is oxidized before regeneration; as a result the regeneration reaction is between $Ca(OH)_2$ and $(NH_4)_2SO_4$ (Figure 4.9).

The $(NH_4)_2SO_4$ solution is a much less reactive absorbent, of course, than are the alkalies discussed up to this point. However, it has a higher capacity than plain water or calcium sulfate solution because of the equilibrium in the ammonium sulfate solution:

$$H^+ + SO_4^= \rightleftharpoons HSO_4^-$$

FIGURE 4.8: MONSANTO ETHANOLAMINE-LIME PROCESS

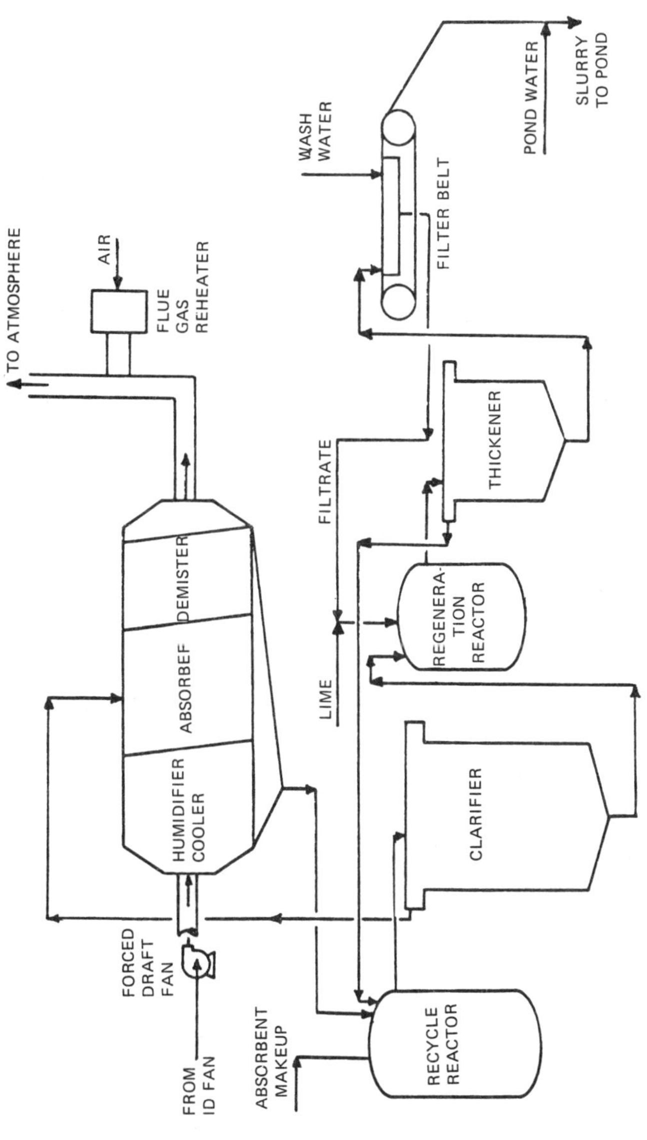

Source: C.R. La Mantia et al (ADL) and N. Kaplan (EPA) at the EPA Symposium, Atlanta, 1974

FIGURE 4.9: KURABO DOUBLE ALKALI PROCESS

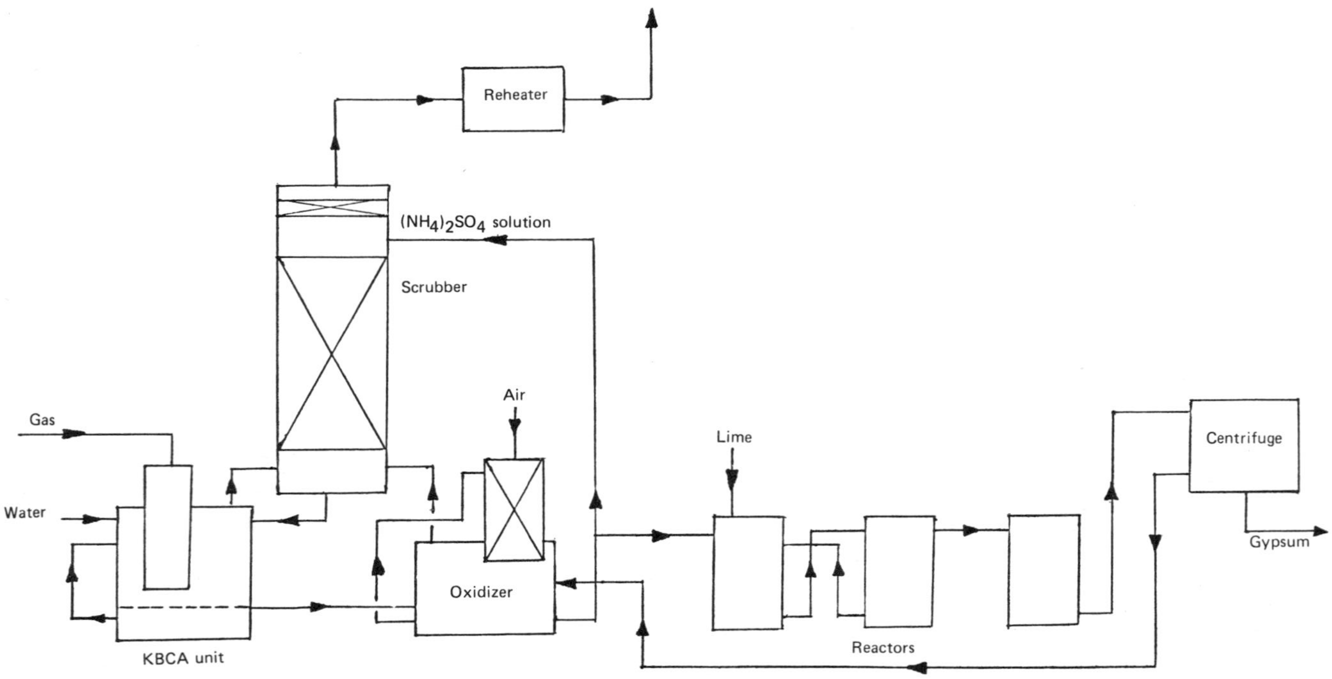

Source: C.R. La Mantia et al (ADL) and N. Kaplan (EPA) at the EPA Symposium, Atlanta, 1974

Thus the L/G requirement (on the order of 30 to 60 gal/mcf) is intermediate between the alkali-type absorbents and the Chiyoda process (see later) which uses weak sulfuric acid as the absorbent.

Gas from the boiler first passes through a KCBA scrubber and then through a packed scrubber in series. The main functions of the KCBA are to cool the gas and concentrate the absorbing solution.

The scrubber loop consists of the two scrubbers plus an oxidizer. Solution flows countercurrently down through the packed scrubber and then into the KCBA where it circulates through the spray-impinger arrangement. From there it flows to the oxidizer where it is circulated through a short tower and contacted with air drawn downward cocurrently through the tower; the air passes to the packed scrubber and flows up through it along with the stack gas. The oxidizer has a level control that controls introduction of water into the KCBA to make up for evaporation loss. About five times the stoichiometric amount of air is used and the retention time is one hour.

Liquor from the oxidizer is circulated to the packed scrubber. A bleed stream is withdrawn from this flow and pumped to a set of three reactors where lime is added to precipitate $CaSO_4{\cdot}2H_2O$. The gypsum is separated in a thickener and centrifuge, washed, and used to make wallboard or other construction materials.

Ammonia is not evolved in gaseous form as in the Kuhlmann process in France, but is rather produced as aqua ammonia in the lime reactors and recycled as such. Apparently the system is so dilute that there is not much ammonia vapor pressure over the reactors. The aqua ammonia is returned to the oxidizer.

The pilot plant (5,000 m^3/hr; 1.5 Mw) has been operated for some time with reportedly good results (over 90% removal). At 1,500 ppm SO_2 in the inlet gas and 0.5 molar $(NH_4)_2SO_4$ in the inlet liquor, the sulfite concentration at the scrubber outlet is 0.01 molar. This indicates the very dilute nature of the system.

The main advantages of the process are (1) no fuming in the scrubber, (2) no scaling, and (3) production of gypsum without use of H_2SO_4 to get calcium sulfite in solution as is common in some of the other Japanese processes. Two full-scale plants are underway, one under construction and the other started up in early 1975.

Dowa Mining

In the Dowa Mining process the absorbent is aluminum sulfate and the regenerant limestone. The process is similar to the Kurabo method in that SO_2 is absorbed by a liquor at low pH and the sulfite is oxidized before reaction with lime or limestone. In the Dowa installation (on a H_2SO_4 plant) the gas is not precooled, although precooling probably would be used in a boiler application. The absorbent is actually basic aluminum sulfate, $Al_2(SO_4)_3{\cdot}Al_2O_3$ made by reacting $Al_2(SO_4)_3$ with limestone.

$$2Al_2(SO_4)_3 + 3CaCO_3 + 6H_2O \longrightarrow Al_2(SO_4)_3{\cdot}Al_2O_3 + 3CaSO_4{\cdot}2H_2O + 3CO_2$$

The aluminum sulfate is quite soluble, about 360 gpl, and has a fair affinity for

SO_2; an 0.7 molar solution will hold about 17 gpl of SO_2 in solution at equilibrium with a gas containing 1,000 ppm SO_2. The various reactions involved are said to be as follows:

Absorption:

$$Al_2(SO_4)_3 \cdot Al_2O_3 + 3SO_2 \longrightarrow Al_2(SO_4)_3 + Al_2(SO_3)_3$$

Oxidation:

$$Al_2(SO_4)_3 + Al_2(SO_3)_3 + 3/2O_2 \longrightarrow 2Al_2(SO_4)_3$$

Neutralization:

$$2Al_2(SO_4)_3 + 3CaCO_3 + 6H_2O \longrightarrow Al_2(SO_4)_3 \cdot Al_2O_3 + 3CaSO_4 \cdot 2H_2O + 3CO_2$$

The pH in the scrubber is in the 3 to 4 range. Other operating data (from operation on H_2SO_4 plant tail gas) are as follows:

L/G: 2.5 l/m^3 (about 18.5 gal/mcf).
SO_2 in exit gas: 5 ppm (99% efficiency).
Al_2O_3 in gypsum: 0.04%.
Moisture in gypsum: 5%.
Gypsum crystal size: 100-200μ.

It should be noted that the scrubber temperature in these tests was relatively low as compared to operation on boiler gas, and that this allows a lower L/G. Dowa estimates that an L/G of 55 gal/mcf or higher would be required for boiler gas. However, a small amount of an inorganic catalyst reduces the required L/G significantly; in the pilot plant, tests have given 95% removal from boiler gas (800–1,700 ppm SO_2) at an L/G of about 42 when catalyst was used.

Twice the stoichiometric amount of air is used in the oxidation step. The air is injected through a special nozzle developed by Dowa. Pressure drop across the nozzle is less than 3 atm.

Since the solution is not neutralized completely in each pass, excellent $CaCO_3$ utilization is obtained; the gypsum contains less than 0.1% $CaCO_3$ (expressed as CaO). Two neutralizers are used in series to increase crystal size.

The process has the advantages that (1) there is no need for a special step to remove calcium from the recycle solution, (2) the product is gypsum, and (3) limestone can be used as the regenerant. Operation in the commercial plant (see Table 4.1) has been very smooth and reliable.

WATER SCRUBBING

The process developed by Chiyoda Engineering in Japan, in which water (actually a very weak acid) is the absorbent and limestone the regenerant, is one of the more important indirect methods for using limestone. It has been installed widely

FIGURE 4.10: CHIYODA WEAK ACID—LIMESTONE PROCESS

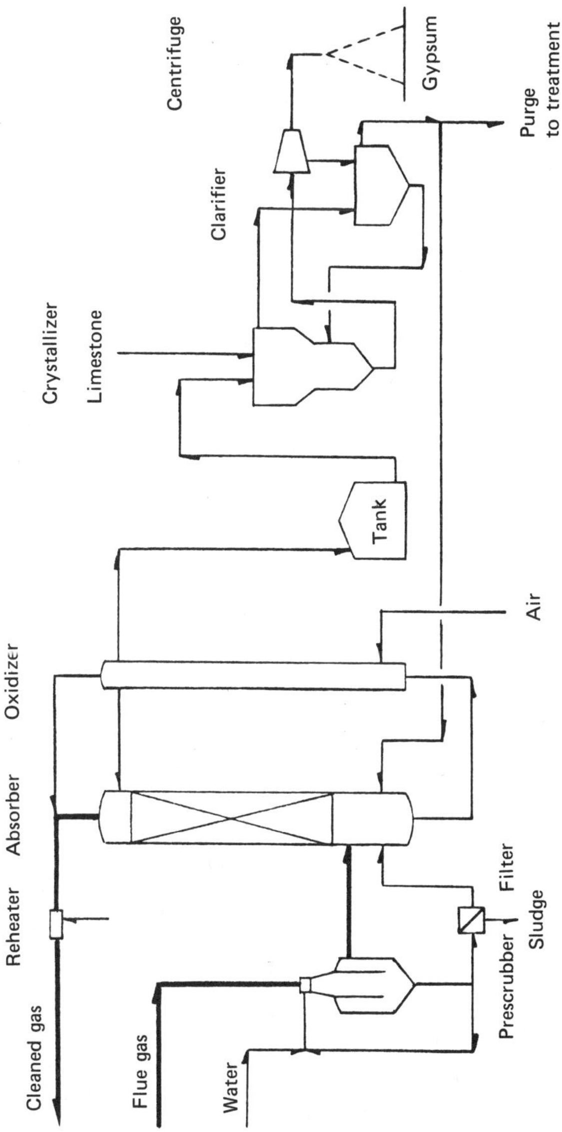

Source: C.R. LaMantia et al (ADL) and N. Kaplan (EPA) at the EPA Symposium, Atlanta, 1974

in Japan, both on industrial and utility boilers, and is being tested in a 20 Mw unit by Southern Services in the U.S.

A simplified flow diagram is given in Figure 4.10. Flue gas is introduced into a venturi type prescrubber for quenching and cooling. The gas then passes to the absorber where SO_2 is absorbed by weak H_2SO_4. The scrubber effluent solution flows to the oxidizer tower where air bubbled through the liquor converts all the absorbed SO_2 to H_2SO_4, with the help of ferric ion dissolved as a catalyst. Part of the liquor leaving the oxidizer passes to gypsum production and the remainder is recycled to the absorber. Thus the recycling absorbent is dilute sulfuric acid solution with a concentration of about 2 to 5%.

In the absorber, oxygen in the recycling liquor and in the flue gas partially oxidizes the absorbed SO_2, thus improving the absorption capacity of the liquor.

The sidestream from the oxidizer is neutralized with lime or limestone to crystallize gypsum. Chiyoda has developed a special crystallizer capable of producing large crystals. The mother liquor leaving the crystal is recycled to the absorber.

The process has several major advantages, including (1) relative simplicity, (2) a high degree of reliability (as demonstrated on oil fired boilers in Japan), and (3) a superior gypsum product. One of the major drawbacks is the very large L/G required because the weak acid will not hold much SO_2 per unit volume. The L/G varies depending on inlet SO_2 concentration and the desired degree of removal. For 1,000 ppm SO_2 and 90% removal, the requirement is on the order of 300 gal/mcf.

The Southern Services test program at the Scholz station is aimed at determining whether use of gas from a coal-fired boiler introduces any problems.

Carbon Adsorption

When SO_2 is adsorbed by carbon, oxidation and water condensation occur so that the product collected in the carbon pores is sulfuric acid. Several ways of removing the acid have been used, one of which is water washing. The product from this is a weak sulfuric acid, usually not more than about 15% H_2SO_4.

Since concentrating the acid is quite expensive, companies such as Lurgi in Germany and Hitachi in Japan have neutralized the acid with limestone to make gypsum. A large Hitachi installation is operating at Tokyo Power's Kashima station in Japan.

The main drawback is the large size and operating complexity of the adsorption step. It seems unlikely that further systems of this type will be constructed.

ECONOMIC FACTORS IN RECOVERY

It is obviously preferable to recover the sulfur oxides in stack gases rather than convert them to a solid form and discard the solid; recovery conserves the sulfur (a valuable natural resource), avoids production of a solid waste and, although it was conceived earlier that recovery processes may reduce the net cost of SO_2 control when compared to throwaway systems, such a situation in reality does not usually exist. Notwithstanding these good points, there are several drawbacks to recovery that have kept it from making significant commercial headway.

Each of the recovery processes has its own peculiar economic characteristics, good and bad; these will be explored in the later process descriptions. In this chapter, the economic factors that apply generally to all recovery methods will be discussed.

GENERAL CONSIDERATIONS

The ideal sulfur oxide recovery process is one that gives adequate emission control without increasing the cost of the main product (e.g., electric power, metals, or sulfuric acid), avoids cooling of the stack gas and consequent loss of plume buoyancy, and gives an easily storable, concentrated product such as elemental sulfur that minimizes marketing problems. It should be said at the outset that such an ideal is probably impossible to attain.

In the first place, the sulfur oxide content of the gas is so low (except in some smelter operations) that recovery ordinarily would not be considered at all; the profitability level necessary to attract investment under commercial conditions is obviously unattainable for a raw material containing only 0.15% or so of a product that sells for only a cent or so per pound.

The problem can be more easily visualized by considering that for 60 lb of sulfur recovered in a power plant from a ton of coal (85% recovery from coal containing 3.5% S), the return from sale of the sulfur at today's prices is less than $1.00. This is a small return for processing the large amount (about 12 tons) of wet, dirty gas formed in burning the ton of coal. In industry, processing of such lean raw materials ordinarily is restricted to the ores of metals such as copper, gold, and uranium—products that command a much higher sales price than does sulfur.

Cooling of the gas can be avoided by using a dry type of recovery process. Such processes have major limitations, however, as will be discussed later. Finally, recovery of the SO_2 as elemental sulfur, although a worthy objective, incurs a considerably higher cost than recovery in an oxidized form. It is not very logical to oxidize the sulfur in the coal to SO_2 (by burning in the boiler or roasting in a smelter), reduce it back to sulfur, and finally oxidize it again for ultimate use as H_2SO_4 (the form most used in industry).

Although it is highly unlikely that SO_2 recovery from dilute gases will ever be a profit-making operation in the usual commercial sense, the recovery route is still preferable if it loses less money than throwaway operation costs. A standoff in this respect may not be enough to justify choice of recovery; since the throwaway approach frees the producer from the problems of higher capital outlay and of product marketing, recovery must have some degree of financial advantage to compete with throwaway.

Published accounts of recovery economics vary widely. There was a tendency for process developers to predict a high profit potential in the early stages of development, followed later by a more conservative evaluation. Some even claimed that their processes could break even, but these cost estimates usually omitted cost of money as a necessary cost item. Moreover, these cost estimates were based on a new undeveloped technology whereby many unknown problems have subsequently emerged which greatly increased the original costs. The estimates have also been difficult to compare because of widely varying basis and scope. The most extensive estimates appear to be those in a series being prepared by TVA for EPA.

The objective in this series is to subject the major processes to a detailed conceptual design analysis and estimate the costs on a uniform basis. Studies completed so far deal with (1) limestone injection (dry system), (2) lime/limestone scrubbing, (3) ammonia scrubbing (with phosphate fertilizer as the end product), and (4) magnesia scrubbing followed by thermal regeneration.

Probably the most accurate and detailed of the TVA studies is the most recent one describing the five most advanced flue gas desulfurization (FGD) processes—limestone slurry scrubbing, lime slurry scrubbing, magnesia slurry scrubbing-regeneration to acid, sodium solution scrubbing-regeneration to sulfur, and catalytic oxidation. The objective of this study is to prepare a set of highly visible, detailed capital and operating cost estimates for comparison on a common, uniform basis.

INVESTMENT AND OPERATING COST FACTORS

The sulfur oxide problem has attracted a large number of inventors, with the result that the number of recovery methods that have been proposed is large—probably well over a hundred if all the variations of major approaches are counted. Most of the emphasis has been on concentrating the SO_2 as a first step, because the main problem in accomplishing removal in an economical manner is the very low concentration in the stack gas. Once the SO_2 is recovered in a concentrated form, either as a gas stream rich in SO_2 or as an undiluted sulfur compound (e.g., Na_2SO_3), further treatment to produce a finished material usually is not a major problem.

Concentration of the SO_2 is usually accomplished by an absorption or adsorption step in which the SO_2 is caught by the sorbent and retained while the gas passes on to the stack. The loaded sorbent then is removed from the gas stream and treated to recover the sulfur. The processes differ mainly in type of sorbent, method of contacting sorbent and gas, and method of recovering sulfur compounds from the sorbent; each of these differences has a major effect on process eco-omics.

Physical Form of Sorbent

From the chemical standpoint, most of the sorbents can be classified as (1) alkali salts, (2) alkaline earth compounds, (3) metal oxides, and (4) adsorbents such as activated carbon. Economic differences between these classes will be treated in later chapters. Another important cost factor is the physical form of the sorbent, which can be either solid, liquid, or, in at least one instance, gaseous.

Solid sorbents can be either (1) granular or (2) nongranular (pulverulent) and liquids can be either (3) melts, (4) compounds that are liquid at ordinary temperature, or (5) aqueous solutions. Chemically, the sorbents cut across these classes. For example, Na_2Co_3 can be used in physical forms (1), (2), (3) and (5). Although sorbent physical form is an important cost factor, there are no good cost comparisons available. Therefore only general considerations as to cost can be given.

Granular Solids: Alkali compounds and metal oxides were used in most of the earlier solid sorption processes. Sodium salts were popular, and manganese, iron, zinc, and copper were the principal metals involved. Sodium aluminate, the sorbent in the U.S. Bureau of Mines alkalized alumina method, is an example of the granular solid class. However, to date, most of these solid sorption processes are abandoned with the exception of the copper oxide process being tested on a fairly large scale in Japan. Universal Oil Products licenses the Shell copper oxide process in the U.S.

The type of contactor used to react the solid absorbent with the gas, an important cost factor, will be discussed in the next chapter. There can also be so much loss in granule strength resulting from the cycle of sorption and regeneration that a large and expensive makeup requirement is needed. Some consider this to be such a major problem as to make other physical forms more promising.

Some of the processes (Lurgi, Chemiebau, Hitachi, Foster Wheeler-Bergbau Forschung) adsorb the sulfur dioxide on carbon granules where it oxidizes and absorbs moisture to give sulfuric acid impregnated in the adsorbent. Different routes are followed in removing the acid. The main drawbacks in the carbon processes are the very large size of equipment required, low concentration of product acid (Lurgi and Hitachi), and high carbon consumption (Chemiebau, Hitachi). These seem likely to be critical problems unless further development brings improvements. The Bergbau Forschung may circumvent some of these problems with a rather novel regeneration scheme. It is to be tested at the Gulf Power's Scholz Plant on a 20-Mw level.

Finely Divided Solids: The problems of attrition and pressure drop can be avoided by using a finely divided solid in the first place. This requires an entrained contactor system and very efficient dust-recovery equipment to recover the sorbent. Thus, in a coal-burning plant the expense of installing two electrostatic precipitators, one to remove fly ash and the other to collect the loaded sorbent, is indicated. Loss of activity over many cycles can also incur heavy makeup cost. The fine solids type of operation was originally tested in the Mitsubishi (Mn_xO_y) and Still [$Ca(OH)_2$] processes. However, both of these processes are now abandoned.

Melts: A few of the proposed methods involve absorption at high temperature in a molten absorbent. Examples are molten iron (Black, Sivalls, and Bryson) and a melt of alkali carbonates (Atomics International). The main advantage is fast reaction rate and resulting small size of contactor. An expensive hot precipitator or other dust removal equipment ahead of the absorption vessel is necessary (or at least highly desirable), however, and retrofit installation would be quite expensive. The BS&B process has not been developed beyond the bench scale level. However, AI's molten carbonate process was tested on a 10-Mw flue gas stream at Consolated Edison's Arthur Kill Station. A combination of process engineering problems, particularly stress corrosion cracking, and Consolidated Edison's financial difficulties forced abandonment of this program before a complete evaluation was available. No future plans for rejuvenating the program appear to be in progress.

Organic Liquids: Sorbents that are liquid at ambient temperature are mainly of the organic type. Dimethylaniline, for example, has been used to scrub stack gas from smelters. The main drawback is volatility of the organic sorbent. The method is generally considered to be applicable only to gases of relatively high SO_2 content.

Aqueous Solutions and Slurries: Most of the problems associated with solid absorbents can be eliminated by using an aqueous solution of a material that has an affinity for sulfur oxides; examples are ammonia, sodium carbonate (or hydroxide), sodium citrate, sodium phosphate, and magnesium oxide. Such solutions or slurries have relatively high capacity, mass transfer is efficient in most cases, oxidation to sulfate is relatively low (which simplifies regeneration in some processes), and transport of the absorbent around the absorption-regeneration loop is easier than for solids.

Offsetting these advantages over solids is the greater difficulty in regeneration (with some exceptions) because of the water present. Moreover, the decrease in thermal lift of the gas caused by cooling and humidification is a major drawback, as discussed earlier. The necessary reheating of the gas is an expensive operation.

Aqueous solution processes include Davy Power Gas, U.S. Bureau of Mines, Stauffer, Stone and Webster Ionics (sodium salt), Chiyoda (dilute acid), and the various ammonia-scrubbing methods (Comico, IFP, TVA-EPA, Czechoslovakian, Romanian, Japanese). Aqueous slurries give certain advantages over solutions, including possible easier separation of the absorption product if fly ash is removed before the scrubber. Examples are Grillo (MgO-MnO_x), Chemico-Basic (MgO), Tsukishima Kikai Company (MgO), and United Engineers (MgO). The slurries are more difficult to handle than clear liquids but not enough so to be a major problem.

A slightly different approach is the Atomics International's Aqueous Carbonate System (ACS). Very concentrated sodium carbonate solution is uniformly sprayed into the wet incoming flue gas, evaporating the water and producing a finely divided powder for regeneration. No reheat is required.

Gases: Ammonia seems to be the only gaseous reactant proposed. The ammonia processes generally involve aqueous solution scrubbing, but a few developers have injected gaseous ammonia into the stack gas stream and collected reaction product downstream. Although equipment requirements are reduced, incomplete reaction and loss of ammonia are major problems. Although ammonia is relatively innocuous to humans, unreacted ammonia could have phytotoxic effects if released in appreciable quantity.

Oxidation and Reduction: Although not sorption processes, the oxidation and reduction methods are included here to complete the process coverage. By use of a catalyst the sulfur dioxide can be oxidized to sulfur trioxide, while still in the gas stream, by the excess oxygen present. Upon cooling, moisture is absorbed and sulfuric acid produced. The leading example of this is the Cat-Ox (Monsanto) process, now being tested on a full scale 100-Mw unit. The main problems are high investment and relatively low acid concentration. Operating cost, however, may be competitive because very little in the way of labor, materials, or utilities is needed. The Tyco process was another example of oxidation in place; nitrogen oxides already present in the gas stream were used to catalyze the oxidation. However, this is another example of a process no longer being developed.

It has also been suggested that the sulfur dioxide can be reduced to sulfur while traveling through the gas system, by adding a reducing agent and using a catalyst that promotes reaction of the reducing agent with sulfur dioxide. Various processes of this type (Peter Spence, Princeton Research, Chevron) have been proposed but all have either failed or do not appear to be very far advanced.

The main drawback to oxidation or reduction in place is high investment, because all units of the system must be large enough to handle the full 12 tons or more of combustion gas per ton of coal burned in a power plant. As discussed later, the required rate of return on investment is a very important factor in evaluating such processes. A further disadvantage of the reduction process is that incomplete reaction of the reducing agent could be expensive and, for some reducing agents (such as H_2S), could give rise to a pollution problem as well.

Fuel Type

Fuel type has a major effect on process economics. For example, FGD systems installed on oil-fired units comparable in fuel sulfur content and plant size to coal-fired units are generally less costly, on the order of 20 to 30%. On the other hand, with the fuel oil market in such a shaky situation, both in respect to cost and supply, it may be more economical from a total cost standpoint to install FGD equipment on coal-fired units. The situation is certainly more complex than this. Some recovery processes require a relatively clean gas input which benefits units burning oil. Others, like the Shell copper oxide process, operate in such a way that dust is not a major problem. Still others may have requirements for clean gas not based on, say, product purity or plugging of absorbers but on the need to reduce harmful side reactions. It is known that fly ash can promote high oxidation rates in FGD systems, an intolerable situation for some sodium scrubbing processes. Political pressure may in the long run dictate what fuels the electric utilities are allowed to burn. If this is the case, coal is likely to be used in increasing quantities regardless of FGD technology.

Sulfur Content of Fuel or Ore

An obvious way to improve recovery economics is to use a raw material (coal, oil, or ore) of higher sulfur content. For example, some smelters, particularly those processing copper ores, have become major producers of sulfuric acid because the high sulfur content of the ore gives a gas of high enough SO_2 content to make sulfuric acid production economical. High-sulfur coal and oil are also available (see Chapter 1), and possibly could be sought out by a utility to make a recovery operation more economical.

High-sulfur coal would give considerable economy in regard to equipment sized on the basis of gas flow rather than sulfur throughput. Thus processes like catalytic oxidation, in which all equipment must handle the full gas stream, are benefited the most. In contrast, wet-scrubbing processes require gas flow only through the scrubber and the size of the rest of the unit must be increased for higher sulfur throughput. There is some economy of scale but much less than for equipment which does not have to be any larger for the higher sulfur loading.

An example of the effect of sulfur content on process economics, taken from the EPA-TVA cost comparisons of flue gas desulfurization study, is given in Figure 5.1. An increase in investment cost of 30 to 40% is seen in changing the sulfur content of coal from 2.0% to 5.0% when using the limestone, lime, sodium, and magnesia processes. No increase is seen for the catalytic oxidation process.

There is the possibility that high-sulfur coal could be obtained at lower cost than for the medium-sulfur type, an advantage that has been assumed in some process analyses. This may be true in situations where high-sulfur coal must be beneficiated at extra cost in order to meet a larger demand for lower sulfur fuel. However, it is conceivable that very high sulfur coals could actually command a premium if recovery processes were generally adopted as the more desirable method for air pollution control.

Moreover, the advantage of high-sulfur content would be lost rapidly if selective mining were required to obtain it. To minimize cost, the coal deposit should average somewhere near the desired content and the coal should be mined as found and handled at the power plant in such a way as to get a fairly uniform boiler feed, as is usually done in large plants.

The use of high-sulfur coal, either naturally occurring or reject material from a deep cleaning process for coal, to improve recovery economics probably would be most feasible in a new plant in which the boiler would be designed for the individual coal

FIGURE 5.1: EFFECT OF SULFUR CONTENT OF COAL ON UNIT INVESTMENT COST*

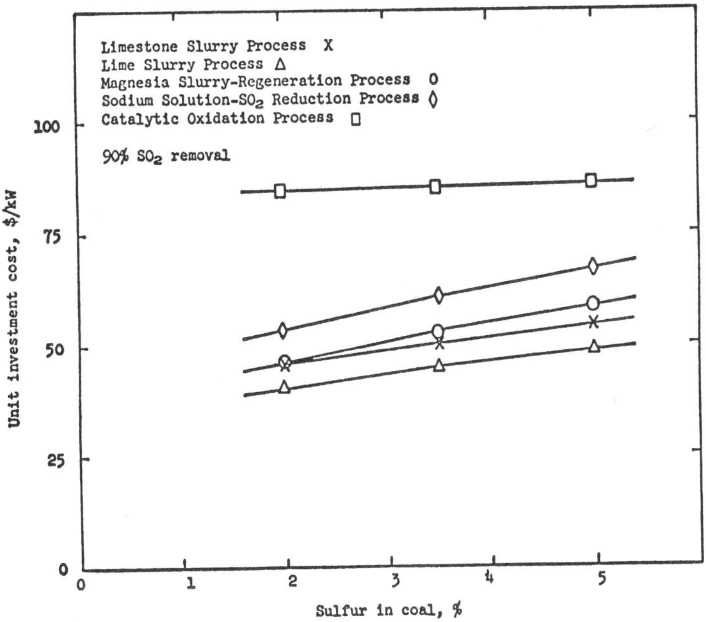

FIGURE 5.2: EFFECT OF SULFUR CONTENT OF COAL ON LEVELIZED UNIT OPERATING COST REGULATED ECONOMICS*

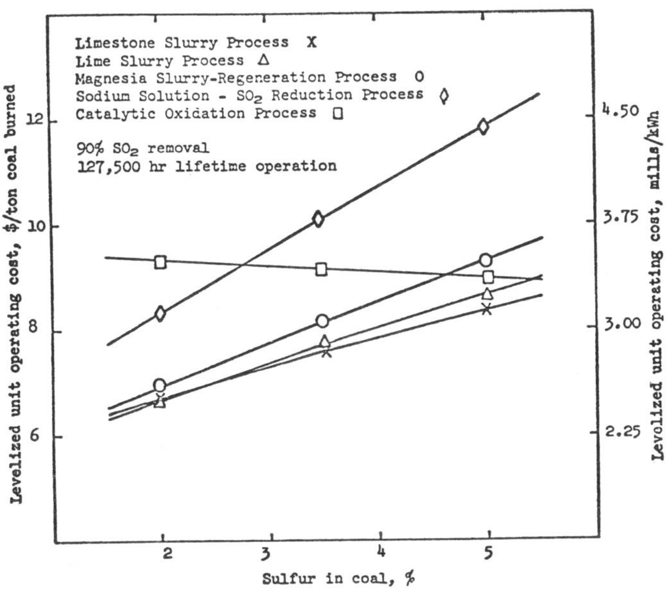

*500-Mw new coal-fired units

characteristics. Another example of the effect of sulfur content on process economics, also taken from the same EPA-TVA study, is given in Figure 5.2. The same relationship as in Figure 5.1 on investment cost is noted in Figure 5.2 which depicts operating cost versus sulfur content in coal.

New Versus Existing Plant

The status of the production unit, whether it is already in existence or only in the planning stage, is an important factor in recovery economics. Some of the processes require major alterations in plant design and all must have space for the equipment, requirements that are much more economically met if they can be provided for in the design before the unit is built.

The resulting saving varies widely, of course, with type of process. The maximum advantage is gained by a process such as catalytic oxidation, which operates in power plants at temperatures higher than air heater exit temperature and therefore requires either installation of process equipment ahead of the air heater or the more expensive operating cost alternative of reheating the stack gas to the required temperature. Processes that operate at air heater exit temperature or lower, and which therefore can be added at the end of the system, have a lower but still significant advantage in new plant construction because the equipment can be located better and cost is generally lower when installation is part of a large power plant construction project.

Some further advantage may accrue from the opportunity of eliminating normal items of plant equipment when a recovery unit is installed in a new plant. For example, wet-scrubbing processes can double in removing dust as well as sulfur oxides. The saving will vary depending on the particular process.

The prospect for taking advantage of new plant economy seems good because most of the new plants being built are large enough for good production economy whereas a large percentage of those presently in operation are not (except where it is feasible to group small units together). The nonrecovery limestone processes appear more suitable for most of the smaller existing units.

Plant Size

Up until the middle 1960s, power plant boiler units were built in relatively small size, typically 200 Mw or so. This is large enough to accommodate the larger scrubber sizes that have been attempted, but the regeneration section would be uneconomically small at this size level (unless very high sulfur coal were burned). Hence to take advantage of economy of scale, larger boiler size is needed or a grouping of small boilers at a large central station.

For smelters, which are generally large operations, the SO_2 output seems large enough for favorable size economy. Sulfuric acid plants, on the other hand, have relatively low SO_2 output. The situation is improving in power plants because of the continuing increase in size of new boilers being installed.

Estimates of future boiler unit size indicate that over 95% of the capacity installed in the 1970s will be in units 600 Mw or larger and about 80% in units of 1000 Mw or higher capacity. By 1980 over half the total capacity in the United States will likely be in units 600 Mw or larger. Practically all of the remainer will be in units 10 years old or more (most of them under 200 Mw) and the capacity factor will be relatively low. Therefore, the pollution problem will be centered in the large, new units because of their major contribution to the power supply.

Because of this there should be ample opportunity to install recovery units in large, new plants, which is not only desirable because of the economy of scale but also because new plants have a better capacity factor and longer life than old ones. The size advantage varies, of course, with type of process. It is possible that some processes will require fairly small modules at a particular point and therefore lose a good part of the size advantage.

In obtaining economy of scale by grouping small boilers, two or more might be ducted into one scrubber, or, if this proved too complicated, the loaded sorbent from two or more boilers could be processed in a single regeneration unit. The same concept could be applied to small plants with total output from all boilers too small for a recovery operation, by collecting loaded sorbent from several plants and regenerating it in a central unit. This approach has been used in this country by the Mystic Station of Boston Edison and the Dickerson Station of Potomac Electric Power Company, both feeding a loaded magnesia absorbent to an acid plant some distance away for regeneration. The small plants served by a single regeneration facility probably would have to be fairly closely located to minimize cost of transporting the sorbent back and forth.

The effect of plant size on investment and operating cost is similar to that resulting from use of high sulfur fuel or ore; in both cases the higher throughput of sulfur reduces unit cost. Curves much like that in Figures 5.1 and 5.2 are generated by relating power plant size to cost. Figure 5.3 and 5.4, taken from the EPA-TVA report, are examples of the change in unit investment and operating cost for varying power unit sizes. Not much economy of scale is seen with units larger than 600 Mw.

Plant Complexity

A process requiring many steps obviously is likely to cost more than one requiring only a few. Direct comparisons of processes in this respect can only be made adequately with a full design and cost study. The EPA-TVA cost comparison study does this for three recovery processes: magnesium oxide, sodium sulffite, and catalytic oxidation. These will be reviewed in detail later but a few generalizations can be made at this point.

For sorption type processes, most of the complexity is in regeneration of the loaded sorbent. The simplest regeneration method is heating a sulfite or bisulfite to evolve SO_2; if the sorbent is recovered without chemical change or undesirable physical alteration, such a process has good possibilities. Normally, however, the sulfate formed by oxidation in the scrubber does not regenerate by heating and a desulfating step must be added. Moreover, the regenerated sorbent may lose activity because of physical deterioration, in which case a sorbent reclaiming step would probably be required.

Some processes are designed to form an intermediate (e.g., ammonium sulfate) which is used as a raw material for some other process such as fertilizer production. This simplifies the sorption part of the process but increases overall com-

FIGURE 5.3: EFFECT OF POWER UNIT SIZE ON UNIT INVESTMENT COST, DOLLARS PER KILOWATT

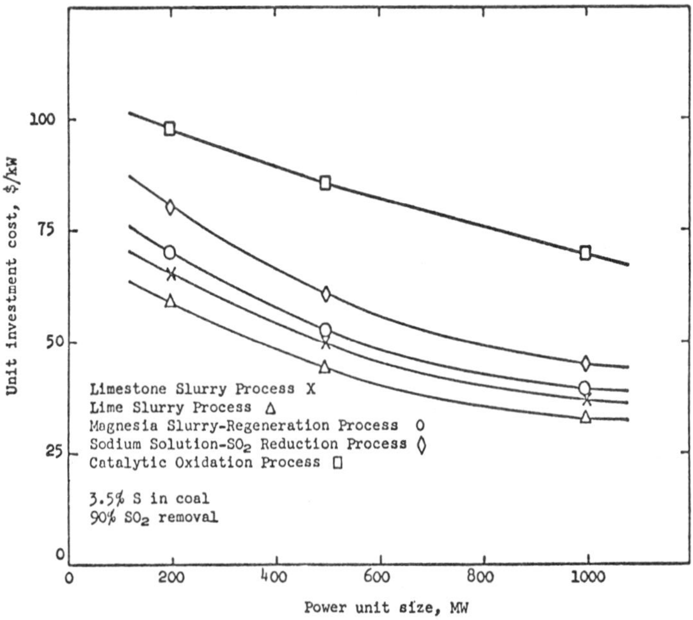

FIGURE 5.4: EFFECT OF POWER UNIT SIZE ON LEVELIZED UNIT OPERATING COST

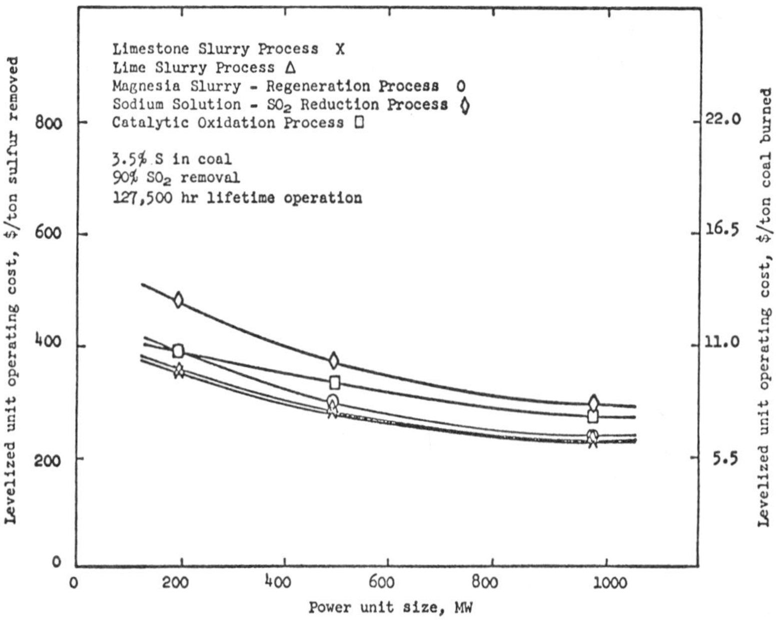

plications. Process complexity is also increased when more than one material is produced. For example, in regeneration of ammonia scrubber effluent solution with nitric acid, both ammonium nitrate and H_2SO_4 are produced.

Probably the least complex of the various methods is catalytic oxidation. Only a catalytic converter, condenser, and mist eliminator are required, and there are no requirements for utilities or labor. However, the very large vessel sizes for this process are a severe drawback, along with other problems.

SO_2 Removal

An important and somewhat surprising result derived by comparing the investment projections in the EPA-TVA cost comparison study is the relatively minor investment savings realized (3.2 to 4.5%) in designing for 80% SO_2 removal as compared to 90%. This may be an important criteria in designing present scrubbing facilities while trying to project into the future what the SO_2 emission standards may be then. The moving target (e.g., particulate emission standards) is always a problem for any industry, and may be worse for the electric utility industry where large gas volumes and varying load factors are involved.

Dust Removal

In coal-burning power plants and smelters, the dust load in the gas must be considered. Since it causes trouble if allowed to mix with the sorbent, most processes include prescrubbing or some other method to remove the dust before SO_2 sorption. The cost of this varies with process type, but generally desulfurization capital investment cost will increase by 20% if dust removal is included.

Capacity Factor

Power plants generally have a low capacity factor (actual production as percentage of production at full design capacity) as compared to chemical plants. The power plant cannot store its product as can most chemical operations, and since the demand for electricity is quite variable the plant must be designed for rapid change in production rate and must operate at less than full capacity for a good part of the time. Smelters and sulfuric acid plants do better in this respect.

Moreover, power plants are usually operated on a system basis; several plants distributed over a service area are operated as an integrated system to supply the power demand, peak as well as average, in the most economical way. In most cases a computerized control system is used to analyze the demand pattern and to assign the load to those plants that can most economically supply it to the points of use.

Thus some plants are operated at full capacity a good part of the time, some operate only at part capacity much of the time, some operate only at peak system load, and some are standbys that operate only in case of primary plant breakdown or other emergency. The range in individual plant capacity factor depends on the variation in system load, which differs between systems because of geographical variations in use pattern. A typical weekly load curve for the TVA system is shown in Figure 5.5 and an annual curve in Figure 5.6.

FIGURE 5.5: TYPICAL WEEKLY LOAD CURVE FOR TVA POWER SYSTEM

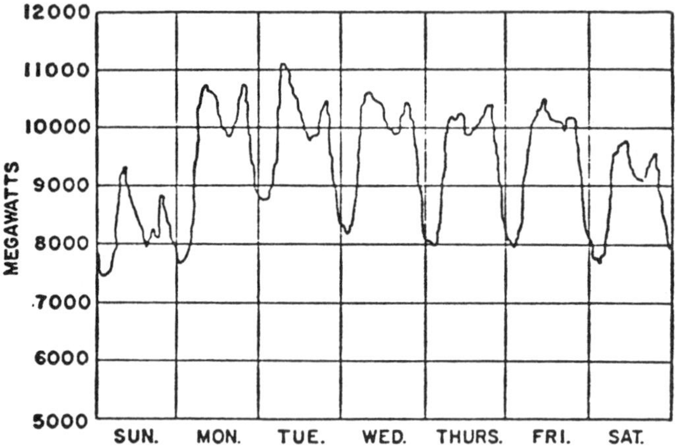

FIGURE 5.6: ANNUAL VARIATION OF LOAD IN TVA POWER SYSTEM

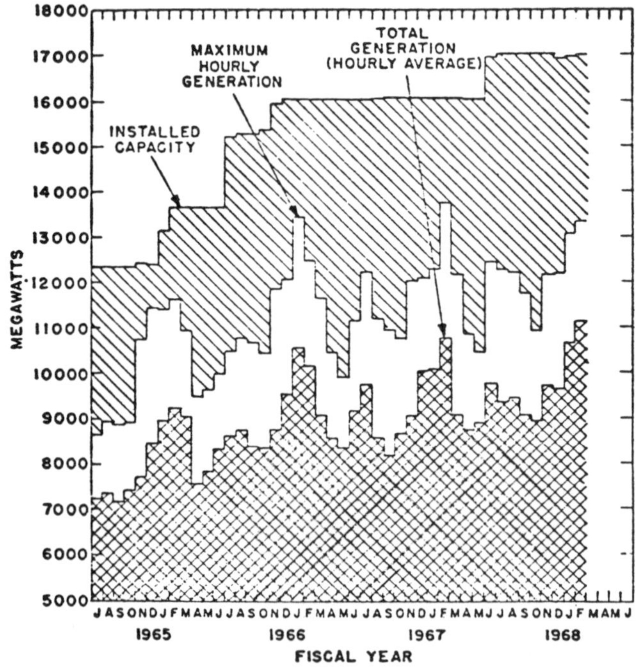

Historically, new generating units added to a power system have been larger, more efficient, and more economical to operate than units already in service.

It follows that the newer units get the most use and that the dispatch system operates the less efficient plants only when necessary. Then as the new units grow older they in turn are pushed into peak load or standby status by even newer and more efficient units.

Even though new generating units are usually assigned a base load status, they do not necessarily have high load factors because at times other factors such as location and cost of fuel may favor shifting of load to some of the older plants. Moreover, problems that develop during initial operation of new units often require additional outage time for modification.

All this is quite undesirable for a chemical operation; chemical products can be stored and the usual practice is to operate the production unit as continuously and uniformly as possible, with swings in product demand taken care of by using storage as a buffer. The plant may be slowed or shut down completely for a period if storage becomes a problem. Operation in conjunction with a power boiler, with wide swings in production rate, would be expensive and for some processes quite difficult to accomplish.

In some sulfur oxide recovery situations it may be practical to use a holding tank for loaded sorbent as a buffer between sorption and regeneration, so that only the sorbing operation would have to vary with the power plant load factor. Or the recovery unit might be designed for only partial sulfur oxide removal, where pollution regulations permit, so that even at reduced boiler load there would be enough sulfur throughput to keep the recovery unit running at capacity.

Buffer storage of loaded sorbent is not feasible for processes such as catalytic oxidation or reduction but may be applicable to scrubber-type methods. Although the scrubber would have to be operated with the boiler, the scrubber effluent, or some other intermediate material, might be held in surge storage so that the main recovery equipment could operate continuously. It appears that the alkaline earth scrubbing methods [Grillo (MgO-MnO_x), and Chemico-Basic (MgO)], would be the most appropriate for this; both produce a storable solid intermediate ($MgSO_3$, and/or $MgSO_4$) in the scrubbing step. Similarly, the Atomics International's ACS system produces a dry storable intermediate (Na_2CO_3-Na_2SO_3) in the scrubbing step. Other sodium processes currently available must regenerate a liquid form of the loaded absorbent.

Some data is available regarding the effect of load factor on operating cost for specific recovery processes. From estimates made in the TVA-EPA study of the reduction in load factor, there was found to be a serious economic deterrent; the operating cost was increased by about 55% by a drop in load factor to 60%, a typical level in power plants. Processes that have relatively high investment incur the most disadvantage from low load factor because of the continuing capital charges when the unit is out of operation or not running at capacity.

Product Type

All the major recovery methods produce either elemental sulfur, sulfuric acid, or a fertilizer material. Some are limited to one product, others make two, and in some the product type is optional.

Sulfur: For most situations, sulfur is the most desirable product, it can be stored at relatively low cost, its high concentration minimizes shipping cost, and it probably would command a higher price in relation to going market price than the other materials. The cost of reducing agent is a significant cost item, however, and a relatively complex plant would be required.

Of the various reducing agents, coal is preferred because of relatively low cost and the simpler supply situation; most power plants do not have natural gas readily available in the quantities required. The impurities in coal are troublesome, however; this and other factors complicate the economics. Producer gas could be made from the coal and used as the reducing agent, of course, but this would add substantially to the cost.

Most of the processes that produce sulfur have an intermediate step in which either a rich stream of sulfur dioxide or a metal sulfate or sulfite is formed. These concentrated materials are then reduced in most processes to hydrogen sulfide which is converted to sulfur by the Claus process. There is some economic advantage for sulfur dioxide or a sulfite because less reducing agent is required than for sulfate. This penalizes the dry absorbent methods, where oxidation to sulfate is much higher than in the wet processes.

It can be said that although sulfur is a desirable product, it will probably cost considerably more to produce than the oxidized products. Since most of the sulfur is oxidized before use, it seems desirable to keep it in the H_2SO_4 or finished fertilizer form except under especially difficult marketing conditions.

Sulfuric Acid: In some of the processes that make H_2SO_4, the loaded sorbent is heated or otherwise treated to release a rich stream of sulfur dioxide that can be converted to sulfuric acid in a plant similar to the type used for making H_2SO_4 from pyrites. The economics of using the loaded absorbent as the H_2SO_4 plant raw material can be compared to use of pyrites; this replacement actually can occur in the Grillo method, for the loaded sorbent (a metal sulfite-sulfate mixture) would be regenerated in pyrites-type sulfuric acid plants in one version of the process.

The main difference is that fuel is required in regenerating the sorbent (in which the sulfur is in the oxidized form) and producing the stream of sulfur dioxide, whereas in use of pyrites (FeS_2), combustion of the iron and sulfur is a heat-producing operation. The cost of the fuel plus the cost of the sorption step must be weighed against the cost of mining and transporting pyrites less the value of the heat produced in burning pyrites. If the latter costs less, some subsidy to the stack gas process must be provided.

Processes such as catalytic oxidation and some versions of carbon absorption have the advantage that no separate sulfuric acid plant is needed; reaction with the oxygen and the moisture present in the gas stream produces the acid directly. Sulfuric acid is attractive in that most of the processes are simpler than those for making other products. The main problems are high storage and shipping costs, plus marketing problems of low grade acid as discussed later.

Phosphate Fertilizers: A promising use for sulfuric acid or for an intermediate such as ammonium sulfate is in production of phosphatic fertilizer at or near the power plant. If phosphoric acid, which is a high-volume material in the fertilizer industry, were the final product there should be much less difficulty in obtaining an acceptable price for it. One of the advantages would be acceptability of acid strength lower than the commercial level; by making 60 to 70% acid, which is adequate for treating phosphate ore to make phosphoric acid, investment and operating costs could be reduced.

Other process variations give finished fertilizers rather than phosphoric acid, which is an intermediate material. A solution of ammonium sulfate can be produced and used along with phosphate ore and nitric acid to make a high-grade nitrogen-phosphate fertilizer. Ammonium phosphate and ammonium nitrate are made in other versions of ammonia scrubbing. All of these have the advantage that the product is a concentrated, widely-used, high-volume material that can be shipped economically.

BASIS OF FINANCING

Plants for producing materials such as sulfur, sulfuric acid, and fertilizers are normally financed as risk-type enterprises. If in the planning stage the project shows promise of realizing a return on investment higher than that obtainable by loaning money out at interest, investors may supply capital on the basis that the chance of getting a return higher than interest adequately offsets the risk that something adverse will happen after the money is invested to reduce the return to a level below the interest rate.

Thus the cost of money is a legitimate project cost in evaluating a process, and it must be set at the level necessary to attract investment. This varies widely with specific conditions, of course, but it seems generally accepted that the projected cash flow (depreciation plus net profit) must be enough to pay out the investment in about 5 days. The economic promise can also be expressed as interest rate of return, which takes into account the time value of money; this can be defined as the interest rate a bank would have to pay to accept and return money on the same schedule as the proposal. An interest rate of return on the order of 15% seems to be a good average for the level adequate to attract investment.

It can also be argued that the cost of capital should be set at the going interest rate, on the basis that funds can be obtained from banks on this basis, and that no profit should be expected from pollution abatement. The drawback to this is that the company building the plant must take the risk that the actual return realized may not be enough to pay the interest, in which case the company must take up the difference as a net loss or go bankrupt. To offset this risk, a projected profit is justifiable even for pollution abatement investment.

Because of the high investment required for SO_2 recovery methods, generally higher than for throwaway operation, the cost of money item becomes a major consideration in evaluating processes. The EPA-TVA estimates and actual real case situations show that there will be no situations in which the projected return will be enough to attract capital on the usual basis.

It should be noted that the term break even has been publicized by some as a worthy objective in developing recovery processes. The term usually means nothing because the cost items considered are not specified. If it is meant that the income is just sufficient to cover operating costs, the process obviously does not break even, as someone must pay the capital costs. There seems little likelihood, unless new unforeseeable methods are developed, that any recovery process will break even on a realistic costing basis.

One possibility for improving (least loss) recovery economics in the power industry comes from the fact that the power producer presumably can treat the cost of pollution abatement as an operating cost and pass it on to the power consumer in the form of a rate increase for the electric power sold. Thus there could be situations in which a chemical company supplied the capital and operated the project, with the power producer paying to the chemical company a fee for the service of SO_2 control. With the two sources of income, sale of product and the service fee, the project would have a better chance of attracting capital.

The size of the service fee would be a matter of negotiation. The limiting consideration would be the cost that would be incurred by the power company if it took care of the problem on its own by installing a throwaway limestone process. The utility possibly would go a little higher than this to avoid getting involved in a chemical process, but probably not much higher.

The effects of the service fee on projected rate of return from a recovery process are shown in Table 5.1. As would be expected, the results depend on the size of the fee charged; for a fee equivalent to a high cost limestone scrubbing process, desirable profitability could be achieved in some cases and for a smaller fee equivalent to a low cost limestone process, low profitability would result in all cases. Funding under this concept will probably be limited.

It may also develop (as the situation now appears) that utilities will finance recovery installations, in which case the investment would be merged, if approved by the state regulatory commission, into the total power plant investment (the rate base) as is the practice for dust-removal equipment. Since the utility is allowed to earn on the rate base at a rate set by the regulatory commission, the SO_2 recovery unit could therefore be installed and operated at little or no risk.

Under this situation there should be no more difficulty in attracting capital for the recovery unit, even if it is a major chemical operation, than for the power plant proper. Any loss on the recovery operation could be offset by an increase in the price charged for the main product, electric power. Such a situation could very well develop into a political nightmare with extreme environmentalists fighting for pristine air, rate payers raging over increases in electric bills, and the electric utility getting the blame from both sides.

This would make the overall cost of recovery processes much lower because the return on investment allowed by regulatory commissions is lower than that which investors must have (as a reasonable prospect) before they can justify putting their money in a nonregulated commercial enterprise. The lower return in the regulated power plant situation is acceptable to investors because of the low risk.

The return allowed on the power plant rate base varies with the situation, but appears to average about 10% (assuming 50% debt and 50% equity). Depreciation is also much lower than that usually applied in nonregulated situations, again because of the differing magnitude of the risk factor. It should be noted, however, that because of governmental regulation there is not as much flexibility in pricing power as there is for chemical products. Although the utility is, in

TABLE 5.1: PROFITABILITY OF SULFUR OXIDE RECOVERY WITH SUPPLEMENTARY INCOME AS PAYMENT FOR POLLUTION ABATEMENT[a]

	Low equivalent payment[c]		High equivalent payment[d]	
Conditions[b]	Payout, yrs.	Interest rate of return, %	Payout, yrs.	Interest rate of return, %
Base case - coal-fired units (500 MW, net unit, 3.5% S in coal, 48.5% avg. capacity factor over 30 yrs., reheat to 175° F)	7.6	8.8	5.6	14.9
Exceptions to base case (coal fired)				
Existing unit (27 yr. life)	7.7	8.4	5.7	13.6
5% sulfur	7.7	8.5	5.4	15.7
200 MW	8.3	7.4	6.7	11.0
1000 MW	7.1	10.0	4.8	18.1
Base case - oil-fired units (500 MW, net unit, 2.5% S in oil, 48.5% avg. capacity factor over 30 yrs., reheat to 175° F)	7.2	9.8	6.1	13.0
Exceptions to base case (oil fired)				
Existing unit (27 yr. life)	7.0	10.0	6.0	12.7
4% sulfur	7.5	9.0	5.8	14.0
200 MW	7.6	8.8	7.1	9.9
1000 MW	6.8	10.9	5.3	15.8

[a] Process: Magnesium oxide scrubbing with sulfuric acid production.

[b] Net sales revenue assumed at $8.00/ton of acid.

[c] Equivalent to limestone-wet scrubbing cost assuming low limestone price, on-site pond disposal of solids.

[d] Equivalent to limestone-wet scrubbing cost assuming high limestone price, off-site disposal of solids.

TABLE 5.2: COST OF MAGNESIA SCRUBBING vs LIMESTONE-WET SCRUBBING UNDER REGULATED ECONOMICS

Conditions	Cumulative net annual costs, $ millions		
		Limestone-Wet scrubbing	
	Magnesia scrubbing[a]	Low limestone process cost[b]	High limestone process cost[c]
Base case - coal-fired units (500 MW, new unit, 3.5% S in coal, 48.5% avg. capacity factor over 30 yrs., reheat to 175° F)	162.1	136.2	170.6
Exceptions to base case (coal-fired units)			
Existing unit (27 yr. life)	160.2	133.4	162.3
5% sulfur	180.5	148.5	194.3
200 MW	91.1	72.7	82.6
1000 MW	237.9	208.3	283.2
Base case - oil-fired units (500 MW, new unit, 2.5% S in oil, 48.5% avg. capacity factor over 30 yrs., reheat to 175° F)	95.9	84.2	94.2
Exceptions to base case (oil-fired units			
Existing unit (27 yr. life)	93.7	84.1	92.3
4% sulfur	111.0	93.8	112.1
200 MW	54.4	46.3	47.7
1000 MW	141.6	129.5	154.3

[a] Net sales revenue assumed at $8.00/ton of acid.

[b] Limestone cost - $2.05/ton; on-site pond disposal of solids.

[c] Limestone cost - $6.00/ton; off-site solids disposal cost, $6.00/ton.

general, allowed to pass on justifiable costs to the consumer, it cannot do so without authorization, which may require an extended period and considerable effort on the part of the utility.

The effect of the lower overall rate of return required on the investment under regulated financing is shown in Table 5.2. The values shown are the cumulative net annual costs over the life of the power unit; thus the values represent the total bill including return on investment and income taxes over the power plant life. Only under very high limestone process costs can the magnesia process be economically competitive.

The situation is not so favorable in the smelting and sulfuric acid industries, where product price is controlled by the market rather than by regulation. The cost of SO_2 removal must be passed on to the customer, of course, but only within the framework of competitive enterprise. The operation is of the risk type rather than protected, and investment therefore is more important than in the utility industry.

The investment required for several recovery processes is fairly well defined. Yet some published figures for new processes (and also for some more developed ones) are so low as to seem unrealistic; others are more reasonable, on the order of \$60/kw or higher, and these likely will continue to rise until inflation weakens and the technology is more fully developed.

PRODUCT MARKETING

A significant economic consideration in sulfur oxide recovery is the income that can be obtained from sale of product. Since the recovered material is a byproduct, the marketing situation is not good in the first place because the material cannot be produced in accordance with the market demand. If the market is poor, the product must either be sold for what it will bring or stored until the market will absorb it, both of which reduce net income. No matter what the case may be, it appears that even under favorable market conditions a net reduction in operating cost of 10% may be the most optimistic situation.

Plant Location

One of the more important economic considerations is the location of the power plant with respect to customers for the sulfur byproduct. The very large amounts of product involved make it almost essential to arrange large-quantity sales to single customers if high sales cost is to be avoided. The sulfur-consuming industry most amenable to this is the phosphate fertilizer industry, which uses almost half the sulfur consumed in the United States. Modern phosphate plants are quite large and provide individual points of high consumption such as needed for marketing power plant byproduct sulfur; for example, a plant producing 1,000 tons per day of phosphate (tons of P_2O_5), which is in the current upper size range, would use 2,800 tons per day of sulfuric acid, the output from about 3,000 Mw of power-generating capacity (at full load).

In order to keep shipping costs from reducing net income too much, the phosphate plant should not be very far from the power plant. Unfortunately, much of the phosphate production is in Florida and along the Gulf Coast, where power production is relatively small, whereas a major part of the sulfur emission is in the Northeast where there is little phosphate production.

The best area appears to be the East North Central region, where the two are more nearly in balance and much of the recovered sulfur might be used in phosphate plants without shipping very far. This would be particularly true for recovery processes in which sulfuric acid is the product, whereas location would be less important for recovery plants producing elemental sulfur.

The upper Midwest is a heavy consumer of fertilizer phosphate (54% of the United States total in 1967) and the high-sulfur coal is there, about 90% of the coal containing over 3.5% sulfur is mined in the block of states comprised of Ohio, Illinois, Indiana, Iowa, western Kentucky, and Missouri. Most of the coal mined in the eastern states contains less sulfur and therefore recovery economics would not be as good. The nonrecovery limestone processes may be more applicable for this area. This depends on land availability, which may not be conducive to throwaway systems, particularly in the northeastern section of the country.

Barge shipping of the product may be feasible from those power plants located on navigable water. Modern barge transport is quite economical; anhydrous ammonia, for example, which requires special refrigeration equipment in transit, is moved on the central rivers for 4 to 6 mills per ton-mile, indicating that sulfuric acid could be shipped a thousand miles for less than $4 per ton. Fortunately, about 80% of the sulfur dioxide emission is from plants located on navigable rivers.

Smelters are not generally located as favorably as power plants in relation to either fertilizer plants, areas of high sulfur product consumption, or navigable waterways. Although several smelters are installing or increasing capacity of byproduct sulfuric acid plants to abate atmospheric pollution, it is not clear where the acid can be sold or what price it will bring.

Product Quality

The price obtainable for the byproduct sulfur will depend in part on the quality of the particular material produced. For sulfur, it may be reasonable to assume that the going market price for primary sulfur (the mined sulfur product) can be realized, which usually is a rash assumption for a byproduct. However, the large market for sulfur, plus the fact that it can be stored at a low cost, makes the assumption seem justifiable. For this to be true, the color and impurity content must be acceptable. If the particular recovery process gave a different color or more impurity than for normal sulfur, some sales price reduction would very likely have to be accepted.

Product quality may be even more important if sulfuric acid is the product. The processes that give a rich stream of sulfur dioxide have an advantage in this respect because any dust in the gas stream with the sulfur dioxide must be removed to protect the catalyst. For catalytic oxidation and carbon adsorption, however, the necessity for cleaning the entire gas stream presents a problem. Even 99% removal leaves enough dust in the gas to give up to a percent or so of dust in the product, depending on how much of the residual dust is caught during condensation of the acid. Another problem is moisture condensed from the gas, which limits the concentration to 80% sulfuric acid or so for catalytic oxidation. In carbon adsorption (Lurgi and Hitachi processes), the dilution is even more pronounced because water is used to wash the acid out of the carbon pores. Most commercial sulfuric acid contains practically no dust and not more than about 7% water. Byproduct acid differing from this probably would have to be priced lower than normal acid to make it salable.

Market Potential

Assuming that the byproduct resulting from sulfur oxide recovery is of standard quality, a further problem is that the market must absorb the production over and above the existing production capacity. The main obstacle to this is the tremendous amount of sulfur product involved. Emission of sulfur oxides from all sources in 1970 was estimated to be equivalent to 18.3 million short tons of sulfur in the United States (see Chapter 1), and the projection for 2000 is 62.9 million short tons (assuming adoption of the breeder reactor in power production). In comparison, United States consumption of sulfur is estimated at 11.6 million short tons in 1970 and 32.4 million short tons in 2000 (on the basis of 3.5% increase per year).

These figures emphasize the problem in developing an adequate market for recovered sulfur oxides. Not all of the tonnage emitted would be recovered, of course, even under the most favorable production cost and market situation. Many of the emitting sources are so small that some method other than recovery would be more economical for control. But any significant fraction recovered would place such a strain on the market that the price structure might well collapse.

Sulfur would again be the favored form in which to recover the sulfur oxides, the high concentration, broad spectrum of use, and ease of storage all give sulfur a leading position in marketing flexibility and support the possibility that a price somewhere near the going level could be obtained. The price of sulfur has varied widely over the past two or three decades, however, in accordance with successive periods of shortage and oversupply. In the periods of shortage, large quantities of byproduct sulfur probably could be absorbed by the market without any large price decline. However, during the years of oversupply, the flow of recovered sulfur would continue and force the price down.

The future market situation for sulfur depends to a considerable extent on the supply of mined sulfur and of sulfur recovered from sour gas. Mining reserves are apparently declining and may approach economic exhaustion in the next two or three decades; it is reported that new reserves are not being found as fast as old ones are mined out, and that mining costs are increasing as the industry moves to reserves of lower quality and accessibility.

Offsetting this is the growing supply of sulfur recovered from sour natural gas (containing H_2S) in various parts of the world, particularly Canada. This is also a byproduct (or at least a coproduct) which is not produced in accordance with the market demand and which therefore tends to force the price down.

Sulfuric acid presents a much more difficult marketing problem than does sulfur, even though over 90% of the sulfur is used in making acid. Most of the sulfuric acid is made by sulfur consumers either for captive use or small-to-medium-lot sale. Hence, selling byproduct sulfuric acid on anything other than a small-lot basis requires that it be sold to established producers as a replacement for acid they would otherwise make themselves (except for new acid-consuming plants that might be built).

The price obtainable cannot be much above the cash outflow for production from sulfur, or the price of the contained sulfur plus a dollar or two for operating cost. Even at this low price (less than half the usual published price), marketing large quantities of byproduct acid may be difficult.

In one study it was concluded that smelters in the western part of the United States possibly could make and sell 5.0 to 5.3 million tons of H_2SO_4 per year, if they were willing to sell the acid at $4 per ton. This would take up the market and leave no room for recovery from other sources such as power plants, and would reduce the smelter emissions only 60 to 65%.

The market potential may be considerably better if the recovered sulfur oxides are used to make fertilizers at or near the sulfur-emitting plant. This approach has been used with some success by smelters in the western United States, although the practice is limited by the low fertilizer consumption in that region. In the East, where most of power plant capacity is located, it would appear promising to locate a fertilizer plant at the power station and integrate sulfur oxide recovery with fertilizer production to minimize operating cost.

Over half of the sulfuric acid produced in the United States is used in making fertilizers. Therefore producing acid or its equivalent at the power plant and using it there in manufacture of fertilizers seems logical. Fertilizer consumption is growing rapidly and should continue to be the major end product from use of sulfur. World consumption of fertilizer is expected to double in the 1970 to 1980 decade.

The results of a recent TVA study for EPA on the marketability of abatement sulfur products indicate that for the approximately 2 million tons of sulfuric acid that could be produced annually by TVA, the net sales revenue above distribution costs would be approximately $8.75 per ton.

However, once a sulfur oxide recovery facility is installed, the continuing necessity of operating it, whether there is a market for the product or not, reduces the market potential severely. Because of this, a combination recovery-throwaway process may be very desirable. The recovery facility could then be operated in a more effective marketing mode by shifting to throwaway during periods of oversupply rather than dumping product on the market, depressing the price, and incurring a loss greater than for throwaway. This would also help in the problem that might arise for a particular plant, that of a major customer stopping or reducing purchase of the product; shifting to throwaway would bridge the gap until a new customer could be found.

RECOVERY PROCESSES—GENERAL PROBLEMS

Each of the many recovery processes that have been tested (full scale, prototype, pilot plant, and bench) or proposed differs from the others, of course, but there are some general problems and considerations that can better be discussed before the processes are considered separately. Several of these were treated in the preceding chapter from the standpoint of effect on process economics; in the present chapter, the emphasis will be on what can be done to reduce the severity of each problem and on how important each may be for the various types of processes. To recapitulate, recovery processes can be grouped as follows.

Absorption by Solid Reactants: A generalized flowsheet for absorption by solids is given in Figure 6.1. Major variations in such a system are (1) location of the dust equipment in the boiler system (perhaps omitted if oil is the fuel), (2) type of dust removal equipment, (3) degree of dust removal, (4) particle size of absorbent, (5) location of contactor in the boiler system, (6) type of contactor, (7) type and location of regenerator (may be integral with the contactor), and (8) concentration of SO_2 or H_2S in the gaseous effluent from the regenerator.

For use in smelters, the flowsheet may be much the same except that there probably is not as much latitude in location of the dust removal and contacting equipment in the overall plant flowsheet. Sulfuric acid plants, of course, would not require dust removal and the location of the contactor (after the acid absorbers) is fixed.

The emphasis on absorption by solids as a method for SO_2 removal has been somewhat cyclical. Up until about 1960, liquid-scrubbing was the approach generally explored by investigators but growing emphasis on the necessity of reheating the gas to maintain plume buoyancy (and the difficulty in doing so) turned attention to the solids-type processes. The emphasis has now shifted back again, although work is still proceeding on solids processes and one of the commercial units is presently in operation (Shell copper oxide system at Showa Yokkaichi in Japan). The reheat problem, although still a difficult one, does not now appear to concern investigators as much as it once did—even though there is a severe energy penalty in reheating.

FIGURE 6.1: GENERALIZED FLOWSHEET FOR SO_2 ABSORPTION BY SOLID MATERIALS

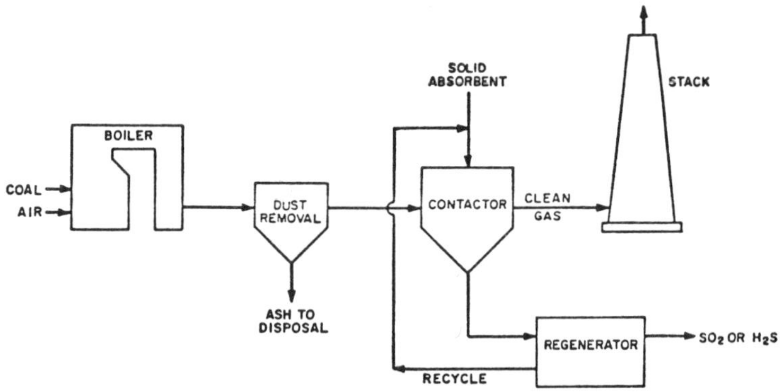

Adsorption on Solid Materials: The activated carbon and other adsorbent processes also use a flowsheet similar to Figure 6.1. Major variations are in dust removal, type of contactor, separate or integral regenerator, and type of product from the regenerator. Some of the commercial and prototype units are of the carbon-adsorption type, including the Sumitomo-Kansai carbon system at Kansai's Sakai Power Station, Hitachi carbon system at Tokyo Electric's Kashima Station, and the Foster Wheeler-Berbau Forschung prototype carbon system at Gulf Power's Scholz Station.

Solution or Slurry Scrubbing: The major division in recovery processes is between sorption by solid reactants and absorption in liquids. Figure 6.2 gives a generalized flowsheet for processes based on absorption in aqueous solutions or slurries. Major variations are (1) type and location of dust removal equipment, (2) degree of dust removal, (3) degree of cooling before scrubber, (4) separation or integration of dust removal, gas cooling and scrubbing functions, (5) type of scrubber, (6) type of regeneration system and (7) type and degree of gas reheating. Since the location of the scrubber is more or less fixed by the need to have as low a gas temperature as possible, flowsheets for smelter and sulfuric acid plant installations would be much the same as for power plants.

Research on wet scrubbing processes goes back for more than half a century. Probably the earliest analogous work was on making ammonium sulfate without need for a sulfuric acid plant—by burning sulfur, absorbing the SO_2 in ammonia solution, oxidizing the solution with air, and crystallizing ammonium sulfate. Later, in Japan, this method was applied to stack gases.

Early development work on liquid scrubbing of stack gases took place in the period bounded roughly by 1930 and 1950. Most of this had to do with smelter gases; ammonia scrubbing was installed at a smelter in Canada and research on absorbents such as aluminum sulfate, magnesia, and sodium carbonate was carried out in various parts of the world. Ammonia scrubbing was later installed on sulfuric acid plants in the United States, Czechoslovakia, Austria, Japan, and Romania and on a power plant in the USSR. Many liquid scrubbing recovery processes are now in commercial use. In addition, several other liquid systems are being vigorously piloted in an effort to reach full scale testing as early as possible.

FIGURE 6.2: GENERALIZED FLOWSHEET FOR SO_2 ABSORPTION BY AQUEOUS SOLUTIONS OR SLURRIES

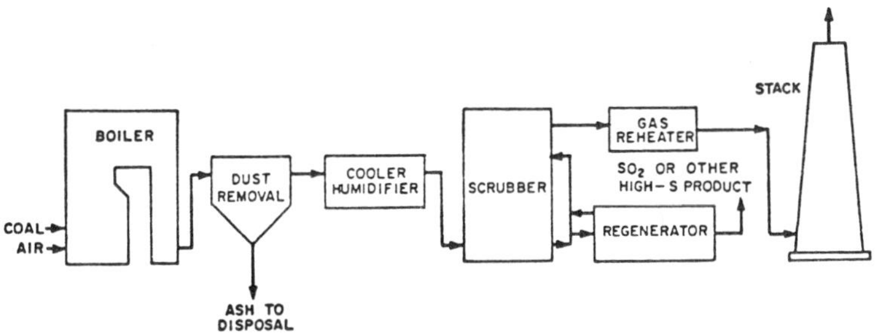

Oxidation or Reduction of SO_2 in Main Gas Stream: In smelter operation, where the SO_2 content of the offgas is usually relatively high, direct oxidation or reduction of the SO_2, without an intermediate step of absorption, has been a general commercial practice for some time. Many smelters make H_2SO_4 directly from the gas and reduction to sulfur has been carried out in a few instances. The low SO_2 content in power plant and sulfuric acid plant stack gases make oxidation or reduction in place much less promising but one commercial process is based on this approach. Figure 6.3 is a flowsheet of the process.

FIGURE 6.3: THE REHEAT CAT-OX SYSTEM

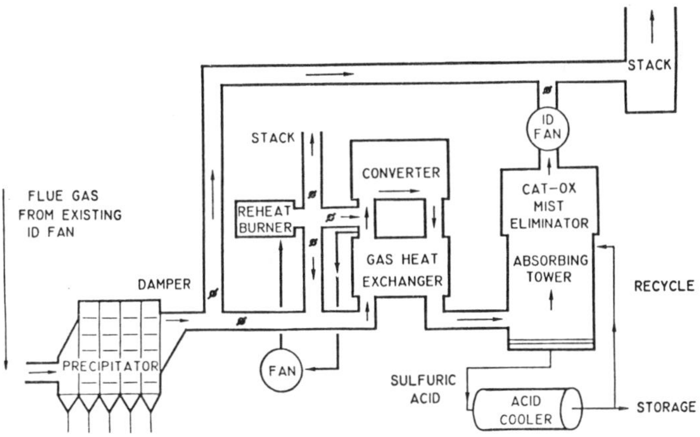

This installation is at the 100 Mw Wood River Station of Illinois Power Company. Very high dust removal efficiency ahead of the reactor is needed for protection of the catalyst.

Thus there are several process problems and considerations that are common to several process types. Dust removal before SO_2 removal is common to all; contactor type and location, regenerator type, and treatment of evolved SO_2 or H_2S are common to most of the solid sorbent methods; precooling, scrubber type and configuration, gas reheating, and treatment of intermediates apply to most of the scrubbing processes; and the other process types have some problems in common with each other and with the sorption methods.

DUST REMOVAL

The removal of dust and other particulate matter from gas streams is a large and complicated subject—much too extensive to cover fully herein. The problem of removing dust, however, cannot be separated from that of removing sulfur oxides in coal-burning power plants and smelters; therefore it must be considered at least in a general way.

General considerations in particulate removal, for both dry and wet SO_2 sorption systems, were covered under the lime/limestone discussions in Chapters 2, 3, and 4. The main differences in regard to recovery processes is that the dust preferably should be removed before the SO_2 removal step, whereas in the limestone methods the two can be removed together, or, if removed in separate units, carryover of dust into the SO_2 absorber perhaps can be tolerated. Recovery processes require either essentially complete prior removal of dust or a subsequent step for separating dust from product. Otherwise, product contamination would likely reduce the sales price. Moreover, in some processes even small amounts of dust in the incoming gas interfere with process operation.

High Temperature Processes

Some recovery processes, mainly the solid and melt absorbent types, must operate at elevated temperature, which, in a power plant, means that they must be installed ahead of the air heater (temperature above 300°F). For the high degree of dust removal required, a hot electrostatic precipitator is the only type of equipment suitable at these elevated temperatures. In recent years the hot side ESP has been shown to be reasonably reliable and effective in removing particulate for general application to power plant flue gas.

Although the hot ESP has had excellent performance in its early development, there are some economic drawbacks. Assuming that the ash resistivity is such as to give good precipitator operation at the usual temperature (about 300°F), the high temperature process must incur the penalty of increased cost because the gas volume is larger and therefore, a larger precipitator is required. This would not apply however, if the coal had such low sulfur content that the precipitator did not perform well at low temperature (see discussion in Chapters 2 and 3).

Moreover, even 0.02 gr/scf of dust remaining in the gas can be a problem, as it would amount to about 1.6 tons/day of dust from a 1,000 Mw boiler—or about 4 lb/ton of equivalent H_2SO_4. Thus processes that can tolerate some dust have

the problem of gas distribution that may occur in configurations which require the gas to make a turn between scrubbers; gas distribution in the venturi is not affected by a duct turn before it, of course, and the distribution into the following scrubber unit is relatively uniform. Even so, type (3) is currently the most popular choice among the full scale installations now in operation.

Carryover of dust-laden liquor as mist into the SO_2 removal stages could be a problem, particularly for processes in which regeneration of the absorbent is a completely closed-loop operation. Either an effective mist elimination system between stages or a means for bleeding ash from the closed loop (without loss of absorbent) would be necessary. The general trend seems to be toward venturis for wet collection of dust. Their proven performance in other industries, both for dust removal and gas humidification, is a major advantage.

SORPTION BY SOLIDS

The use of solid sorbents, both of the absorbent and adsorbent type, presents several problems in design and operation. As compared to absorption in liquids there is not nearly as much engineering experience on which to base design and predict operation difficulties. The method also has the drawback that the SO_2 must find its way some distance into the sorbent granule or particle if an adequate loading during each pass through the sorption vessel is to be attained, whereas, in a liquid the SO_2 can reach any sorbent molecule without difficulty. Thus the entire solid sorbent system may be designed for promoting SO_2 penetration into the granule or particle interior.

The sorbent loading attainable also depends to some extent on ability of the SO_2 to penetrate through reacted material, on the granule surface or on the internal surfaces in pores, which is affected by the ratio of reaction product unit volume to that of the sorbent. If the fraction product volume is larger, the reacted layer will tend to be dense and nonporous, like an oxide film on aluminum. If it is smaller, a porous or discontinuous film is formed through which the SO_2 can penetrate with relative ease.

Contactor Design

One of the major problems is design of the gas-solid contactor in which sorption takes place. The design objective is to maximize contact of gas and sorbent with a minimum of (1) pressure drop in the gas, (2) contactor size, and (3) stress (mechanical or otherwise) on the sorbent. Design considerations differ considerably depending on whether the sorbent is granular or finely divided.

Packed Bed: The contactor that puts the least mechanical stress on a granular sorbent is the packed bed type. Since the sorbent remains in place, the operation must be cyclic; that is, gas is passed through the bed until the sorbent has taken up SO_2 to near its limit and sorption begins to drop off, after which the gas flow is cut off and the sorbent regenerated by treatment to remove the absorbed SO_2. A parallel set of contactors is required so that the gas flow can be switched from the saturated bed to a fresh one.

A sketch of a packed bed arrangement is shown in Figure 6.4. Although this system gives maximum protection to the absorbent, it has some major drawbacks.

The major one probably is the cost and operating difficulty in shifting the very high tonnage of gas from one vessel to the other at frequent intervals. The dampers required are not only expensive but also subject to various operating and maintenance problems.

FIGURE 6.4: FIXED BED CONTACTOR ARRANGEMENT FOR SO_2 ABSORPTION

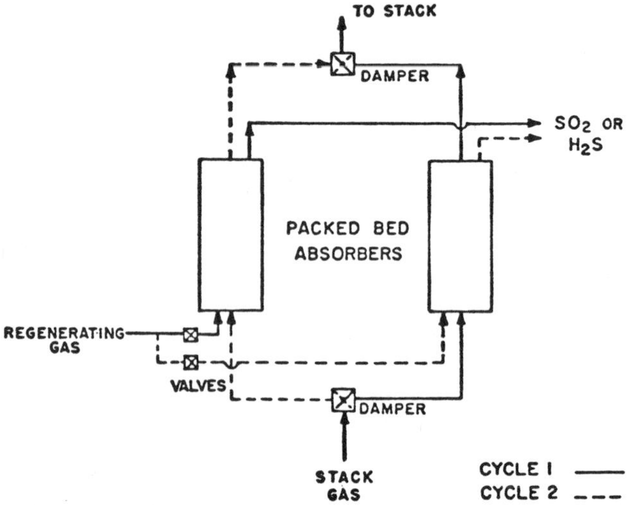

Other problems are (1) the need to remove dust almost completely before the absorber in order to avoid the necessity of frequent removal and cleaning of the absorbent, and (2) necessity for very tight dampering when a reducing gas is used for regeneration (to avoid explosion resulting from mixing reducing gas with the oxygen-bearing stack gas). Finally, the design problem is made difficult when the absorbent must be at higher temperature during absorption than during regeneration, as it must be in many of the processes. This requires a heating and cooling step in each cycle, with resulting delay in the cycle and danger of excessive thermal stress on the equipment.

The alternation of steps in the packed bed system is much like the operation of making semiwater gas from coke, which once was a popular way of making synthesis gas (for production of ammonia, methanol, and other basic chemicals) but today is regarded as nearly obsolete. Continuous, noncyclic methods have been developed that are cheaper to install and easier to operate.

The high investment and cumbersome operation of the cyclic fixed bed system are such major drawbacks that many developers have ruled it out on the basis that better contactors are available. Hitachi, however, has installed such a process on Tokyo Electric's 150 Mw Kashima Station. The loaded sorbent is regenerated by acid washing to produce a 20% acid. One of the departures is that developed by Shell in the Netherlands. The reactor is built in such a way that the gas does not pass through the sorbent but through channels alongside it.

The sorbent is held in thin, vertical beds by a mesh-type material so that the gas can diffuse into the sorbent. Good operation on a 40 Mwe (gas flow equivalent) refinery in Japan is claimed, with low pressure drop and no plugging of sorbent by dust.

Moving Bed: Continuous noncyclic operation can be attained by use of a contactor in which absorbent granules, usually in plug flow, move downward through the vessel countercurrent to the upward flow of gas. A second vessel is used for regeneration.

A generalized version of the moving bed type is shown in Figure 6.5. The sorbent moves down through the absorber vessel, is removed and transported to the top of the regenerator (by a bucket elevator or pneumatic conveyor), moves down through the regenerator in contact with the regeneration gas, and is elevated back to the top of the absorber.

The arrangement is quite like the moving bed technique of catalytic cracking in the petroleum industry, which made its debut during World War II. In this process the catalyst promotes cracking in one stage but collects a carbonaceous deposit that must be removed at short intervals by burning in another stage. The moving bed technique allows economical operation with cycles as short as from 10 to 20 minutes. The process has been simplified somewhat by performing both operations (catalysis and regeneration) in the same vessel. The catalytic step takes place in the upper section of the tower and the catalyst then moves downward through the regeneration zone in the lower section. The process gas enters at the top and flows out through a side exit at the midsection, while air enters at the bottom and leaves at another side exit; a purge stream of steam is introduced at the midsection to keep the two gas streams from mixing.

One of the main problems is mechanical attrition of the catalyst, which was increased when bucket elevators were generally replaced with air lifts. Major improvements have been made in the pellet hardness but the catalyst make-up required because of attrition loss is still a major operating cost. For SO_2 absorption the Chemiebau process (updated version of the Reinluft process) is an example of the moving bed contactor system. No commercial demonstration is in operation or planned.

Fluidized Bed: One or both of the moving beds in Figure 6.5 can be replaced by a fluidized bed with some advantage. The transport of absorbent from one vessel to the other is considerably simpler as the material behaves like a fluid and therefore can overflow from one bed to the other. Moreover, a more uniform temperature can be attained in the bed, which is important in the regeneration step of some processes.

Offsetting these good points are the drawbacks of (1) nonuniform residence time of absorbent in the bed, (2) greater likelihood of solids entrainment, and (3) poor turndown capability. However, the fluid bed technique is used successfully in several industries, including catalytic cracking, limestone and ore calcination, pyrites burning, and solvent vapor recovery (adsorption on carbon). For SO_2 recovery, it was originally proposed for the absorption step in the alkalized alumina process (now abandoned due to severe attrition problems) and the Westvaco carbon adsorption process now ready for commercial development.

FIGURE 6.5: MOVING BED CONTACTOR SYSTEM FOR SO_2 ABSORPTION

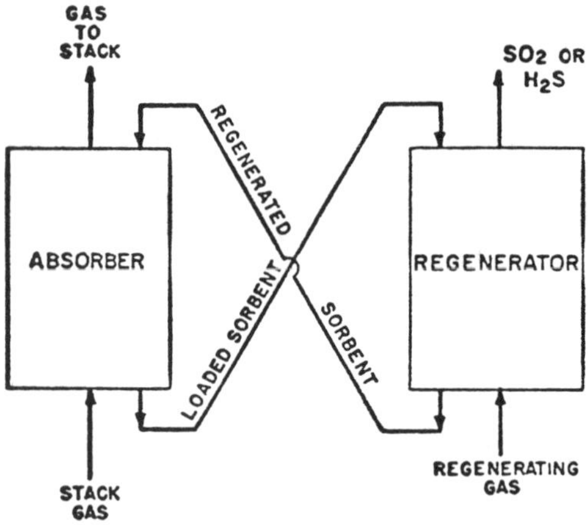

Entrained Reactor: The problem of solids attrition can be avoided or minimized by use of an entrained contacting system, in which finely divided absorbent is introduced into the gas, carried along with the gas stream for some distance, and then removed in a dust collector. Regeneration can then be carried out in any type of system amenable to treatment of fine solids. The attrition problem is reduced because the absorbent is finely divided in the first place and any finer material formed by attrition will be captured in the dust collection unit (although formation of extremely fine particles may require larger dust collection equipment). A relatively long retention time can also be obtained, which is helpful in some processes.

One disadvantage for a coal-burning power plant is the need for two dust collection units, one for ash collection and the other to recover the absorbent. Erosion of the contactor surface can also be a problem, and there is some loss of absorbent even from an efficient dust collector. The leading example of the entrained reactor is the Mitsubishi manganese oxide process tested at the Chubu Electric Power Station in Yokkaichi, Japan. However, because of the troubles encountered with slurries in the regeneration section and the difficulties of selling ammonium sulfate, it is no longer in operation.

Crossflow Granular Bed: In another type of system, ash and SO_2 are removed simultaneously by use of a thin bed of granules placed at right angles to the gas flow. The granules can move downward either continuously or intermittently and, for ash removal alone, they can remain in place with occasional back flow of gas to remove the dust. For SO_2 removal the granules would be made of an SO_2 sorbent material and would be regenerated after shaking off the ash. Both the Foster Wheeler Berbau Forschung and the Sumitomo-Kansai carbon processes are examples of this design.

Granule Resistance to Cycling

The main problem encountered thus far in use of solid sorbents is loss of sorbent effectiveness, either from structural deterioration or loss of porosity, after extended cycling (through absorption and regeneration). In addition to mechanical attrition caused by granular particles colloding with or rubbing against each other or equipment surfaces, there is the problem of structural stress resulting from the chemical change during the cycle of sorption and regeneration.

Since this involves a change in crystal structure, there is likely to be weakening of the granule structure and eventual collapse. The situation is somewhat different from that in catalytic operations, where cycles may also be involved, because the catalyst merely provides surface for a transient reaction whereas the sorbent must react throughout the granule, to the extent feasible, so that it can carry the maximum amount of SO_2 out of the sorbent vessel.

Those processes in which the absorbent is subjected to a high temperature during regeneration are subject to the hazard of sorbent sintering, or loss of porosity due to closing up of pores. Although sintering during each pass can be reduced to a low level, the cumulative effect over an extended period of cycling can severely limit absorbent life. The problem applies to both granular and finely divided absorbents but there is not enough experience as yet for adequate evaluation. Adsorbents such as activated carbon are not so much subject to this because the use of carbon normally does not involve high temperature.

ABSORPTION BY LIQUIDS

The problem in using aqueous solutions, aqueous slurries, nonaqueous liquids, and melts as SO_2 absorbents are quite important because the leading processes, both for recovery and throwaway, are of the liquid absorption type. There are several common problems, including gas cooling, mist elimination and gas reheat, that are unique to liquids and are not encountered when solids are used.

Gas Humidification and Cooling

The advantages, problems and techniques of gas cooling are mainly about the same in recovery scrubbing processes as in lime/limestone scrubbing (see Chapter 3). One exception is that inlet gas temperature can be more important in recovery because in some processes the SO_2 vapor pressure over the absorbent solution is much more sensitive to temperature than when lime/limestone slurry is used.

In some of the earlier ammonia scrubbing test projects, for example, direct water cooling has been used to bring the gas temperature down to below 100°F before entering the scrubber—much below the wet-bulb temperature of the power plant stack gas being scrubbed. Although this benefits SO_2 absorption, a very large cooling tower and a large volume of cooling water are required. Cooling to such a low temperature level does not appear to be justified economically except in special situations.

Indirect precooling of the gas, however, may be especially desirable for temperature sensitive processes. Partial cooling by indirect means (e.g., use of liquid circulating between liquid-gas heat exchangers located before and after the scrubber) lowers the wet-bulb temperature and allows a lower scrubber temperature without necessity for condensing moisture from the gas with cooling water. Although the temperature reduction by this means is limited to not more than about 10°F, even this much could be a major advantage in some processes.

Gases from smelters and sulfuric acid plants have the advantage that the wet-bulb temperature is usually lower than for power plant stack gas. For this reason processes that are relatively sensitive to scrubber temperature do better when used in smelters and acid plants. Conversely, methods that are not temperature sensitive are advantageous for power plants. Some processes are so insensitive to gas temperature that they are operated purposely to give a scrubber outlet temperature higher than the wet-bulb, the purpose being to reduce or eliminate the gas reheating requirement. This is done by operating with such a low gas residence time in the scrubber that complete humidification does not occur. The Atomics International Aqueous Carbonate System operates in this manner by using a spray drier type absorber.

Scrubber System Design

There are some important differences in regard to scrubber design between recovery processes and lime/limestone scrubbing (see Chapter 3). In the first place, the main limiting mechanism in limestone scrubbing—rate of limestone dissolution—is seldom a factor in recovery scrubbing operations because most of the absorbents used are soluble in water. Since these usually have a high affinity for SO_2, the limiting resistance is transfer in the gas phase and scrubbers can be designed accordingly. The design objective is to get adequate gas phase transfer with minimum pressure drop, liquid rate, and scrubber size.

Various scrubber types have been used in pilot plants and in full scale scrubbing operations. Since another major difference between recovery and limestone scrubbing is that scaling is seldom if ever a problem in recovery, the moving bed type of scrubber has not found favor. The favored types appear to be packed bed, sieve tray and venturi.

Most of the earlier work was with packed beds, the packing usually being rings or wood grids. Since the usual types of dumped packing have a pressure drop that is quite costly for the large gas volumes involved, stacked or vertically arranged packing such as used in cooling towers has some advantage; the wood grid packing is of this type. Another is a special type that has been developed in Czechoslovakia and is being used there in a commercial scrubbing operation on a sulfuric acid plant; the packing is made up of spirals shaped like long springs, arranged in a vertical array and made of glass or plastic covered metal.

The design objective is to allow high gas velocity without excessive entrainment. A gas velocity of 16 ft/sec with practically no entrainment has been claimed, along with high rate of mass transfer and low pressure drop. The Mitsubishi Heavy Industry grid (packed type) is a very popular absorber in Japan for both throwaway and recovery systems.

Venturi scrubbers have been adopted for the few large scale MgO slurry absorption processes now in operation. It appears that the contact time in a venturi, even though short, is adequate for good SO_2 absorption efficiency. For oil-fired boilers where little dust enters the absorber, a single stage venturi may be sufficient, whereas dual stage venturis will be necessary where moderate to high levels of dust enters the absorber (for example, coal-fired boilers with no ESP or where the existing collector has poor removal efficiency).

Tray type scrubbers have given good results in Japanese full scale FGD systems using sodium scrubbing. This type has a special advantage, over and above the usual considerations, because of the need in some recovery processes to get a high bisulfite to sulfite ratio in the scrubber effluent liquor—a requirement that does not enter in lime/limestone scrubbing. To do this requires two or more separate scrubbing stages with the composition of the recycled liquor in each stage adjusted to give the proper scrubbing effect; for example, in ammonia scrubbing the liquor in the last gas stage should be relatively dilute and low in pH to scrub out any ammonia escaping from the first stages, whereas in sodium scrubbing no volatilization occurs on the top stage and all efforts are devoted to operating the bottom or first stage highly concentrated and at low pH—in the bisulfite region.

Such separation of stages requires that the liquor from each stage be collected and withdrawn for recirculating, a requirement that is difficult to arrange for packed scrubbers if the stages are to be stacked in a single scrubber shell for economy. As noted earlier, a scrubber such as the sieve tray type already has the liquor withdrawal capability; liquor is fed onto the tray from above, gas rushing up through holes in the supporting grid holds the liquor on the tray, and a weir at the periphery withdraws the liquor. Carryover of liquor from one stage to the next can be a problem in such a staged scrubber arrangement.

A disk-and-doughnut type scrubber has also been used in SO_2 recovery. In this type, curtains of liquor fall from disk to doughnut to disk down the tower, so that something resembling spray tower operation is obtained. Other aspects of scrubbing with liquids, such as materials of construction, oxidation of sulfite in the scrubber, mist elimination and gas reheat are much the same as in lime/limestone scrubbing (Chapter 3).

A more recent design is the Showa Denko vertical cone scrubber, which is somewhat like the Zurn and Bahco scrubbers in that liquor is entrained in the entering gas and is carried upward with the gas through a draft tube. The gas then passes through a disengaging section where the liquor separates and falls back to the bottom of the scrubber. Figure 6.6 illustrates this design for one of four identical scrubbers serving the 150 Mw boiler at Chiba, Japan.

Oxidation is a more severe problem in recovery however, because it affects not only scrubber operation but also the regeneration step. In most processes the regeneration method does not work for sulfate, so a special step, always expensive, must be inserted in the flowsheet to take care of the sulfate. For many recovery processes oxidation can be considered to be the major process problem.

CONVERSION OF INTERMEDIATES

In most of the sorption type processes, the regeneration step produces an intermediate sulfur-containing material rather than a finished product. Treatment of this intermediate to give a salable final product is thus a common problem in several processes.

Sulfur Dioxide

The intermediate most often involved is SO_2. Here the absorption and regeneration steps serve the function only of concentrating the SO_2, which is a useful function nevertheless because a concentrated stream of SO_2 can be treated in several ways, and with reasonably good economy, to give a useful and salable product. (The SO_2 is a useful product in itself but the market is too small to absorb any significant amount of by-product from stack gas recovery operations.)

FIGURE 6.6: VERTICAL CONE SCRUBBERS AT SHOWA DENKO

In most processes, the intermediate SO_2 is converted to sulfuric acid. This is a much more expensive operation than making acid by burning sulfur, the general practice in the United States. In this respect it is analogous to producing acid from pyrites (FeS_2), which involves an expensive gas cleaning system to remove dust, catalyst poisons, and moisture from the pyrites roaster gas.

Design of wet type sulfuric acid plants is well developed and will not be discussed here. Many such plants are already in use for recovering SO_2 from smelter stack gas, where the SO_2 content is ten times or more higher than in power plant stack gas. The absorption and regeneration steps in power plant recovery processes bring the SO_2 content up to a level high enough for such an acid plant but there are still some differences as compared to smelter gas or gas from a pyrites roaster.

The main difference is in the properties of the rich SO_2 stream produced in the regeneration step, mainly in regard to the content of diluent (usually combustion gas), dust, and moisture. The diluent content can range from very little (in steam stripping of bisulfite solution) to near the critical maximum for contact sulfuric acid plant operation (in thermal decomposition of wet or dry sulfite precipitates by direct heating).

The dust content depends on the type of operation producing the rich SO_2 stream—near zero for steam stripping to a possibly high content for thermal decomposition of sulfites in a kiln. Moisture content in some processes—particularly those involving steam stripping of bisulfite solution—is fairly high, mainly because the loaded absorbent from the scrubber contains moisture. For the magnesia process, the crystals are fairly well dewatered in a rotary drier before being sent to a kiln. Hence little water from the scrubbing liquor appears with the concentrated SO_2 stream.

In a pyrites or smelter gas plant, one or more wet purification stages are usually included to remove metal and halogen compounds that might damage the catalyst. Presumably these could be omitted in SO_2 recovery and the gas purification operated in a dry mode, if desired, by using dry dust collection equipment. If the usual commercial grade of sulfuric acid is to be made, however, most of the moisture in the SO_2 stream must be removed by cooling (usually to 100°F in standard wet plants). This could amount to a considerable cooling load because moisture can come from several sources including free water in the sulfite or sulfate decomposed to give the SO_2, water of hydration, moisture in combustion air, and moisture formed by combustion of hydrocarbon fuel.

It may be possible to eliminate moisture removal from the gas if the resulting dilution of the sulfuric acid can be tolerated on the basis that the acid is to be used in some operation adjacent to or near the power plant. An example is production of phosphoric acid for the fertilizer industry. However, producing the more dilute acid aggravates corrosion.

In comparing production of sulfuric acid by a recovery process with production from sulfur, recovery must bear the economic penalty of higher investment and operating cost. The magnitude of the penalty depends on the cost of purifying the SO_2 stream and on the acid concentration that is acceptable.

One possibility for reducing the economic penalty is to reduce the size of the acid plant as compared to a normal plant. The resulting overloading would produce a tail gas higher in SO_2 content than could be released to the atmosphere, but this should not be a problem because the gas could be recycled to the main scrubber in the power plant. Some credit can also be claimed for the investment and operating cost that would be required in any event for removing SO_2 from the tail gas in a normal acid plant.

It has also been suggested that the gas drying tower normally incorporated in sulfuric acid plants (using product acid as the drying agent) could be eliminated if the tail gas were recycled to the power plant scrubber—on the basis that sulfuric acid mist generated in the converter and coolers from reaction of the moisture with SO_3 would be removed in the scrubber system. The mist is known to be quite difficult to remove however, and might not be collected adequately even with an alkaline scrubbing medium.

The drying tower is also eliminated in the Russian CO process announced several years ago. In this method, ore roaster gases are passed directly to the converter without any treatment other than dust removal, which is claimed to reduce investment and operating costs. The problem of mist formation is taken care of by two stages of mist elimination, including an electrostatic precipitator.

As an alternate to H_2SO_4 production, the SO_2 from the regeneration step can be reduced to elemental sulfur, a much more desirable product than sulfuric acid. The technology for producing sulfur, however, is not nearly so well developed as that for making sulfuric acid. The subject has been investigated widely but there has been only limited commercial use, apparently because of the reducing agent cost and operating complexity.

Various reducing agents can be used, including hydrogen, carbon monoxide, producer gas, carbon and natural gas (methane). Carbon monoxide was used in the Bolidens process, operated for several years in Sweden and discontinued in 1943 because of increase in cost of coke. The CO was produced by blowing air, plus part of the stack gas (containing 5% SO_2), through red-hot coke; the remainder of the stack gas was then mixed with the CO and passed over a catalyst to produce sulfur.

Reduction with carbon (usually supplied as coke) produces not only sulfur but also COS, CS_2 and H_2S. In the Trail process (Cominco, Trail, B.C.), these side products were reduced with CO over an alumina catalyst. The method was tested on a plant scale. Recently, Foster Wheeler has developed a proprietary process (Resox) where SO_2 rich gas is passed through a vessel containing crushed coal. Here, SO_2 is reduced to gaseous elemental sulfur and the liberated oxygen combines with a portion of the coal to form carbon dioxide. Reduction in the Resox reactor is carried out at 1200° to 1500°F. The reduction process is planned for demonstration on a 20 Mw prototype unit at Gulf Power's Scholz plant early in 1975.

Methane and other hydrocarbon gases seem to have given the best results, although formation of COS and H_2S are problems here also. The Thiogen process, tested at a smelter in California around 1910, used the following reactions to avoid side products

$$CaSO_4 + CH_4 \longrightarrow CaS + CO_2 + 2H_2O$$
$$CaS + 2SO_2 \longrightarrow CaSO_4 + 2S$$

The process was never fully developed. Gas preheating was required, as well as at least 7% SO_2 in the gas.

The Asarco porcess, developed by the American Smelting and Refining Co. in Utah in the 1940s, also used methane. Preheated smelter gas (500°C, 5 to 8% SO_2) was heated to 1200°C with methane, cooled to 450°C, passed over a bauxite catalyst to decompose COS, heated to 210°C, passed over an activated alumina catalyst (after addition of the proper amount of SO_2) to convert H_2S to sulfur, and cooled again to condense the sulfur. The method was tested in a 5 ton per day unit. International Nickel and Texas Gulf Sulfur joined in development of another process in the mid-1950s for reduction of SO_2 with natural gas. The method was never commercialized.

The most important venture thus far in SO_2 reduction appears to be the project by Allied Chemical to reduce SO_2 to sulfur at the pyrrhotite roasting plant of Falconbridge Nickel Mines near Sudbury, Ontario. The plant was started up in early 1971. Little difficulty has been reported in the operation, which produces 180,000 tons/year of sulfur using natural gas as a raw material. This process will be used at the Northern Indiana Public Service Company's Mitchell Station where a Davy Power Gas FGD system will remove SO_2 and concentrate it to approximately 95%. The Allied process will reduce the concentrated SO_2 stream to elemental sulfur.

A somewhat unusual approach to reducing SO_2 to elemental sulfur is being taken by the Bureau of Mines in the citric acid FGD process. Here, H_2S is added to the scrubber solution to precipitate pure sulfur. Two-thirds of the sulfur produced is reduced back to H_2S for reuse and the other one-third is product. Another process uses a similar approach with phosphate as the buffering solvent.

The various methods for reducing SO_2 to S were evaluated in an early study made by Allied Chemical for EPA. Reducing agents considered were coke, methane, H_2S (made by reducing part of the product sulfur with methane), and reformed or producer gas. The Asarco process with methane was the most economical for the higher SO_2 concentrations (15.8 and 90% SO_2 evaluated). For 4.5 to 0.3% SO_2, use of H_2S in a low temperature Claus process was best (conversion on a bauxite catalyst in a moving bed converter at a temperature level at which the sulfur condenses on the catalyst).

However, the costs for the low concentrations were so high as to be probably unacceptable. High concentration reduced cost markedly; operating cost of less than $10/ton of sulfur was estimated for a 90% SO_2 stream. Thus, SO_2 absorption processes that produce a highly concentrated SO_2 stream in the regeneration step have a major advantage.

Hydrogen Sulfide

Some SO_2 recovery processes produce H_2S as an intermediate by reducing the SO_2 absorption product. A further step, common to all such processes, is conversion of the H_2S to elemental sulfur. The method normally used is the Claus process, in which a third of the H_2S is burned to SO_2 and the remainder reacted with the SO_2 over a catalyst (bauxite) to form sulfur and water. The technology is complicated, with several variations, but is well developed. Its use has grown rapidly in the past 10 years with the increase in recovery of sulfur from sour natural gas; the H_2S is separated from the methane by one of several methods and then converted to sulfur by the Claus process.

Use of the process on H_2S from an SO_2 recovery operation does not appear to present any problem. If dust is generated in the process for making the H_2S, it must be removed. In some cases it may be feasible to economize on reducing agent by reducing only two-thirds of the absorption product and evolving the remainder of the sulfur as SO_2. The two streams should then serve as suitable feed to a Claus plant. If the temperature of the mixture is below 320° to 380°F, the usual temperature of Claus gas entering the catalytic converter system when heat exchangers are used to bring the gas to converter temperature (about 475°F), some additional heat may be necessary.

Claus plants have a pollution problem of their own because conversion of the H_2S to sulfur is not complete. The usual practice is to install enough converter capacity to meet pollution regulations on sulfur emission, burn the residual H_2S, and vent the SO_2. With the present restrictions on sulfur emissions from Claus plants, a considerable cost will be incurred for either more converter capacity or a tail gas scrubbing system. In a power plant the SO_2 from H_2S combustion could be recycled to the scrubber. There is also the possibility, in some processes, of avoiding the waste of H_2S by using excess SO_2 to drive H_2S conversion to completion and then recycling the unused SO_2.

Other Intermediates

Various other intermediates occur in the voluminous SO_2 recovery technology. An example is ammonium sulfate, formed by oxidizing the effluent solution from an ammonia scrubber. It can be used as one of the raw materials in a nitric phosphate fertilizer process. This and most of the other intermediates are specific to a single recovery process and will be discussed in later chapters.

NITROGEN OXIDES

The nitrogen oxides present in power plant stack gas are important both because they are pollutants that should be controlled and because they may have an effect on SO_2 removal processes. In a study made by Esso Research and Engineering for EPA, it was concluded that almost 60% of the NO_x emitted in the United States comes from the combustion of fossil fuels, and that the power generating industry contributes about 38% of this, or roughly 25% of the total.

Most of the coal-fired boilers produce an NO_x content in the gas leaving the system in the range of 500 to 1,000 ppm, although values lower and higher have been reported. From 90 to 95% of the NO_x is NO and the rest NO_2. Many factors affect the amount of NO_x formed in the boiler, including excess air, air preheat temperature, heat release rate, heat removal rate, degree of gas backmixing, fuel type (coal gives the most NO_x), and fuel nitrogen content.

The main effect of NO_x in SO_2 removal processes is the well-known reaction of NO_2 with SO_2 or $SO_3^=$ to give SO_3 or $SO_4^=$. This is the basis of the Tyco process (no longer being developed) which will be discussed later; the NO_2 in the gas stream is used to oxidize SO_2 and make sulfuric acid.

Although the oxidizing effect of NO_2 is desirable in the Tyco process, it is objectionable in most of the recovery processes based on wet scrubbing because it oxidizes sulfite to sulfate. The regeneration step is usually based on sulfite decomposition and is either not effective on sulfate or is effective only at added cost. The NO_x may also be responsible for part of the SO_3 in the gas entering the scrubber, which also forms objectionable sulfate. It seems likely that part of the NO_2 formed in the boiler oxidizes SO_2 before the gas reaches the scrubber. No data appear to be available on the point.

Thus it would be desirable to either remove NO_x from the gas or reduce NO_2 to NO before the scrubbing step in recovery processes. The latter has been accomplished by use of ferrous compounds in a prescrubber to reduce the NO_2; this work was done by Aerojet General in a study made for EPA. No other efforts in this direction appear to have been reported.

The NO_x can be decomposed to elemental nitrogen by several methods, the main one being reduction over a catalyst with a reducing agent such as H_2, CO, CH_4, or NH_3. The main application of this has been in treating tail gas from nitric acid plants, where the NO_x content is some ten times higher than in power plant stack gas. Since even at this high concentration full success has not been attained, there is some question whether reduction would be practicable for power plant stack gas. However, reduction processes are under development in Japan.

Aside from the need for NO_x removal to reduce sulfite oxidation, it is desirable to reduce emission of NO_x as a means of pollution abatement. The ideal of course, is to remove NO_x along with SO_2 in the same scrubbing step. Numerous proposals of this type have been made. Most such processes involve addition of a recycled stream of NO_2 to the gas in order to produce an equimolar mixture of NO and NO_2, a step that is necessary because NO is quite inert and can be absorbed most easily by conversion first to N_2O_3:

$$NO + NO_2 \longrightarrow N_2O_3$$

$$2CH_3OH + N_2O_3 \longrightarrow 2CH_3NO_2 + H_2O$$

This method has been used to treat nitric acid plant tail gas with some success.

The difficulty in applying this approach to power plant stack gas is that the recycled NO_2 prefers to oxidize sulfite in the absorbing solution rather than form N_2O_3 with the NO. Hence a large amount of recycled NO_2 probably would be necessary, first enough to oxidize all the sulfite to sulfate and then enough to match all the NO (that originally present plus that formed by reduction of NO_2). This is the procedure in the Tyco process, where oxidation of SO_2 is the main consideration, whereas in scrubbing processes oxidation of sulfite is quite objectionable. It has been suggested that the oxidation of sulfite can be prevented if the scrubber solution is kept alkaline. No data appear to be available on the point.

Recently, Chiyoda Chemical Engineering announced the development of a simultaneous NO_x and SO_2 removal process (Chiyoda 102). Nitric oxide is partially oxidized by a small quantity of ozone after the flue gas has been prescrubbed and precooled. The partially oxidized flue gas is led into the flue gas desulfurization unit where NO_x is absorbed simultaneously with SO_2 in an absorbent where redox reactions occur with an oxidation catalyst. Part of the absorbed NO_x becomes nitric acid, while the rest becomes N_2 and other harmless gases which are released into the atmosphere with the flue gas. In pilot plant tests an NO_x removal rate of more than 60% was attained.

The chemistry of the Chiyoda system is somewhat complicated but appears feasible. The following reaction sequence is postulated from the data available:

$$3O_3(g) + 3NO(g) \longrightarrow 3O_2(g) + 3NO_2(g)$$

$$NO_2(g) + HNO_3/H_2SO_4(aq) \longrightarrow HNO_3 + HNO_3/H_2SO_4(aq)$$

$$2NO_2(g) + 2NO(g) \longrightarrow 2N_2O_3(l)$$

$$N_2O_3(l) + 2H_2SO_3 \longrightarrow N_2O(g) + 2H_2SO_4$$
$$N_2O_3(l) + 3H_2SO_3 \longrightarrow N_2(g) + 3H_2SO_4$$

The net reaction is,

$$3O_3(g) + 5NO(g) + 5H_2SO_3(l) + HNO_3/H_2SO_4(aq) \longrightarrow 3O_2 + 5H_2SO_4(l) + HNO_3(l) + N_2O(g) + N_2(g) + HNO_3/H_2SO_4(aq)$$

A somewhat more complicated SO_2-NO_x removal system is that being developed by Sumitomo Metal Industries, Ltd. and Fugi Kasui. Here, ClO_2 (chlorine dioxide) is added to the precooled gas stream to oxidize NO to NO_2. The gases are then absorbed via the following reaction sequence:

$$2NO(g) + ClO_2(g) + H_2O(g) \longrightarrow NO_2(g) + HCl(g) + HNO_3(g)$$
$$HCl(g) + NaOH(aq) \longrightarrow NaCl(aq) + H_2O$$
$$HNO_3(g) + NaOH(aq) \longrightarrow NaNO_3(aq) + H_2O$$
$$SO_2(g) + 2NaOH(aq) \longrightarrow Na_2SO_3(aq) + H_2O$$
$$NO_2(g) + 2Na_2SO_3(aq) \longrightarrow \tfrac{1}{2}N_2 + 2Na_2SO_4(aq)$$

It is obvious that aqueous scrubbing to remove NO_x can present formidable water quality problems. More work is needed in both processes to develop a more complete regeneration mode.

Shell also has an NO_x removal process, a dry type. Ammonia is added ahead of absorbent in the Shell copper oxide FGD system which purportedly reduces NO to very low levels. No further data are available.

Currently, the best solution to the NO_x problem is to reduce formation in the first place. The Esso study and more recently TVA's work indicate that various combustion modifications can reduce formation to 10 to 50% of the usual amount for existing units. This is accomplished by using low excess air and two-stage combustion.

ALKALI ABSORBENTS

Alkali compounds have probably received more attention than any other type of absorbent over the history of SO_2 removal from stack gases. The reasons for this include relatively high affinity for SO_2, applicability of sulfite-bisulfite chemistry to the absorption-regeneration cycle, and ability to keep all the compounds involved in solution—thereby avoiding scaling and silting problems in scrubbers.

The alkalis that have been used are ammonium, sodium, potassium, lithium, hydrogen, and aluminum. The various name processes employing these are listed in Table 7.1; many others have been proposed, mainly in the patent literature, that are not as well-known. The absorbent anion is listed generally as $SO_3^=$ in the table, although in some processes the regenerated form of the alkali is the oxide or carbonate.

In practice the amount of scrubber liquor recirculated to give an adequate liquor-gas ratio is so much larger than the regenerated absorbent stream that sulfite becomes the major sorbent. Also, oxide or carbonate is rapidly converted to sulfite in the scrubber.

The table shows that both the ammonium and sodium systems have been investigated much more fully and have been used more than any of the others.

Most of the alkali absorbent methods involve use of an aqueous solution in a scrubber. One of the major efforts, however, has been the alkalized alumina ($NaAlO_2$) process, in which the absorbent is a granular solid.

In another, sodium carbonate is used in a spray-dryer absorber producing a dry solid product. An alkali melt at high temperature has also been used, but, unfortunately, both it and the alkalized alumina process have had many problems and are not currently being developed.

TABLE 7.1: SULFUR OXIDE ABSORPTION PROCESSES OF THE ALKALI TYPE

Process	Active Absorbent Cation	Active Absorbent Anion	Development Status
Cominco	NH_4^+	$SO_3^=$	Commercial
Institute of Inorganic Chemistry, Czechoslovakia	NH_4^+	$SO_3^=$	Commercial
IPRAN, Romania	NH_4^+	$SO_3^=$	Commercial
Showa Denko (Japan)	NH_4^+	$SO_3^=$	Prototype
NIIOGAZ (U.S.S.R.)	NH_4^+	$SO_3^=$	Commercial
Kuhlmann-Électricité de France	NH_4^+	$SO_3^=$	Prototype
Institut Francaise du Petrole	NH_4^+	$SO_3^=$	Prototype
TVA-EPA	NH_4^+	$SO_3^=$	Pilot plant
Wellman-Lord	Na^+	$SO_3^=$	Commercial
Universal Oil Products (Sulfoxel)	Na^+	$SO_3^=$	Prototype
Johnstone	Na^+	$SO_3^=$	Pilot plant
Ionics-Stone and Webster	Na^+	$SO_3^=$	Pilot plant
Bureau of Mines (alkalized alumina)	Na^+	AlO_2^-	Pilot plant
Bureau of Mines (citric acid)	Na^+	$SO_3^=$	Pilot plant
Pfizer-Peabody-McKee (citric acid)	Na^+	$SO_3^=$	Pilot plant
Stauffer (phosphoric acid)	Na^+	$SO_3^=$	Pilot plant
Wellman-Lord	K^+	$SO_3^=$	Pilot plant
TVA	K^+	$P_2O_7^{-4}$	Small scale
Consolidation Coal	K^+	$OOCH^-$	Pilot plant
Atomics International (molten salt)	Na^+, K^+, Li^+	$CO_3^=$	Prototype
Atomics International (aqueous carbonate)	Na^+	$CO_3^=$	Pilot plant

SULFITE-BISULFITE EQUILIBRIUM

Scrubbing with alkali solutions gives the unique advantage that the loaded absorbent can be regenerated by heating at a temperature level not very far above absorption temperature. For most absorbents this is not feasible, because if the absorbent reacts with SO_2 to give a product with low SO_2 vapor pressure, which is necessary if the SO_2 content of the gas is to be reduced to a low level, then the product is too stable to release the SO_2 in regeneration, except at high temperature. Much of the work on absorbents has been on identifying a sulfite (reaction product of absorbent and SO_2) that has a high temperature coefficient of vapor pressure.

Alkali solutions have the advantage that the sulfite-bisulfite equilibrium can be used to decrease the temperature differential between absorption and regeneration. Bisulfite ($MHSO_3$, M is alkali cation) is relatively unstable and SO_2 can be desorbed from it easily; it can be formed in a countercurrent scrubber because the relatively high SO_2 vapor pressure at the bottom of the scrubber drives the sulfite-bisulfite reaction toward bisulfite:

$$M_2SO_3 + SO_2 + H_2O \rightleftharpoons 2MHSO_3$$

Scrubbing efficiency is not good at this stage of the scrubbing because the SO_2 vapor pressure over the bisulfite is too high; in the upper part of the scrubber, however, the incoming sulfite has little vapor pressure and therefore removes the SO_2 to a low level. A generalized reaction scheme is as follows:

(1) In the upper part of a countercurrent scrubber (or in the first liquor stage of two or three back-mixed stages), a recycled sulfite solution (at about 120°F) removes SO_2 to a low level. With enough scrubber volume and contact surface, almost any degree of removal can be attained.

(2) The resulting solution flows to the bottom of the countercurrent scrubber (or to the further liquor stages of a staged back-mixed scrubber) where the high SO_2 content of the incoming gas builds up the bisulfite content of the solution to a high level; unabsorbed SO_2 passes on up through the scrubber and is absorbed in the top section or stages. With enough scrubber capacity, the bisulfite/sulfite mol ratio can be pushed to 4:1 or higher. However, there will be an economic balance that determines the optimum level.

(3) In the thermal regeneration step (stripping), reaction (1) is reversed by heating the loaded solution to 200°F or so. The heating drives off a rich stream of SO_2 and regenerates M_2SO_3 for recycling.

The solution composition at various points in the absorption-regenerator system is usually expressed as S/C ratio, where S is the total mol/l of sulfur and C is the effective molar concentration of alkali, that is, the amount tied up with sulfite and bisulfite. Since part of the alkali is combined with sulfate formed incidentally in the system, it is not available to contribute to the sulfite-bisulfite equilibrium.

Thus, for complete conversion to bisulfite, S/C is 1.0. The operating level, both in the scrubber and the regenerator, is usually somewhere between this and 0.5, the S/C for all sulfite. At 0.9, for example, the solution contains 80% bisulfite (molar basis) and 20% sulfite.

A major problem in the thermal regeneration route is the high vapor pressure of water at the temperature required to strip out SO_2. As a result, much of the heat energy supplied must be expended in evaporation of water. The main difficulty is the rapid drop in SO_2 partial pressure as S/C decreases. At an S/C of 1.0 the SO_2 pressure is quite high and stripping is economical.

However, as SO_2 is evolved the bisulfite is converted to sulfite and the SO_2 partial pressure drops rapidly, requiring an increasing amount of water evaporation if SO_2 stripping is to be continued to an economical level before return of the stripped solution to the scrubber. As a result it may be necessary to evaporate 10 lb or so of water per pound of SO_2, which is likely to make the process uneconomical.

The unique chemistry of the sulfite-bisulfite system again affords an opportunity for improving the situation. For ammonium and sodium, the sulfite is much less soluble than the bisulfite, therefore, if the circulating system is operated at as high a sulfite concentration as possible, it may be possible to crystallize sulfite in the regenerator when bisulfite is converted to sulfite. Thus a steady state situation can be reached in which sulfite formed by bisulfite decomposition immediately crystallizes out, plus incoming sulfite, with the result that the S/C in the solution remains at a steady level and further drop in SO_2 partial pressure does not occur, except from decline in bisulfite concentration which has much less effect than drop in S/C. Thus, although the sulfite-bisulfite chemistry gives the opportunity for thermal regeneration at low temperature, there are some serious drawbacks that have limited its use and directed the technology more to chemical regeneration. Even here, however, formation of bisulfite is important in several of the processes because sulfite is either not reactive with the chemical regenerant, produces more side product than does bisulfite, or requires large regeneration systems.

In all processes, bisulfite formation has the disadvantage that the accompanying reduction in pH aggravates corrosion. Another drawback is that an additional scrubbing stage is required as compared to processes that can operate with sulfite. Chemical methods of regeneration can be summarized as follows:

(1) Oxidation of Sulfite and Bisulfite to Sulfate – This applies only to the ammonium system, as the sulfate can be used as such as a fertilizer or as an intermediate in production of phosphate fertilizers. However, in Japan, abatement sodium sulfate is being used in the pulp and paper industry.

(2) Acidification – Treatment of the scrubber effluent with a strong acid evolves SO_2 and forms the alkali salt of the acid. Various acids can be used, including sulfuric, nitric, and phosphoric, as well as alkali bisulfate.

(3) Precipitation of Another Sulfite – If the oxide or carbonate of a metal whose sulfite is more stable than the alkali sulfite is added to the scrubber effluent, the metal sulfite crystallizes and the alkali is regenerated. The metal sulfite is then separated and regenerated by one of several methods.

(4) Reduction – In some processes, sulfite and bisulfite are reduced, while still in solution, to the sulfide or to H_2S, either of which can be treated in a fairly simple way to produce sulfur. Disproportionation of sulfite to sulfate and sulfur is another way of converting at least part of the sulfur to a reduced form.

AMMONIA SCRUBBING

Ammonia has some major advantages over the other alkalis as an absorbent for SO_2, including lowest cost per equivalent, more versatility in regeneration, and feasibility of leaving it in the final product. At a delivered cost of $130/ton, the cost per pound equivalent is about $1.11; in contrast, Na_2CO_3 at $52/ton costs $1.37. Although the absorbent is recycled in regeneration processes, there are always losses, which gives ammonia an advantage because of its lower cost.

Moreover, the cost of regeneration can be avoided because the ammonia can be left in a product of the fertilizer type to supply nitrogen, a major plant nutrient. This is not feasible in sodium system scrubbing because the sodium has no commercial fertilizer value. Potassium is a plant nutrient, but the form in which it might be supplied to an SO_2 scrubbing system (hydroxide or carbonate) is too expensive for fertilizer use. However, ammonia scrubbing has an inherent fume problem which until recently has negated many of its good features.

Absorption Chemistry and Kinetics

The ammonia scrubbing system is relatively simple from the chemical standpoint. In the pH range 4.2 to 7.0, the main dissolved species are HSO_3^-, $SO_3^=$, and NH_4^+. Sulfate will appear depending on the amount of SO_3 in the flue gas and the degree of scrubber oxidation. Other materials will be introduced with the gas (ash, Cl) and may accumulate if the scrubber circuit is of the closed loop type. These do not appear to effect scrubber efficiency.

The SO_2 and NH_3 partial pressures over $(NH_4)_2SO_3$-NH_4HSO_3-H_2O solutions are given in Figure 7.1 for a solution containing 22 mol percent NH_3 (not counting NH_3 tied up with $SO_4^=$). The curves show the difficulty in getting high bisulfite (high S/C) unless the SO_2 partial pressure in the incoming gas is relatively high. Hence high-sulfur coal could not only improve economics by increasing sulfur throughput but also by allowing a higher bisulfite-sulfite ratio for those processes that are helped by high S/C.

The SO_2 partial pressures shown in Figure 7.1 can be reduced, of course, by using more dilute solutions; equations for calculating the resulting pressures are available in the literature. Although dilution would help in getting higher bisulfite, there would be disadvantages such as higher oxidation rate, more water to evaporate, and large regeneration equipment.

The sulfate content of the solution also affects SO_2 partial pressure. An increase from 0.2 mol $(NH_4)_2SO_4$/liter to 2.2 mols raises the SO_2 partial pressure at 50°C from 3.0 to about 3.8 mm at an S/C of 0.91 to 0.92 (with C at 10.2 to 11.4 mols/liter.

Oxidation of sulfite in ammonia scrubbing is a very important consideration, as it is in most recovery processes; not only is the SO_2 partial pressure increased but the sulfate is nonregenerable in many of the processes. The oxidation mechanism has been reported to be the same for alkali sulfites as for calcium sulfite, which was discussed in detail in Chapter 3. Only a few differences have been reported; for example, thiosulfate ($S_2O_3^=$) and trithionate ($S_3O_6^=$), side products that form and accumulate in a closed loop system, are said to be effective catalysts. In other work, Co^{++}, Fe^{++}, and Ni^{++} were found to be quite effective, in the order named. Organic compounds such as pyridine have also been proposed as catalysts for processes in which sulfate is desired as a product.

Mass transfer in SO_2 absorption by ammoniacal solutions is controlled mainly by the gas phase resistance. The chemical reactions involved are rapid; hydration of SO_2 is the slowest but it is sufficiently fast to be nonlimiting up to a concentration of 3 to 4% SO_2 in the gas.

Liquid film resistance is also quite low except at low pH such as would be encountered in the first stage of a multistage scrubber; the liquid phase resistance becomes equal to the gas phase resistance at an S/C of 0.92 to 0.96. The transfer rate falls rapidly with increase in temperature from a relative rate of 10.2 at 23°C to 2.3 at 52.5°C. Thus, cooling below the wet bulb temperature, although expensive, would give much better SO_2 transfer to the solution.

FIGURE 7.1: EQUILIBRIUM PARTIAL PRESSURES OF NH_3 AND SO_2 IN THE SYSTEM NH_3-SO_2-H_2O

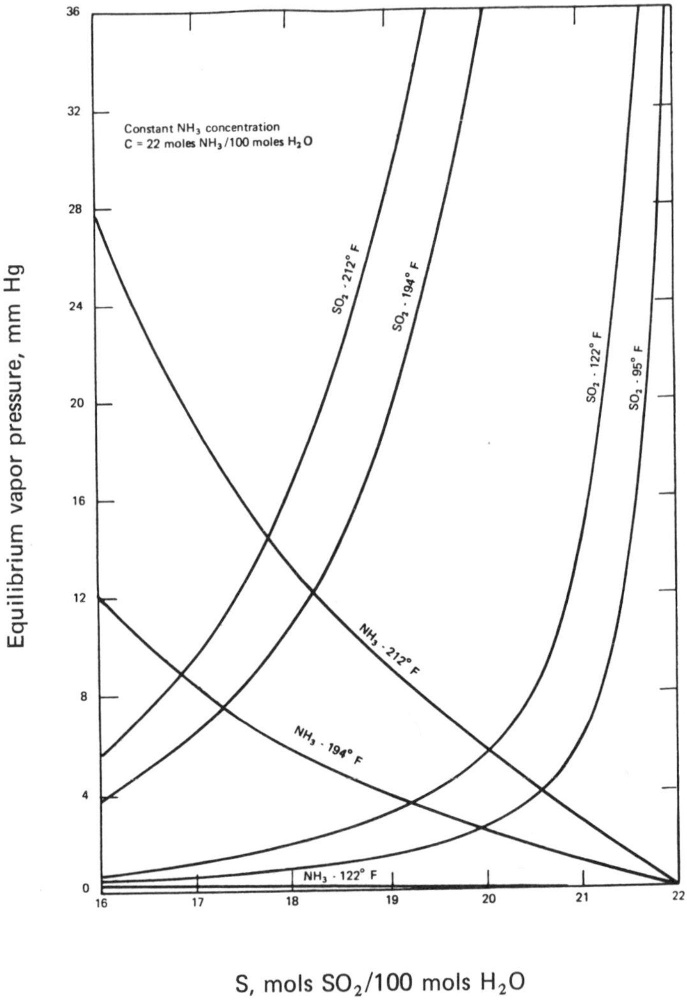

Adapted from West, W.E., Jr., Thesis, *Evaluation of Sulfur Dioxide Recovery Processes,* University of Illinois, Urbana, Illinois (1953)

Scrubber Design

Ammonia scrubbing differs from use of the other alkalis in that the cation as well as the anion is volatile and the scrubber therefore must be designed to recover both.

For processes that require high bisulfite content in the scrubber effluent liquor, there must be an enriching stage in which the driving force of the SO_2 partial pressure in the incoming gas produces a high bisulfite-sulfite ratio. Thus at least three scrubber stages are needed: (1) the SO_2 enriching stage; (2) a central stage in which most of the absorption takes place; and (3) an ammonia recovery stage to prevent ammonia loss to the gas leaving the scrubber (Figure 7.2). In practice it has been found that four or more stages are preferable for maximizing bisulfite content. The composition of the solution circulated in each stage is controlled in accordance with the function of the particular state.

FIGURE 7.2: AMMONIA SCRUBBER WITH THREE SO_2 ABSORPTION STAGES

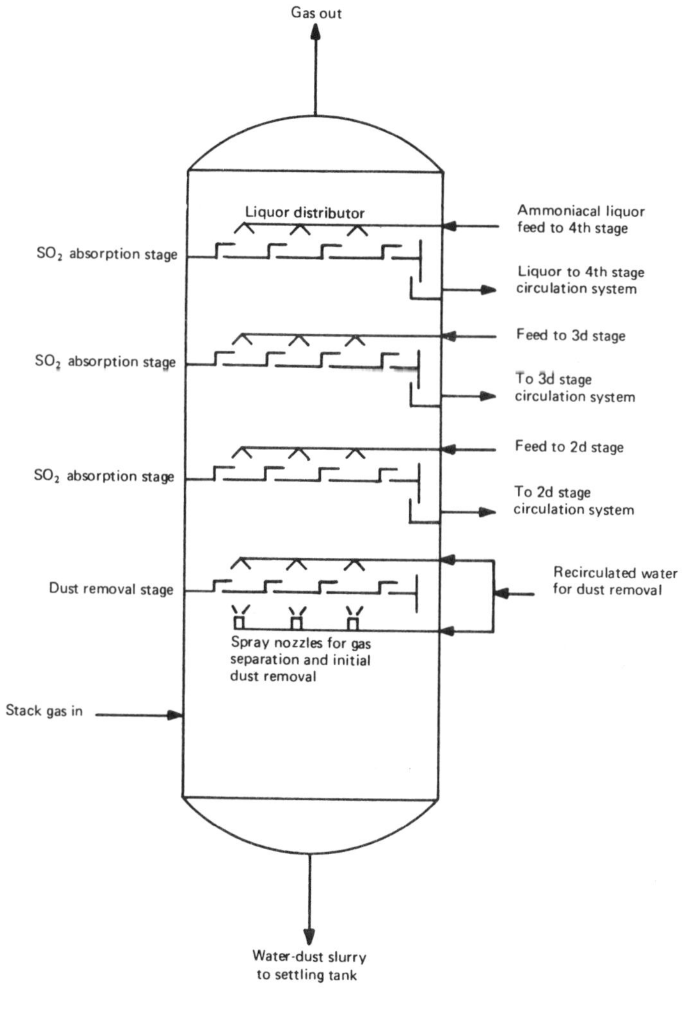

A few special types of scrubbers have been applied in ammonia scrubbing. One is a packed scrubber in which the packing is an assembly of helixes (spirals) arranged in a vertical array. This device was developed by the Research Institute of Inorganic Chemistry in Czechoslovakia and is used there in a commercial plant. The Linac Separator (Wade Co.) is made up of polypropylene rings held apart by spacers, forming a series of venturi-like openings through which the gas accelerates and then decelerates. Good removal of SO_2 is claimed.

In Japan, several different absorber types are being used. Nippon Kokan KK employs a simple grid-type packing. Kurabo uses a packed absorber and Sumitomo-Fuii Kasui a tray type. All seem to be very good SO_2 removal devices.

Although several different scrubber types have been used in ammonia scrubbing, the performance data are difficult to compare because of the widely varying conditions in the various test programs. In an effort to get comparative data, EPA and TVA have installed a 4,000-scfm pilot plant at TVA's Colbert station in north Alabama and a test program is under way.

Various scrubbers have been tested, including tray, plate, and mobile bed types. So far, each absorber has given good SO_2 removal results but some appear better from an operational standpoint. In addition, operating parameters are being optimized to give the best combination of bisulfite-sulfite ratio, degree of oxidation, and solution concentration for various regeneration systems.

In a few processes the ammonia has been injected directly into the stack gas, followed either by a scrubber or a solids collection device. The objective in both cases was to form solid ammonium sulfite in the gas phase and thereby either eliminate the scrubber or reduce the size requirement. The drawback to this is incomplete reaction of NH_3 and SO_2 in the gas phase; calculations indicate that even at a temperature as low as 150°F, only about half of the SO_2 would react to form a solid.

Injection of part or all of the ammonia into the gas may be desirable to protect equipment upstream from the scrubber against corrosion. For example, the corrosion of a heat exchanger-economizer used to pick up heat ahead of the scrubber for use in reheating the stack gas, caused by condensation of water from the gas and absorption of SO_3 to form H_2SO_4, should be prevented by injection of NH_3 before the exchanger. However, deposits can form and cause plugging. Ammonia has also been injected into the gas in Poland as a means of SO_2 emission control, on an episode basis.

Scrubbers have been built of aluminum and of mild steel lined with lead or coated with a resin. Type 316 stainless steel and neoprene-lined steel have also been used at critical points in scrubber systems.

Fume

One of the more difficult problems in ammonia-scrubbing is fume formation. Figure 7.3 shows a typical ammonia scrubber plume (including steam) emitted during a routine TVA pilot plant operation with no gas reheat (exit temperature is approximately 125°F). The plume was formed during operation with three scrubbing stages, no prior humidification, and while producing a liquor with a

FIGURE 7.3: TYPICAL PLUME DURING ROUTINE OPERATION; HIGH SCRUBBER LIQUOR CONCENTRATION (C = 10); NO REHEAT

C of about 10 and an S/C of 0.8. Sulfur dioxide removal efficiency under these conditions is typically about 90%. TVA efforts to reduce or eliminate the plume through manipulation of the pH of the scrubber liquor and the addition of a humidification step ahead of the scrubber did not reduce the plume opacity significantly. However, under these test conditions the steam plume may have masked any reduction in the ammonia-based plume. Adding a wet electrostatic precipitator after the scrubber failed to produce significant improvement at the normal gas flow rate of 2,600 scfm. (The precipitator was found to be only 50% efficient in removing the particulate at this flow rate, 15 fps shell velocity through the precipitator.)

Chemical and petrographic analyses of the plume collected in an impaction sampler (Brinks mass sampler) indicate that the major fraction of the ammonia-sulfur salt was ammonium sulfate. It is probable that the particulate was formed in the vapor phase as ammonium sulfite and then oxidized to the sulfate form in the sampler. A portion of the particulate has been analyzed as ammonium chloride. (The coal used during these tests normally contained 0.1 to 0.2% chlorides.)

Tests showed that a water wash ahead of the scrubber materially reduced the chloride content of the gas entering the scrubber. The fume particle size was determined to be 0.25 micron with 10^7 particles per cubic centimeter in the size range of 0.005 to 0.5 micron. Reheating the scrubbed gas was effective in reducing plume opacity.

There are differing schools of thought on how fume formation occurs. Probably the most complete evaluation is that developed somewhat concurrently by Air Products, Inc.-Catalytic Inc. on the one hand and by TVA-EPA on the other. Both approaches were different, one being theoretical with limited pilot plant data and the other based primarily on pilot plant results. Regardless of the bases, it is now apparent that control of the partial pressures of gaseous NH_3, SO_2, and H_2O vapor in each scrubbing stage is very critical to avoiding fume.

In addition, a prewash section ahead of the first ammoniacal scrubbing stage to remove chlorides and reduce or eliminate stripping of ammoniacal liquor in the first stage is necessary. Finally, some means to reduce ammonia losses on the top stage (i.e., low pH liquor) is also required. An easy way to accomplish the above objectives is to use very dilute ammoniacal solution, on the order of C = 1 or less. At these low solution concentrations, the partial pressures of NH_3 and SO_2 are also low and fume formation is kept at a minimum (assuming a prewash section to remove chlorides).

Air Products-Catalytic has developed and recently patented a thermodynamic formula based on solubility products which predicts fume formation using the following criteria: inlet gas composition, scrubber solution concentration, prewash, and afterwash.

The formula assumes that the solution is in thermal equilibrium and that any excursions in temperature (particularly downward) can accelerate or initiate fume formation. Hence, it is required that all duct work and equipment from the scrubber to the reheater section be insulated. In a short series of runs using the TVA pilot plant, several of these theories were tested. Although it was impossible statistically to elucidate the effect of scrubber insulation in these runs, overall fume formation was apparently depressed. The effect of reheat on fume formation is a complicated phenomenon. Originally, the Japanese believed that reheat would eliminate the fume; however, electrostatic precipitators are now being used. Reheat can have the following effects: decompose fume (salt) into its gaseous components, except for ammonium sulfate; evaporate any mist carry-over from the top stage of the scrubber; and eliminate the condensed steam portion of the saturated exit gas.

In addition, the higher stack temperature will allow higher partial pressures of NH_3, SO_3, and water to exist before exceeding the solubility product, an obvious advantage. Evaporation of mist, on the other hand, may aggravate the situation by putting more NH_3, SO_2, and H_2O into the gas phase. Taking all this into consideration, moderate reheat at least to dissipate the condensed steam portion of the plume appears beneficial.

Another approach is to allow fuming to occur and simply remove it in a particulate collector downstream of the absorber. This is possible but considering the small size of the particulates, it may be more economical to retard initial formation than try to collect it after formation. It is obvious that more work is needed in this area, both in avoiding fume formation in the scrubber and removing it in a particulate collector.

Regeneration Process

The large amount of research carried out on ammonia as an absorbent for SO_2 has been coupled with study of a wide variety of regeneration processes.

Thermal Stripping: The earliest work on thermal stripping of ammonia scrubber effluent solution appears to be that of Cominco (formerly Consolidated Mining and Smelting Co.) in Trail, B.C., during the 1930s. Operation of zinc and lead smelters by the company caused a pollution problem because of the sulfur oxides evolved. The usual smelter method for controlling such a problem, installation of a sulfuric acid plant, was applied in the zinc smelter but since the gases from the lead smelter were too dilute for this, ammonia scrubbing was developed as an alternate.

In the course of this work several ways of regenerating the scrubber solution were tested, including thermal stripping, called the exorption process. A commercial plant was built and operated for some time. Scrubber solution was heated to 130°C to evolve SO_2. The process apparently was not considered economical because an acidification method was adopted as the standard operational process.

During this general period Johnstone and coworkers at the University of Illinois worked on ammonia scrubbing, measuring vapor pressure and evaluating thermal stripping, acidification, and oxidation. The group concluded that stripping was the most expensive of the three methods.

Also in the mid-1930s, the U.S. Bureau of Mines did fundamental work on the process, including measurement of vapor pressures. American Smelting and Refining Co. did pilot plant work on the Guggenheim process, which was similar to the Cominco exorption method. Extensive work on thermal stripping, as well as on other regeneration processes, has been done by the State Research Institute of Industrial and Sanitary Gas Cleaning (NIIOGAZ) in the U.S.S.R.

In 1952 a power boiler of about 65-Mw capacity was fitted with ammonia scrubbing and thermal stripping equipment. The plant was operated until 1961, when the power plant was converted to natural gas. Pertinent data from this operation are as follows.

SO_2 in gas	0.3 %
Scrubber type	3-stage sieve tray
Gas temperature	150°-160°C
Scrubber temperature	35° C
SO_2 removal	90 %
S/C at scrubber inlet	0.78
Solution capacity	40 kg SO_2/m^3
SO_2 oxidation	10-15%
Solution concentration,	
SO_2	3.2 mol/l
NH_3	4.1 mol/l
$(NH_4)_2SO_4$	2.0 mol/l
Scrubber pressure drop	22 mm H_2O
S/C at scrubber outlet	0.92-0.93
Regeneration temperature	90°-95°C
Regeneration pressure	300-350 mm Hg
Consumption per ton SO_2	
NH_3 [in $(NH_4)_2SO_4$]	0.14 ton
CaO	0.12 ton
Inhibitor (to reduce oxidation)	0.57 kg
Steam	7.7 ton
Power	700 kwh
Daily production,	
SO_2	15-16 ton
$(NH_4)_2SO_4$	2.9-3.0 ton

The NIIOGAZ process is shown in Figure 7.4. The gas is cooled to 35°C by use of a very large amount of cooling water in the first section of the scrubber; about 10% of the SO_2 is absorbed in this stage, requiring use of lime for neutralization before the water is discarded. Three stages of scrubbing are used for SO_2 absorption in order to build up a high S/C and still minimize ammonia loss.

The final stage in the scrubber is an electrostatic precipitator serving as a mist eliminator. The ammonium sulfate formed by sulfite oxidation and the thiosulfate and polythionate formed by solution decomposition in the regenerator are removed by autoclaving a side stream of solution, thereby producing ammonium sulfate and elemental sulfur (very small amount) as coproducts with the SO_2. The stripping is carried out by indirect heating with steam.

Early cost estimates by NIIOGAZ indicate that ammonia scrubbing-stripping is preferable to lime or magnesia scrubbing. A full-scale demonstration is planned for a 300-Mw coal-fired boiler at the Ryazan power station. Two different absorber types, each 500,000 m^3/hr, will be used. One will be a sieve-plate type and the other will have the helical packing developed in Czechoslovakia. Regeneration will be by steam stripping in a plate column and the stripped solution will be cooled before recycle to the absorber. Wet ESPs will be installed between the absorber and the stack for fume control. A conceptual design and economic study for this project was scheduled for completion in March 1975, work drawings will be completed by mid-1976 and operation started late 1978.

In 1967-1968 a 30-Mw test system was installed at the St. Ouen station (in a Paris suburb) to test the Kuhlmann ammonia scrubbing method. Gas from the oil-fired boiler (2,000 ppm SO_2) was scrubbed with ammoniacal solution in two scrubbers (in series) designed by Weiritam, a French equipment designer and manufacturer. In the recovery process, the scrubber effluent was heated to evolve a rich stream of SO_2. No facilities were installed to convert the SO_2 to an end product such as sulfur or sulfuric acid.

In the early Cominco work, the main problems had been high energy consumption for volatilizing the SO_2 and formation of side products such as elemental sulfur and thiosulfate in the heating step. Kuhlmann, however, claimed that in their small-scale work the energy consumption was much less than that reported by Cominco.

When the St. Ouen test unit was operated in the recovery mode, it was found that the energy balance did not work out as indicated by the small-scale work. About 10 lb of steam was required per pound of SO_2 produced rather than the 1.5 lb predicted. Oxidation of sulfite to nonregenerable sulfate was also higher than expected.

Kuhlmann became discouraged with marketability of both the process and the product (sulfuric acid or sulfur) in France. Tests of the recovery process therefore were discontinued.

More recent development involves the Institute Francaise du Petrole's (IFP) wet Claus process in which an H_2S-SO_2 stream is reacted at relatively low temperature in an organic solvent to produce molten elemental sulfur.

The basic IFP process has been extended to flue gas desulfurization by use of ammonia scrubbing to produce a solution that can be evaporated to give a gaseous mixture of SO_2, NH_3, and H_2O. H_2S is then introduced or formed in situ to give a composition that is suitable for conversion to elemental sulfur in the wet Claus unit. The ammonia does not react and thus can be recycled back to the scrubber. Sulfate formed in the scrubber is reduced in a separate reactor to SO_2.

In the IFP regeneration system, the scrubber solution first goes to a scrubber through which the tail gas from the regeneration process is passed to absorb H_2S as thiosulfate by reaction with sulfite. The solution then passes to two heat exchangers, each capable of handling 60% of the flow, and the first stage evaporator; the solution is concentrated to 15% H_2O but about 40% of the sulfur salts still remain in solution. In the second evaporator, the heat is supplied directly by burning natural gas under reducing conditions (about 5% less than the stoichiometric amount of air). Too much reduction at this point will crack ammonia but some H_2 and CO are needed. One of the important reactions in this step is thiosulfate with bisulfate to give SO_2.

The second evaporator was originally a submerged combustion unit. Because of various problems, natural gas is now burned and the combustion gases introduced into the melt through a graphite dip pipe lined with refractory to protect the graphite from the hot gases. The second evaporator operates at 350°C and volatilizes practically everything. Since part of the sulfur is volatilized as SO_3 from this evaporator, the next step is reaction of the hydrogen in the gas with the SO_3 to give SO_2. This requires a catalyst.

FIGURE 7.4: NIIOGAZ PROCESS FOR SO_2 REMOVAL FROM STACK GASES

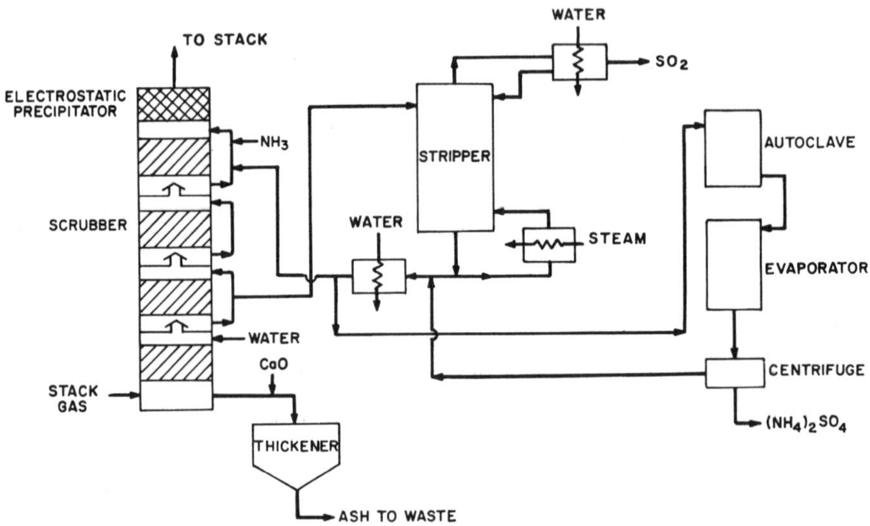

From the catalytic SO_3 reducer, the gas passes to two reducer vessels in series in which synthesis gas is introduced to produce enough H_2S for the later Claus reaction. The synthesis gas (H_2 plus CO) is made in a separate unit from natural gas and steam. The product gas is passed through a waste heat boiler to get it as cool as possible so as to offset the exothermic reaction of SO_2 plus H_2 to form H_2S. Part of the synthesis gas is introduced into the first reducer vessel and the mixed gas then cooled to about 400°C. The rest of the synthesis gas goes to the second reducer, from which the gas stream goes to a waste heat boiler.

The gas then contains 2 mols of H_2S per mol of SO_2 and is ready for the Claus reaction. The solvent in the Claus scrubber-reactor is recirculated and cooled between cycles. The molten sulfur collects in the bottom of the vessel and is drained off, a high purity grade of sulfur being produced.

The off-gas, which contains some unreacted H_2S, all the NH_3, and a little SO_2, is quenched in a water scrubber. Entrained solvent is separated and returned to the Claus reactor. The gas is then subjected to further cooling to condense water as NH_4OH plus small quantities of $(NH_4)_2S$ and $(NH_4)_2SO_3$; the resulting solution is returned to the main scrubber. The final tail gas is then scrubbed with effluent solution from the main scrubber to remove residual H_2S.

The final tail gas from the regeneration system contains a small amount of ammonia and a little hydrogen (used in small excess in the regeneration train). As the system gets older it may be necessary to use more excess reducing gas, in which case the tail gas will be flared. There is provision to add some elemental sulfur in the second evaporator to promote full reduction of sulfate. This is expensive so it is hoped that the addition will not be necessary.

IFP has negotiated with Électricité de France (EDF) to test this process at Champagne, using the ammonia scrubbers that were installed for the Kuhlmann double alkali project. The IFP regeneration equipment has been installed and was in the first phases of start-up in early 1975. The project is being funded by EDF, IFP, and the French government.

In Japan, The IFP process is licensed by Mitsubishi Heavy Industries and Toyo Engineering. Two small units (14 and 3-Mw equivalent) are in the start-up stage at refineries. Various problems are being encountered. It is interesting that neither the Russians, French, nor Japanese have much concern for reducing or eliminating the fume problem in the scrubber. They all accept the fact of its occurrence and try to remove it afterwards. In the U.S. the opposite approach is taken with some success.

The current status of thermal stripping is somewhat difficult to analyze. Many investigators appear to agree with Johnstone that the heat requirement is so high that chemical methods of regeneration are more economical. However, there are several possibilities for improving the economics, such as those listed below. These possibilities need further evaluation to assess their application.

Use high-sulfur coal in order to gain the advantage of the high S/C that can be generated when the SO_2 content of the gas is high. The resulting higher SO_2 vapor pressure in the stripper should reduce the heat requirement.

Use multiple effect evaporation. For a large power plant, the investment in multiple effect might be justified as it is in other processes involving evaporation of large amounts of water.

Use waste heat or low-value heat available in a power plant. This could be low-pressure steam extracted from the power cycle or waste heat recovered from the stack gas after the air heater. It has been proposed in regard to the latter that the gas be cooled with mineral oil, the hot oil mixed with the solution to the stripper, and the oil separated after stripping and recycled.

Crystallize $(NH_4)_2SO_3$ in the stripper to maintain a high SO_2 vapor pressure. The relatively low solubility of $(NH_4)_2SO_3$ in the system $(NH_4)_2SO_3$-NH_4HSO_3-H_2O (Table 7.2) favors such a method; although data are available only at low temperatures the trend of the invariant composition with temperature indicates a bisulfite/sulfite weight ratio at saturation of 4:1 to 5:1 at stripper temperature.

TABLE 7.2: SOLUBILITY IN THE SYSTEM $(NH_4)_2SO_3$-NH_4HSO_3-H_2O

	Composition of Saturated Solution	
Temperature (°C)	$(NH_4)_2SO_3$	NH_4HSO_3
0	10	60
20	12	65
30	13	67

Thermal stripping should be more effective for sulfuric acid plant tail gas, and possibly for smelter gases, because the lower wet bulb temperature allows a higher S/C without excessive consumption of cooling water. Such applications, however, have not been reported.

Oxidation to Sulfate: The problems in thermal stripping, particularly the high heat requirement and the difficulty with formation of sulfate and other side products in the cycled liquor, can be avoided by purposely oxidizing all the sulfite and bisulfite to sulfate so that ammonium sulfate is the end product of the scrubbing operation. This involves no recycling of absorbent at all and is possible only because the ammonia has end-product value of its own (as a nitrogen fertilizer).

A flowsheet of the method is given in Figure 7.5. As compared to other methods, the process has the advantage that the scrubbing step is simpler because there is no need for high S/C and oxidation in the scrubber is desirable rather than a process problem.

FIGURE 7.5: PRODUCTION OF AMMONIUM SULFATE BY OXIDIZING AMMONIA

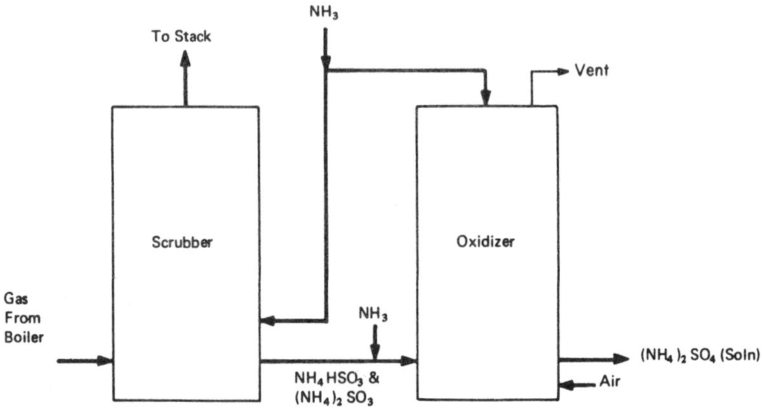

Oxidation in the scrubber, to the extent that it can be accomplished, is a relatively inexpensive way of carrying out the process. Several investigators have studied the factors affecting oxidation, but the emphasis has been on preventing oxidation rather than promoting it. The large excess of oxygen in the stack gas, about 30 times the stoichiometric amount for oxidizing all of the sulfite to sulfate, favors oxidation but it is not known to what degree oxidation could be promoted in the scrubber without interfering with absorption of sulfur dioxide. The net effect of oxidation is to tie up ammonia and thus reduce solution capacity.

Thus, all the sulfite could not be oxidized to sulfate in the scrubber and still retain absorptive capacity, unless perhaps an additional prescrubbing step were added in which conditions would be adjusted to give oxidation rather than sulfur oxide absorption. This is in effect adding an oxidizer vessel, and a separate oxidizer may be preferable. A possible way for carrying out all the oxidation in the scrubber is to cool a side stream and crystallize ammonium sulfate from it in an amount sufficient to remove the SO_2 absorbed in the scrubber. The amount of sulfate crystallized would also have to be equal to that formed in the scrubber. Hence the rate of oxidation would have to equal the rate of absorption, which might be difficult to accomplish. Perhaps the system would be self-regulating; for example, if oxidation rate were initially not quite as high as absorption, it seems likely that absorption would decrease to an equilibrium level consistent with the solution composition, or perhaps an operating parameter could be identified that, when varied, would increase absorption and decrease oxidation, or vice versa. Gas temperature or amount of oxidation promoter added might be suitable parameters to vary.

In either case, it would be necessary to promote oxidation to a rate higher than normal. Even if a separate oxidizer were used, it should be desirable to maximize oxidation in the scrubber, to the extent possible without interfering with absorption, so as to reduce the load on the oxidizer. The obvious way to accomplish this is to do the opposite of what researchers have found to be effective ways of inhibiting oxidation. Thus the following would be indicated: low gas velocity; a scrubber type, e.g., packed bed, that maximizes oxygen absorption; low solution concentration; high solution temperature; air injection into the gas before the scrubber, or use of high excess air in the boiler; and ash allowed to mix with the solution.

Unfortunately many of these interfere with absorption or make scrubber operation more expensive. Hence it may be more economical to optimize scrubber operation for absorption and carry out the oxidation in a separate system designed for it. Use of catalysts or oxygen carriers might be an exception. For example, the use of activated carbon recirculated in the scrubber solution has been claimed to be an excellent way of promoting oxygen dissolution and sulfite oxidation. Manganese ions have also been found effective.

Oxidation in a separate vessel, as noted earlier, is an old and fairly well established practice, and must be considered as the basic method of converting the entire scrubber output to sulfate. The main difficulty in separate oxidation is getting an adequate rate of oxygen dissolution in the scrubber effluent solution, a problem that has dictated the use of pressure in most of the systems developed.

The most complete development is represented by the Japan Engineering Consulting Company process in Japan. In this method, solution from the scrubber is neutralized with ammonia (to prevent evolution of sulfur dioxide from NH_4HSO_3 in the scrubber) and then oxidized with compressed air or oxygen injected into a pressurized, water-cooled reactor through a spinning cup oxidizer.

The solution is recycled through the oxidizer and the heat of reaction removed by a cooler in the recycle circuit. Ammonia and sulfur dioxide losses are recovered by introducing the spent air stream into the stack gas before the absorber. There are many installations in Japan using this technology for sulfite oxidation (mainly oxidation of calcium sulfite).

Of the other development efforts in this area, that of Simon-Carves in England appears to be the most significant. During work by this company on the autoclave process, tests were also made on oxidizing the scrubber solution with air, using a special atomizing device developed by Simon-Carves for another purpose.

Although good results were obtained, only small-scale tests were made. Most of the other work is reported only in patents. Bergwerksverband (Germany) describes a process in which the scrubber solution is neutralized with ammonia and oxidized with air or oxygen at 5 to 25 atmospheres and 150°C. Lonza (Switzerland) claims oxidation at 50°C (presumably at atmospheric pressure) with 10 cubic meters per hour of air per kilogram of sulfur dioxide.

Another variation involves oxidation without first neutralizing the scrubber solution, the objective being to evolve sulfur dioxide from the bisulfite present. In effect this is the same as the acidification method except that the oxidation is carried out in the solution rather than in a sulfuric acid plant.

The flowsheet proposed is shown in Figure 7.6. A possible advantage is that the relatively high SO_2:NH_3 ratio would promote oxidation rate. Limited tests of such a method have been made at TVA; scrubber effluent solution was treated in a packed tower with a countercurrent flow of air. Results were poor, possibly because of insufficient pressure or need to atomize the air into the solution. Catalysts might also be used to increase the oxidation rate, as discussed earlier.

Miscellaneous oxidation methods include electrolytic treatment of the solution, and separation of solid ammonium sulfite from solution by crystallization followed by oxidation of the solid with air at atmospheric pressure. The only commercial use of ammonia scrubbing followed by oxidation to sulfate apparently has been by the sulfuric acid industry in Japan. At least three companies used the process at one time but all are reported to have changed to some other method of operation.

A pilot plant designed to treat 26,000 acfm (approximately 9 Mw) gas from an iron-ore sintering plant at the Keihin Works of Nippon Kokan operated for more than three years using ammoniacal liquor scrubbing followed by oxidation to ammonium sulfate. The operation has since been converted to a double alkali (with CaO) type process. However, Nippon Kokan now has a large system under construction in which ammonium sulfate will be produced.

Although oxidation is a relatively simple method of sulfite conversion, use of the ammonium sulfate is a major problem. In Japan the sulfate is a widely used fertilizer (although its popularity is declining), but in other parts of the world, particularly in the United States, sale of the material is sometimes quite difficult. Industries such as caprolactam and coke manufacture produce large quantities of by-product sulfate, so much that the market sometimes cannot absorb it at any reasonable price.

In manufacture of nitric phosphate fertilizer, ammonium sulfate can serve the purpose of tying up and removing calcium from phosphate ore (commonly called phosphate rock), an operation that must be carried out to make the phosphate usable as a fertilizer.

Various methods have been developed in the fertilizer industry for using nitric acid to dissolve phosphate rock so that it can then be converted to an effective fertilizer. Using nitric acid in this way eliminates the need for the sulfuric acid that is normally used, thereby gaining both process and economic advantages. The nitric acid method, however, introduces a problem because calcium nitrate, a very hygroscopic material, is formed and must either be removed or converted to a more acceptable material. One of the methods that seems promising for dealing with the calcium nitrate problem is to treat the phosphate rock-nitric acid reaction slurry with ammonium sulfate, which reacts with the calcium nitrate to form soluble ammonium nitrate (an acceptable fertilizer material) and insoluble calcium sulfate. The precipitated calcium sulfate is then removed from the system and reacted with ammonia and carbon dioxide to convert it back to ammonium sulfate for recycling; the calcium carbonate also formed is discarded. The process shows promise as a way of reducing fertilizer cost.

FIGURE 7.6: AIR OXIDATION OF AMMONIUM SULFITE-BISULFITE SOLUTION

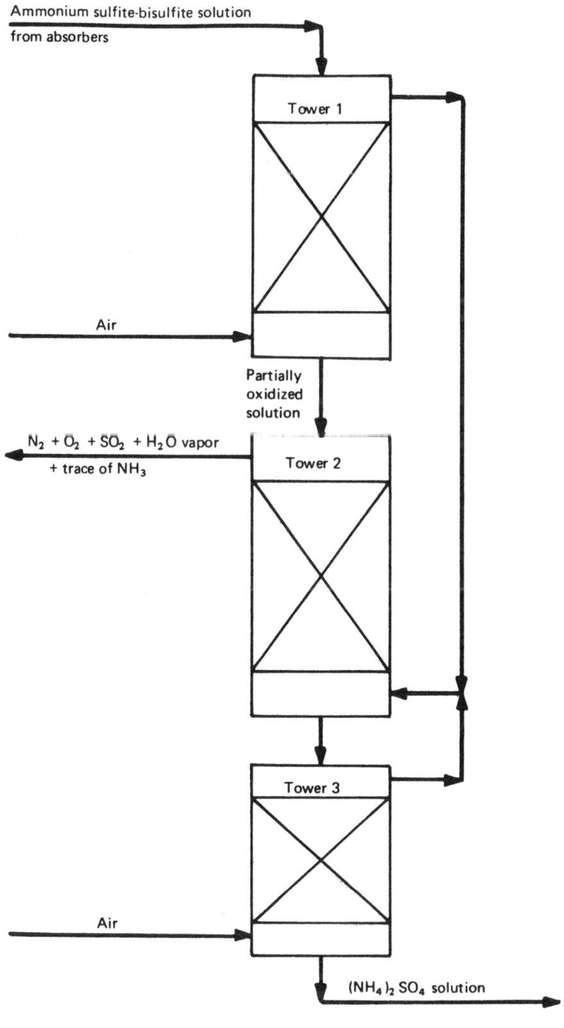

The products are ammonium sulfate and SO_2 (for conversion to H_2SO_4).

If ammonium sulfate were already available from a power plant as the product of a sulfur oxide recovery operation, it could be used in the nitric phosphate process in place of the recycled ammonium sulfate. It would not be necessary then to go through the step of converting calcium sulfate to ammonium sulfate, and the cost of this conversion could be credited to the ammonia scrubbing process. The precipitated calcium sulfate would be discarded and the filtrate, containing ammonium nitrate and phosphoric acid, reacted with ammonia to give an ammonium phosphate-nitrate fertilizer.

A flowsheet for the process is shown in Figure 7.7. An analogous system is employed in a European fertilizer plant, where by-product ammonium sulfate from a caprolactam operation is used to precipitate calcium sulfate in a nitric phosphate plant of the type shown.

FIGURE 7.7: USE OF AMMONIUM SULFATE FROM STACK GAS RECOVERY SYSTEM IN PRODUCTION OF NITRIC PHOSPHATE FERTILIZER

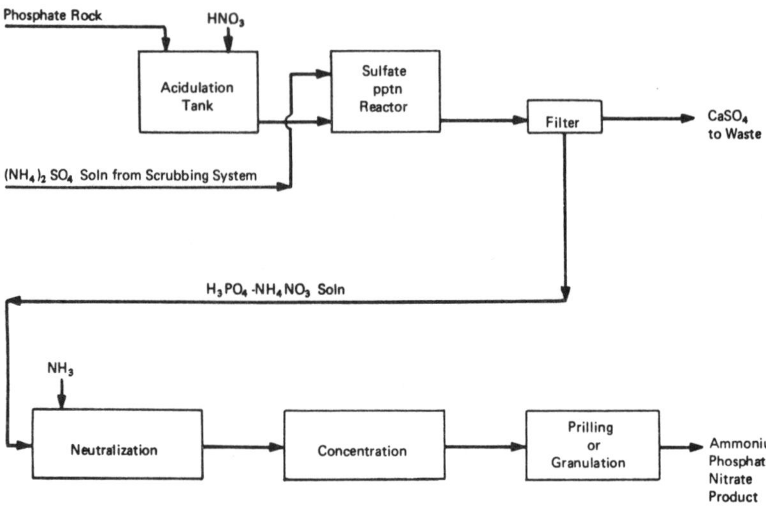

Such a system was evaluated in one of the earlier EPA-TVA conceptual design studies. The conclusions (some of which may not be true today, particularly those regarding relative costs) were as follows.

> Ammonia scrubbing and production of phosphate fertilizer has promise, under certain conditions, as a method for recovering sulfur oxides from stack gases.
>
> The main economic factors are product volume (depending on power plant size and S content of coal), net sales revenue, and basis of financing.
>
> Private industry participation in financing and operation appears unlikely because of the high projected cash flow necessary to attract capital. There is a net profit for plants about 500 Mw and larger but

it is not large enough to attract investment except perhaps under special conditions. The situation is improved if there is supplemental income in the form of a payment for the service of sulfur oxide control; in this case the larger product volumes, e.g., 1,000 Mw at 3.5% S, give a projected total income adequate for financing.

The economics under power industry financing are more promising. For plants 500 to 600 Mw and larger in size the recovery method is favored over limestone wet scrubbing because there is less deficit passed on to the power consumer after payment to investors of the regulated return on investment.

Only a few recovery installations of this type can be accommodated by the fertilizer market.

It may be possible to improve the economics by increasing the lifetime capacity factor, reducing the degree of gas reheat, and development of new departures that may reduce investment and operating cost.

At the present time (early 1975), the fertilizer industry is prosperous and there is a good market. However, expansion of capacity to use power plant sulfur seems unlikely. Moreover, the potential oversupply of sulfur, particularly from by-product sources (mainly from smelters and sour gas sweetening operations), makes sulfuric acid so cheap that nitric acid has lost much of its attraction as an acidulant for phosphate rock.

Conversion of ammonium sulfate to bisulfate is another way of using the sulfate advantageously. Heating the sulfate to about 700°F converts it to bisulfate with evolution of half the ammonia.

$$(NH_4)_2SO_4 \xrightarrow{700°F} NH_4HSO_4 + NH_3\uparrow$$

The ammonia can be recycled to the scrubber. The bisulfate is acidic and can be used in place of sulfuric acid in several ways; examples of application to fertilizer technology are as follows.

Ammonium Phosphate-Sulfate Fertilizer—A solution of the bisulfate will dissolve phosphate rock as other strong acids do. The resulting slurry would be filtered (Figure 7.8) to remove calcium sulfate and the filtrate ammoniated to give ammonium phosphate-sulfate fertilizer. The method has been tested only on a bench scale. It was evaluated in the earlier EPA-TVA study and found to be less attractive than the nitric phosphate route; however, not enough data were available for a good evaluation.

Phosphoric Acid—A major problem in the fertilizer industry is the impure nature of the phosphoric acid produced by the usual method, dissolution of phosphate ore in sulfuric acid. The impurities extracted by this procedure (Fe, Al, Mg) cause precipitates when the phosphoric acid is used in making liquid fertilizers.

It has been found that relatively pure acid can be produced by reacting phosphate ore with ammonium bisulfate and extracting the resulting melt or slurry with an alcohol, e.g., methanol. The alcohol extracts the acid, which can be separated by boiling off the alcohol.

FIGURE 7.8: USE OF AMMONIUM SULFATE RECOVERED BY STACK GAS SCRUBBING IN PRODUCTION OF PHOSPHATE FERTILIZER BY THE BISULFATE METHOD

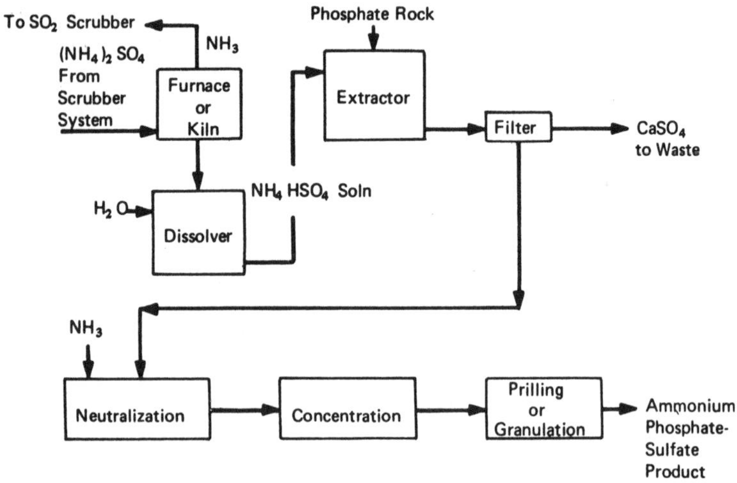

A generalized (and simplified) equation is as follows.

$$Ca_3(PO_4)_2 + 6NH_4HSO_4 \longrightarrow 2H_3PO_4 + 3CaSO_4 + 3(NH_4)_2SO_4$$

There are several variations of the general method. One of the more promising is shown in Figure 7.9. The solid $CaSO_4$ is separated and discarded and the $(NH_4)_2SO_4$ is recycled; thus half the sulfate converted to bisulfate must be supplied, perhaps from a stack gas recovery unit, and half is from recycle.

The general process has been studied by several organizations, including TVA and Bohna Engineering and Research, Inc. Only small-scale or minipilot plant work has been done. Esterification of the alcohol during the extraction step has been a problem in some of the work, and the process is obviously somewhat complicated. Further work must be done before an adequate evaluation can be made.

For all bisulfate processes, the step of converting the sulfate is a difficult one. High temperature is required, the reaction mixture goes through a sticky stage during the heating period, and corrosion is a major problem. A considerable amount of development work remains to be done in this area. Exxon (formerly Esso) did major pioneering work years ago using an electrical decomposer for regenerating ammonium sulfate. Further work, on a fairly large scale, is under way in France using a direct oil-fired decomposer.

Acidulation: Since sulfurous acid (H_2SO_3) is a weak acid, any strong acid added to the scrubber effluent will capture the ammonia and release sulfur dioxide from the sulfite and bisulfite.

FIGURE 7.9: USE OF AMMONIUM SULFATE RECOVERED BY STACK GAS SCRUBBING IN PRODUCTION OF PURE PHOSPHORIC ACID BY THE BISULFATE METHOD

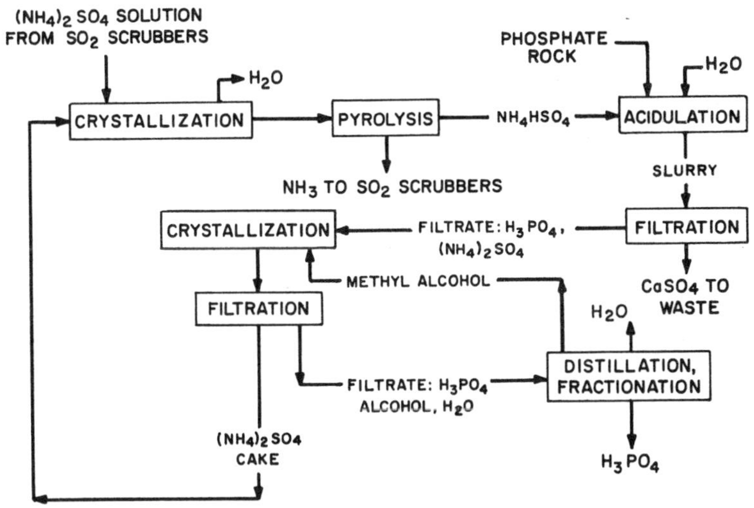

This method gives two products: (1) either sulfuric acid or elemental sulfur made from the evolved sulfur dioxide, and (2) the ammonium salt of the added acid. Since sulfuric, nitric, and phosphoric are the most appropriate acids to use, the coproducts are ammonium sulfate, ammonium nitrate, and ammonium phosphate respectively (Figure 7.10).

FIGURE 7.10: UTILIZATION OF AMMONIA SCRUBBER EFFLUENT BY ACIDIFYING TO DECOMPOSE SULFITE AND BISULFITE

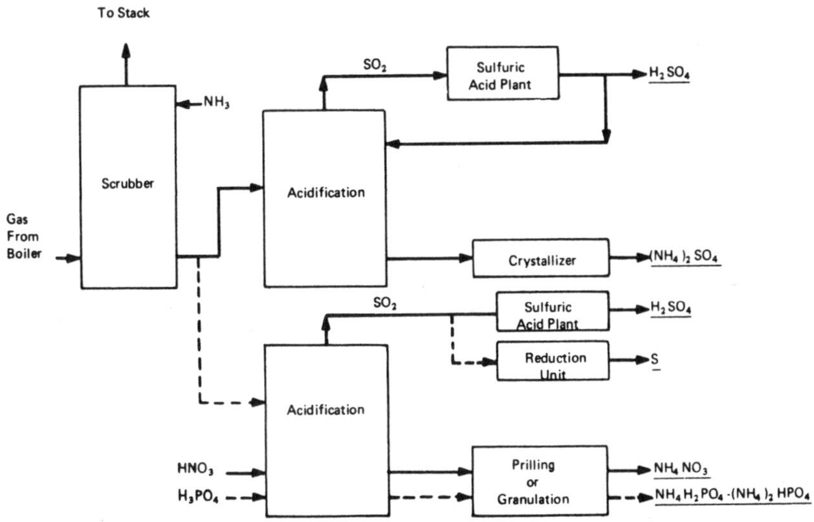

If sulfuric acid is used, an obvious source of the acid is the sulfur dioxide evolved, converted to sulfuric acid in an acid plant installed as part of the recovery installation. More acid will be produced than needed for effluent acidification, so that there is some net production of sulfuric acid for sale. The amount depends on the content of sulfite in the scrubber effluent, as each mol of bisulfite produces one mol of sulfur dioxide in excess of that required to make the sulfuric acid for recycling, whereas all the sulfur dioxide evolved from ammonia sulfite must be recycled.

$$2NH_4HSO_3 + H_2SO_4 \rightarrow (NH_4)_2SO_4 + 2SO_2 + 2H_2O$$

$$(NH_4)_2SO_3 + H_2SO_4 \rightarrow (NH_4)_2SO_4 + SO_2 + H_2O$$

Thus to maximize production of sulfuric acid, the scrubber should be designed and operated for high bisulfite; if ammonium sulfate is preferred then operation to give sulfite is indicated. In most situations sulfuric acid is the more desirable of the two products.

The major work on acidulation with sulfuric acid has been done by Cominco in British Columbia. The process, started up almost 40 years ago and still in operation, involves ammonia scrubbing of smelter gases followed by heating the scrubber effluent solution by heat exchange with the acidifier effluent. The heated solution flows to the acidifier, a brick-lined vessel, where strong H_2SO_4 is added; the evolution of SO_2 gas provides adequate mixing. The acidified solution overflows to a stripper where an air stream strips out most of the residual SO_2, from which the solution passes to a crystallizer; ammonia is added first to neutralize any residual acidity. The evolved SO_2 is converted to H_2SO_4 in a standard contact-type sulfuric acid plant.

Several other organizations have worked on acidulation with sulfuric acid, including RIIC (Czechoslovakia, prototype test unit), Fertilizers and Chemicals, Travancore (India, full-scale), and TVA (pilot plant). The process can be considered to be fully developed because of the large amount of test work and the fact that it is in commercial use in a smelter and in at least four sulfuric acid plants (India, U.S.A., Austria, and Czechoslovakia).

The main drawback is the difficulty in selling or otherwise disposing of the $(NH_4)_2SO_4$ portion of the product, which outweighs the sulfuric acid considerably in percentage of the total sulfur recovered. Depending on the bisulfite/sulfite ratio, the sulfate accounts for a percentage ranging from about 60% to nearly all of the sulfur.

Acidulation with nitric acid gives the advantage that the NH_4NO_3 produced is a popular fertilizer with much better sales potential than $(NH_4)_2SO_4$. The major work with nitric acid has been done by RIIC in Czechoslovakia, where the several process problems have been worked out and the method applied commercially in a sulfuric acid plant. A flowsheet of the process is shown in Figure 7.11.

Tail gas from the sulfuric acid plant (at Lovosice, 300 tpd H_2SO_4) passes through two scrubbers in series, operating at about 25°C. Removal of SO_2 is excellent, about 90%, and a high S/C is obtained. The NH_4NO_3 solution obtained in the reactor-stripper is pumped to an adjacent fertilizer plant where it is used in complex fertilizer production. The SO_2 is returned to the H_2SO_4 plant.

Among the problems in this type of process are: escape of NO_x with the evolved SO_2, causing damage to the catalyst in a contact-type sulfuric acid plant; and presence of sulfate (from SO_3 in the gas and oxidation in the scrubber) in the NH_4NO_3 solution. The NO_x problem can be avoided by careful operation of the stripper, thereby keeping NO_x evolution low enough to avoid catalyst damage. Sulfate formation, if it does not exceed 10 to 15% of the total sulfur, can be tolerated if the NH_4NO_3 is used in a mixed fertilizer (NP or NPK) production. It is also probably acceptable for straight NH_4NO_3 production, although the nitrate will not be the usual commercial grade, which would have some adverse effect on sale of the product.

The nitric acid acidulation route appears to be one of the better prospects among the recovery processes that have been demonstrated commercially. It has not been applied in power plants, however, and there will be problems in doing so. Probably the major one is the higher wet bulb temperature of the power plant gas, which reduces the S/C attainable. The main drawback to low S/C in this case is the higher tonnage ratio of NH_4NO_3 to H_2SO_4.

FIGURE 7.11: UTILIZATION OF AMMONIA SCRUBBING EFFLUENT SOLUTION BY ACIDULATING WITH NITRIC ACID

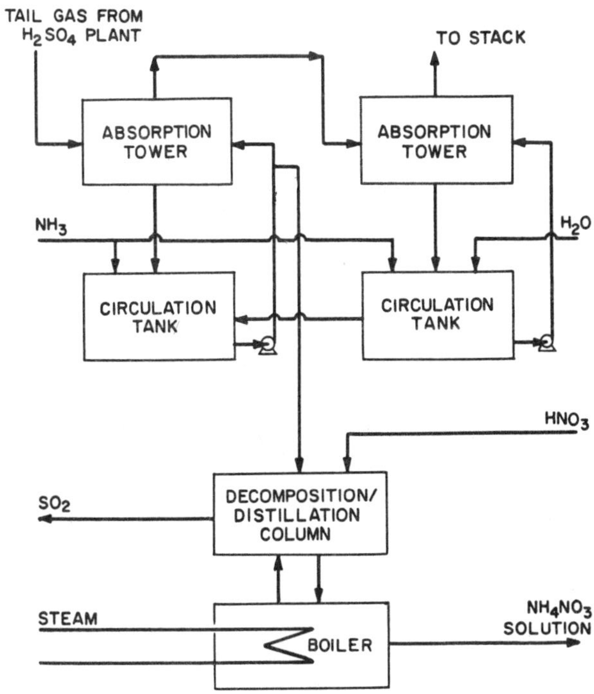

Process developed by the Research Institute of Inorganic Chemistry in Czechoslovakia

Acidulation with phosphoric acid has been studied by several organizations but the major work has been done by IPRAN (Inorganic Chemistry and Fertilizer Engineering Institute) in Romania. Here also the process problems have been worked out and full-scale production achieved. The flowsheet is given in Figure 7.12.

FIGURE 7.12: UTILIZATION OF AMMONIA SCRUBBING EFFLUENT SOLUTION BY ACIDULATING WITH PHOSPHORIC ACID

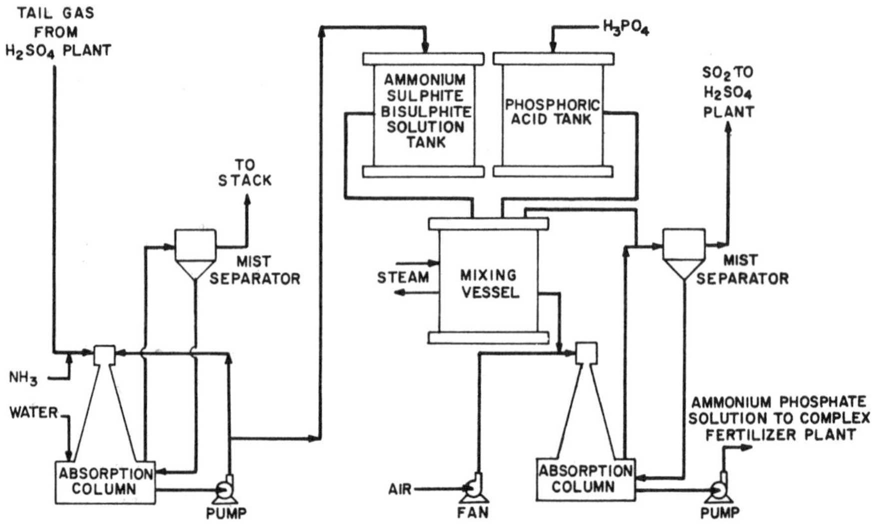

Process developed by the Inorganic Chemistry and Fertilizer Engineering Institute in Romania.

Tail gas from a sulfuric acid plant is scrubbed in a venturi-type scrubber of special design. The resulting solution is mixed with H_3PO_4 (50% P_2O_5) in a heated vessel (95°C) and then stripped of SO_2 in a second venturi-type vessel through which is blown a stream of air. The SO_2 is returned to the acid plant and the ammonium phosphate solution (mainly $NH_4H_2PO_4$) pumped to an associated complex fertilizer plant.

One of the main problems in this process is the relatively low reactivity of phosphoric acid. Whereas it would be desirable to produce diammonium phosphate [$(NH_4)_2HPO_4$] in order to reduce the amount of phosphate produced, the second hydrogen ion of the acid is too weak to react with $(NH_4)_2SO_3$, making the main reaction:

$$(NH_4)_2SO_3 + 2H_3PO_4 \longrightarrow 2NH_4H_2PO_4 + SO_2 + H_2O$$

For a sulfuric acid plant associated with a complex fertilizer plant this is acceptable because much more ammonium phosphate is involved in the total operation than that produced in the SO_2 recovery unit. This situation makes it feasible in the IPRAN process to use only one scrubber stage as a means of reducing scrubber cost. As a result, the S/C is low (about 0.7) and more H_3PO_4 is required, but this is not a problem because there is plenty of H_3PO_4 available in the fertilizer plant operation. For a power plant, however, where the amount of SO_2 recovered is the controlling factor in amount of phosphate produced, it would be highly desirable to minimize the amount of H_3PO_4 required.

Another drawback is that the monoammonium phosphate produced is not a major fertilizer in the United States. Some market development would be required, but the material is intrinsically a good one and probably would be acceptable for large-scale production. The small amount of sulfate in the product should not be a problem.

The process has been installed in three sulfuric acid plants (total of 300,000 tpy H_2SO_4) at the Turnu Magurelle fertilizer complex in Romania. Routine operation was established in October 1970 and performance is reported to be satisfactory. Other installations may be made. The process is promising and could find application in SO_2 recovery from power plants. Acidulation with ammonium bisulfate is also a possibility; it is a strong acid and thus can be used to decompose the scrubber solution.

$$(NH_4)_2SO_3 + 2NH_4HSO_4 \rightarrow 2(NH_4)_2SO_4 + SO_2 + H_2O$$

$$NH_4HSO_3 + NH_4HSO_4 \rightarrow (NH_4)_2SO_4 + SO_2 + H_2O$$

A flowsheet is given in Figure 7.13. Like stripping, the method has the drawback that sulfate formed in the gas and in the scrubber accumulates in the system and must be removed; there is no coproduct to collect it as in acidulation with HNO_3 and H_3PO_4 and no automatic sulfate removal as in the oxidation process.

The only known work on the method (except that indicated in patents) has been at TVA, where an ammonia scrubbing program at TVA's Colbert pilot plant has been under way since 1969. This effort began as a wholly funded EPA program with the objective to fully characterize ammoniacal liquor scrubbing of SO_2-laden flue gases. Since 1971, the program has become a jointly funded effort in which the operational phase of ammonia scrubbing was coupled with the bisulfate regeneration scheme.

The process is promising because it appears to have fewer problems than most of the regeneration processes. The energy requirement is relatively low (much lower than in stripping), side reactions do not appear to be a problem, there is no major coproduct, and the product is a stream of essentially pure SO_2 suitable for conversion to elemental sulfur.

The sulfate problem can be handled either by selling the $(NH_4)_2SO_4$ formed (which should be a relatively small amount) or by decomposing with lime. The $(NH_4)_2SO_4$ is easy to separate from the main flow because all the sulfate in the system passes through the stripper on the way to the decomposer; part of it, enough to bleed off that formed in the scrubber circuit, can be removed at that point. A flowsheet of the absorption section of the TVA ammonia pilot plant is shown in Figure 7.14.

FIGURE 7.13: UTILIZATION OF AMMONIA SCRUBBING EFFLUENT SOLUTION BY ACIDULATING WITH NH_4HSO_4

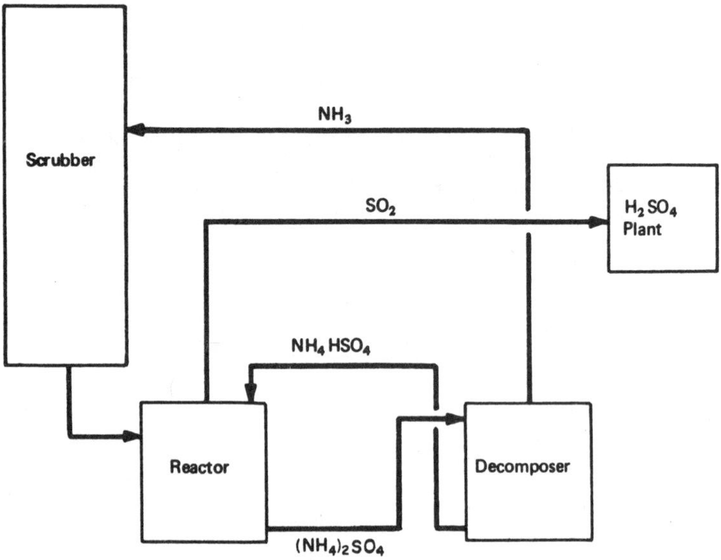

FIGURE 7.14: ABSORPTION SECTION

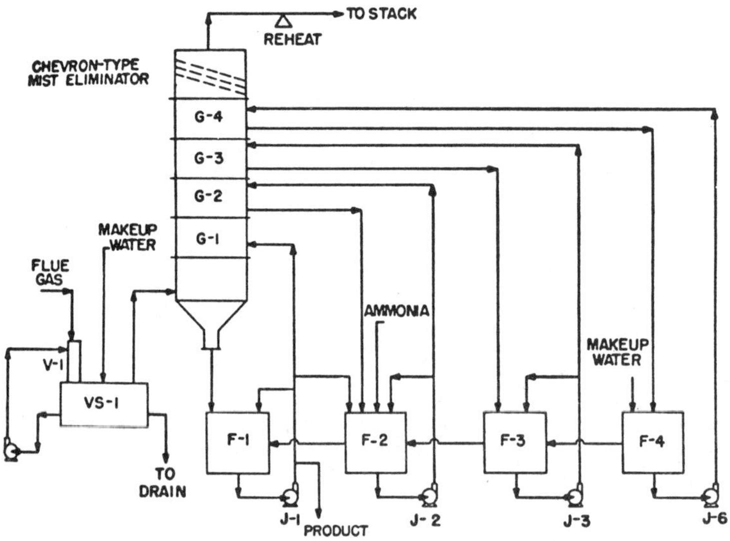

The pilot plant is designed to treat approximately 3,000 cfm of flue gas at absorber conditions (saturated at about 125°F). The gas flows through a prewash section containing a venturi-type element. The conditioned gas is then routed to the absorption tower. The absorber consists of as many as four independent absorbing stages (valve tray elements).

The purpose of the gas pretreatment is to humidify and cool the flue gas and to remove fly ash and chlorides from the gas stream. Humidification, cooling, and chloride removal are considered necessary for reducing the plume emission from the absorber. Fly ash is removed before the absorption section to decrease the amount of solids in the scrubber liquor. These solids could cause major problems in the absorption and regeneration sections.

There were three major problem areas identified during the operation of the prewash section: (1) low pH of sump liquor (corrosion), (2) disposal of low pH purge water from prewash sump, and (3) mist carryover from the prewash section to the absorber section.

During initial operation, the quantity of makeup water added to the prewash section was that consumed in the process for humidification and absorber product liquor bleed-off, approximately 1.2 gpm. At this flow rate, the pH of the wash liquor was 1.0 or less. The makeup water throughput was increased to 15 to 30 gpm to increase the pH of the wash water and to reduce its corrosion potential. This raised the pH of the wash water to approximately 2.5, but corrosion continued.

An inspection of the gas pretreatment section was made after 2,000 hr of operation. The entire pretreatment section, constructed of Type 316L stainless steel, was badly damaged by corrosion. The rods across the throat of the venturi were pitted badly, especially on the ends, and several needed replacing. The liquor sump walls were corroded with the most severe attack above the normal gas-liquor interface. In this area, pits 40 to 60 mils deep were common. The gas duct from the sump to the absorber and the sample probes inside the duct (316 stainless steel) were also severely corroded.

The gas prewash section was operated open loop to minimize corrosion of the equipment. The problem of disposing and/or utilizing low pH water will be common to most SO_2 removal processes that require a prewash. Attempts to close the loop around the prewash section will be studied in future operations.

The absorber shell consists of 32" x 32" x 4' sections. At normal operating conditions, the gas velocity in the tower is approximately 7 fps. Two types of absorption elements, marble beds and valve trays, have been tested.

In the most recent tests, two absorber configurations were used. During the first test period, the absorber consisted of one valve tray followed by two marble beds (Figure 7.15). Tests were made to determine the highest liquor concentration that could be obtained using this configuration while limiting the outlet SO_2 concentration to $\leqslant$250 ppm and the ammonia concentration to $\leqslant$50 ppm. Product liquors with high concentrations require less energy for regeneration and smaller regeneration equipment. The maximum product liquor concentration obtained using this configuration was a C of 14. Factors limiting the C in the initial series were: the diluting effect of the mist carryover from the gas pretreatment section, and the SO_2 and NH_3 vapor pressure over the liquor on the top stage. The second series of tests was made with the fourth stage added to the absorber. Figure 7.16 shows the absorber arrangement after addition of the fourth absorber stage. The two marble-bed absorption elements were replaced with valve trays during the modification because of occasional excessive stage-to-stage weepage which upset the liquor concentrations throughout the absorption tower. Experience showed that this problem does not occur with the valve tray when the gas flow rate remains relatively constant although weepage from the valve trays is a problem when the gas flow varies.

Tests using this absorber arrangement indicated that product liquor could be produced that had C's of 20 and above. Table 7.3 shows data from the absorber operation while producing liquor with a C of 21. No absorber operational problems were encountered during operation at the high concentration. As stated earlier, the most apparent problem in the absorption step is that of fume formation (discussed earlier).

TABLE 7.3: TYPICAL ABSORBER LOOP TEST–AMMONIUM BISULFATE PILOT PLANT

Test Conditions	
Gas to absorber	
Flow rate (at 32°F)	2,350 scfm
Temperature	295 °F
SO_2	2,800 ppm
Gas leaving venturi	
Temperature	120 °F
SO_2	2,720 ppm
Gas leaving first stage	
Temperature	123 °F
SO_2	1,160 ppm
Gas leaving second stage	
Temperature	124 °F
SO_2	430 ppm
Gas leaving third stage	
Temperature	115 °F
SO_2	510 ppm
Gas leaving fourth stage	
Temperature	114 °F
SO_2	390 ppm
SO_2 removal	84 %
Fresh water feed to absorber	0.2 gpm
Ammonia feed to absorber	22.5 lb/hr
Liquor flow from first stage to second stage	1.6 gpm
Liquor to first stage (product)	
C	21.1
S/C	0.73
pH	6.2
Specific gravity	1.286

(continued)

TABLE 7.3: (continued)

Test Conditions	
Liquor to second stage	
C	21.5
S/C	0.67
pH	6.3
Specific gravity	1.282
Liquor to third stage	
C	6.6
S/C	0.78
pH	5.9
Specific gravity	1.138
Liquor to fourth stage	
C	2.5
S/C	0.86
pH	5.7
Specific gravity	1.086

FIGURE 7.15: THREE-STAGE ABSORPTION

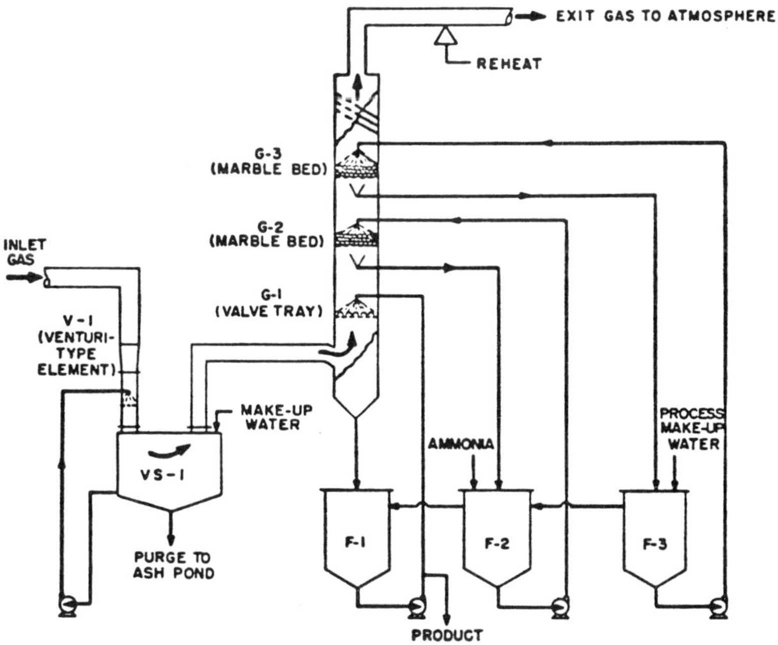

FIGURE 7.16: THREE- AND FOUR-STAGE ABSORBER

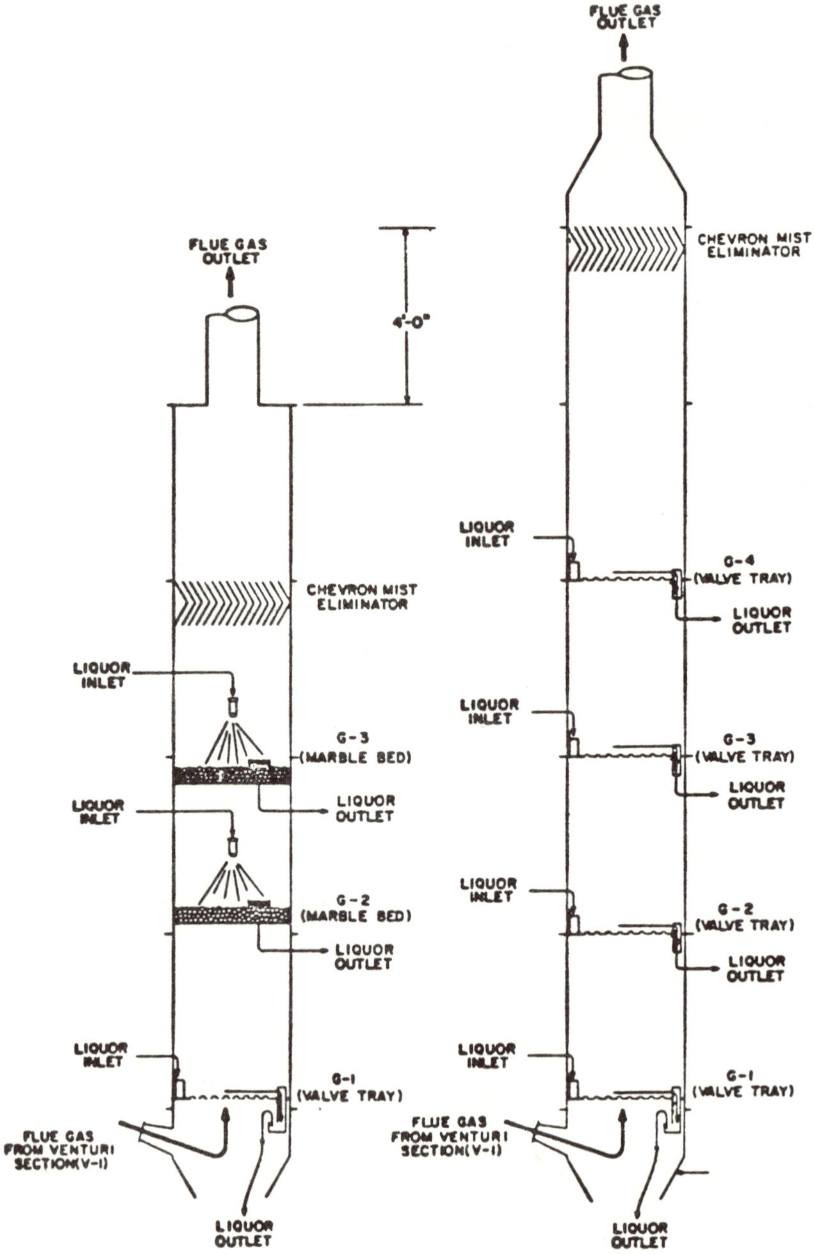

The precipitation of solids in the absorption section may present another major problem. A yellow solid, tentatively identified petrographically as a homogeneous iron-ammonia-sulfur compound was present in all of the absorber liquors. Fly ash is the probable source of iron in the solid. Although most of the fly ash is removed in the prewash section, some passes through the venturi and is removed in the absorber. The mist carry-over from the prewash section also contains dissolved iron along with other metals.

Oxidation of sulfite to sulfate occurred in the absorption section. In the range of liquor concentrations tested, the oxidation ranged from 10 to 15% of the absorbed SO_2. An average value of 13% was obtained over an extended operating period using the four-stage absorber. A flowsheet of the regeneration section of the pilot plant is shown in Figure 7.17. The absorber product liquor is fed to an acidulator where it is mixed with an acid ion to chemically release the SO_2.

A portion of the released SO_2 is evolved in the acidulator. In past operation, the acid ion was supplied by plant-grade sulfuric acid. In the future, the acid ion will be supplied by ammonium bisulfate generated by thermal decomposition of ammonium sulfate (anticipated operation of the electrically heated decomposer is late 1975).

FIGURE 7.17: REGENERATION SECTION

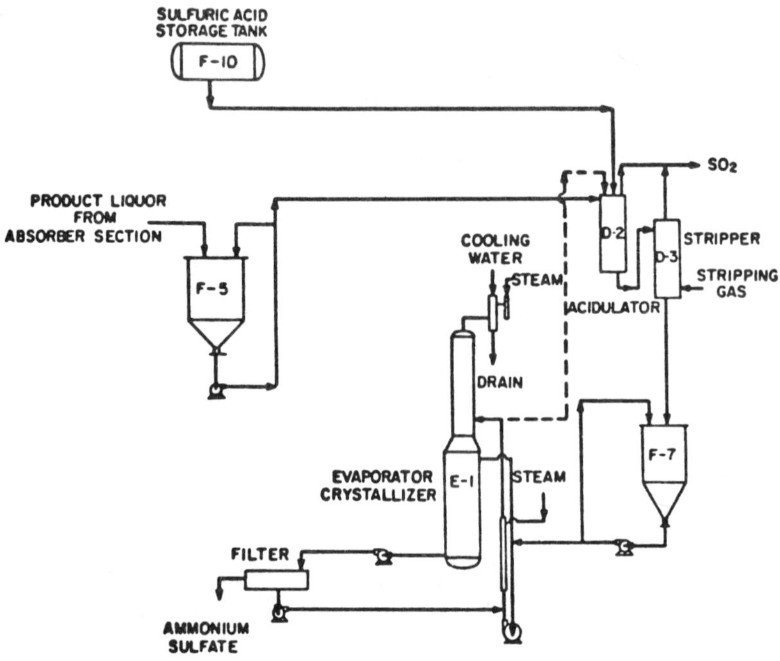

The liquor from the acidulator overflows into a stripper where the remaining SO_2 is stripped from the liquor with a countercurrent flow of air. Under the conditions employed, 96% of the SO_2 in the ammonium sulfite-bisulfite liquor to the acidulator was removed during acidulation and stripping.

The stripped liquor (ammonium sulfate solution) is metered into the evaporator-crystallizer where water is evaporated to produce a slurry of ammonium sulfate crystals. The crystals are separated from the mother liquor in a continuous belt filter and fried. The crystals were dewatered from about 80 to 5% water on the belt and were then fed to a propane gas-fired dryer. Product from the dryer contained less than 1% water and was free-flowing.

Results of these pilot operations have shown that the ammonium bisulfate process is a promising candidate for second generation SO_2 removal systems. Continuing pilot-plant activities will be directed toward long-term operation of an integrated ABS system and demonstrated performance of 90% + SO_2 removal. The approach to this objective will include the following.

Closing the water balance around the gas prewash section and reducing carry-over into the absorber.

Further reducing the plume emissions from the absorber.

Improving the acidification and stripping steps to obtain a complete release of SO_2 using ABS melt for acidulation.

Improving the solids separation from the absorber product liquor.

Studying ammonium sulfate crystallization (energy required, crystal growth, crystal size, etc.).

Reduction: If $(NH_4)_2SO_3$ and NH_4HSO_3 are heated under pressure, disproportionation occurs and $(NH_4)_2SO_4$ and elemental sulfur, plus other decomposition products, are formed. Thus the reduction occurring in the disproportionation reaction can be used to advantage in converting at least part of the recovered SO_2 to the preferred form, elemental sulfur.

A generalized flowsheet for the method is given in Figure 7.18; it has been studied mainly by Cominco, Simon-Carves (U.K.), NIIOGAZ, and TVA. The chemistry appears to be quite complicated, with investigators giving various accounts of the mechanisms involved. It seems generally agreed that both thiosulfate and sulfuric acid must be present for good operation. Some have reported that these are necessary to prevent explosion hazard; others hold that the H_2SO_4 enters into various reactions, one of them being conversion of thiosulfate to $(NH_4)_2SO_4$ and S.

Although the process was developed to a fairly advanced degree by Simon-Carves and Cominco, it has never been used commercially. Pilot plant work (about 15 megawatts) was done on an oil-fired boiler in the U.S.S.R. Although elemental sulfur is produced, two-thirds or more of the SO_2 is converted to $(NH_4)_2SO_4$. Another problem is the high corrosion encountered in the reactor; it has been proposed that this be minimized by operating at higher pH, accepting incomplete reaction in the autoclave, and recirculating unconverted solution to the scrubber. It has also been proposed that reduction to sulfur be carried out by treating the sulfite solution with H_2S or other reducing agent. These methods are described only in patents and it is not known to what extent they have been developed.

FIGURE 7.18: UTILIZATION OF AMMONIA SCRUBBING EFFLUENT SOLUTION BY AUTOCLAVING TO PRODUCE $(NH_4)_2SO_4$ AND ELEMENTAL SULFUR

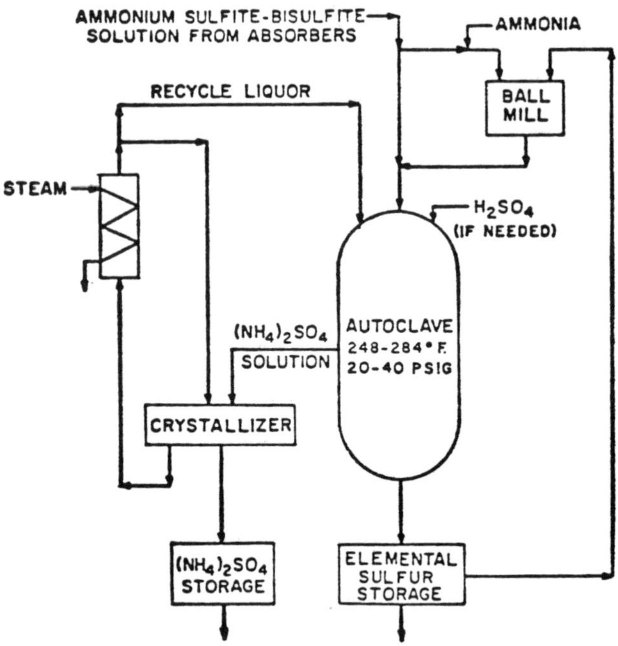

The process has been a favorite one with inventors, not only for treating ammonia scrubbing effluent but also for solutions or slurries obtained with other absorbents (other alkalis, alkaline earths, and metal compounds). The simplicity is appealing, since elemental sulfur is produced in one step and is easily separated. The main drawback is formation of numerous side products that can accumulate in the closed loop system. With one or two exceptions, little progress seems to have been made on these problems. The more promising work has been done in the sodium system, discussed later.

Other Methods: Several other regeneration processes have been proposed but none appears to have been developed very far.

Crystallization of $(NH_4)_2SO_3$ and use as such, for fertilizer or in other industries.

Crystallization of $(NH_4)_2SO_3$ followed by complete oxidation, over a catalyst, to HNO_3 and H_2SO_4. In one version, the $(NH_4)_2SO_3$ is heated to decompose it to NH_3 and SO_2 and the two gases passed through the catalyst. In another, either solid $(NH_4)_2SO_3$ or $(NH_4)_2SO_4$ is blown into the catalyst.

Separation of $(NH_4)_2SO_4$, reaction with ZnO, and decomposition of the $ZnSO_4$ to ZnO and SO_3.

Reaction of ZnO with ammonium sulfite-bisulfite solution to form $ZnSO_3$, which is decomposed thermally to ZnO and SO_2. (This method has been developed mainly for the sodium system, as discussed below.)

SODIUM COMPOUNDS

Sodium compounds have several advantages over other alkali absorbents. As compared to ammonia, they can be used in solid absorbent systems and, since the cation is not volatile, there is no problem with absorbent volatilization in scrubbing processes. As compared to potassium, sodium compounds are generally cheaper and the solubility relationships in scrubbing processes are more favorable. The products of SO_2 absorption in sodium systems are similar to those in ammonia scrubbing; usually Na_2SO_3, $NaHSO_3$, and Na_2SO_4 are obtained. Most of the regeneration systems give Na_2SO_3, Na_2CO_3, or NaOH for recycling but some processes involve a special absorbent, e.g., $NaAlO_2$.

Absorption by Solids

Compounds such as Na_2CO_3 or NaOH are reactive enough that they can be used in solid absorption systems, at somewhat elevated temperature if desired.

Sodium Carbonate: Very little work appears to have been done on absorption by solid Na_2CO_3. Laboratory tests made in the U.S.S.R. indicated good absorption at 40° to 75°C. Similar work in Czechoslovakia indicated that the presence of water vapor in the gas promotes the reaction in the temperature range studied (120° to 160°C). There are a few patents on the method but they give little information.

Sodium Bicarbonate: There has been some interest in nahcolite, a natural form of sodium bicarbonate; a very large deposit of the material occurs in the Piceance basin in northwest Colorado. Various tests have been made in which nahcolite was injected into stack gas. In one of the more successful tests, the gas temperature at the point of injection was 240°C and the product was caught in a bag filter. Removal of SO_2 was in the range of 60 to 75%. The product was presumably Na_2SO_4. Tests have also been made (by Superior Oil) on SO_2 absorption in a moving bed of granular nahcolite.

The cost of the nahcolite is the important consideration, of course, either for the injection method or for solution scrubbing, as it should be usable in either. It is not being mined at present, but the developers in the early 1970s estimated a delivered cost on the East Coast of \$16/ton (72% $NaHCO_3$). This is equivalent, on the basis of the amount required to react with SO_2, to ammonia at about \$110/ton. The delivered cost of nahcolite would certainly be \$25/ton or more at today's prices, or the equivalent of \$175/ton for ammonia, which makes nahcolite quite expensive.

However, if injection of the solid would give adequate SO_2 removal, the reduction in capital costs, as compared to ammonia scrubbing, would go far in offsetting the higher raw material cost. The work so far appears to be aimed at using nahcolite in a throwaway process. For this, the material has an even higher cost handicap, since on an equivalent basis limestone costs only about 15 to 20%

as much as the estimated delivered price of nahcolite (assumed for this reference to be \$25/ton). Again, however, further work might show injection of nahcolite to be adequately effective whereas it is evident that limestone injection is not, in which case the lower investment for injection as compared to scrubbing would tend to offset higher raw material cost.

For recovery processes, nahcolite would have to compete against sodium carbonate as makeup for system losses. Since the equivalent cost for Na_2CO_3, based on \$25/ton for nahcolite, is about \$70/ton, Na_2CO_3 might be more economical since it should be obtainable for less than \$70 in most areas. This might also be true, of course, for thrcwaway operation. The main problem for throwaway might be in finding a way to dispose of soluble Na_2SO_4 (or Na_2SO_3) without incurring a water pollution problem.

Trona ($Na_2CO_3 \cdot NaHCO_3 \cdot 2H_2O$), another mineral obtainable at relatively low cost, has been tested in a dry-type system in which the mineral is held in place by a device such as a bag or panel filter. Good sorption efficiency was obtained but the problems are about the same as for nahcolite.

Sodium Aluminate: Although no longer considered a viable method, the solid absorbent process that has received the most attention is use of sodium aluminate as the absorbent, the so-called alkalized alumina process. The method was developed originally by the U.S. Bureau of Mines at Pittsburgh, Pa., where a pilot-plant test program was carried out. The Central Electricity Generating Board (CEGB) in the U.K. also took an interest and did experimental work at an operating power plant in northern England. Various other groups have contributed, including W.R. Grace (absorbent development), AVCO (kinetics studies), and M.W. Kellogg (process design and evaluation). The process is based on use of $NaAlO_2$ as the absorbent in a solids-absorption system.

$$2NaAlO_2 + SO_2 + \tfrac{1}{2}O_2 \rightarrow Na_2SO_4 + Al_2O_3$$

The alumina is included to assist in regeneration rather than in absorption, although the alumina base for the absorbent provides porosity that is difficult to attain with sodium salts alone. In regeneration the alumina serves a very useful function by providing an anion to combine with the Na^+ and thereby reduce the energy level required. The regeneration involves reduction rather than thermal decomposition.

$$Na_2SO_4 + Al_2O_3 + 4H_2 \rightarrow 2NaAlO_2 + H_2S + 3H_2O$$

Other processes that involve reduction of alkali sulfate or sulfite are plagued with various operating problems, including high reaction temperature, formation of corrosive melts and formation of side products. In contrast, reduction in the alkalized alumina process is relatively smooth and uncomplicated. A major requirement for good regeneration, however, is a small content of iron (1%) in the sorbent; without this the regeneration rate is slow.

Various flowsheets have been proposed for the process. Figure 7.19 gives a simplified version of the one proposed by CEGB. The stack gas is contacted with regenerated absorbent in a fluidized reactor, operated preferably at 300° to 350°C (although lower temperature can be used if necessary for retrofitting). Small-scale tests indicate that 90% removal of SO_2 is attainable with a reasonable amount of absorbent. The loaded absorbent is heated to 650° to 700°C by heat exchange (plus heat from burning unconverted H_2S in a later process step) and then regenerated, in a fluidized bed, by reaction with reformed gas or any other suitable reducing gas. Part of the evolved H_2S is burned to SO_2 and the two gas streams passed through a Claus plant to produce elemental sulfur. Unconverted H_2S in the Claus plant tail gas is burned to SO_2 and recycled.

FIGURE 7.19: ALKALIZED ALUMINA PROCESS FOR RECOVERY OF SO_2 FROM STACK GAS

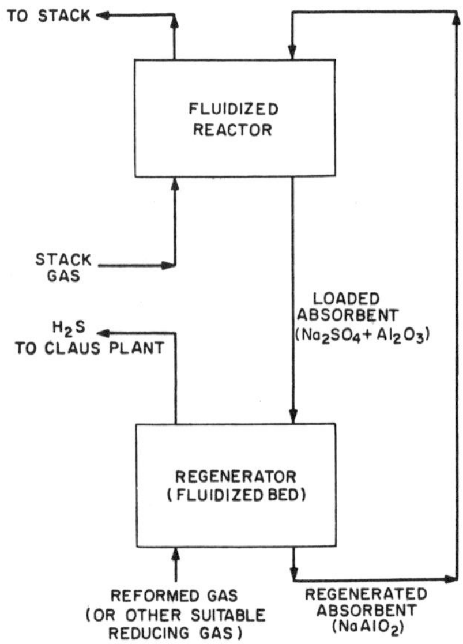

The absorbent contains about 20% Na_2O and is formed into particles in the size range of 10 to 14 mesh. A considerable amount of research effort has been devoted to increasing porosity of the granules in order to increase absorption rate of SO_2. This has been successful, making adequate SO_2 removal possible in the short retention time available and at the same time loading the absorbent enough to avoid excessive recycling tonnage. The preferred method for fabricating the absorbent is to react solutions of Na_2CO_3 and $Al_2(SO_4)_3$ to form $NaAlCO_3(OH)_2$ (dawsonite), which is then heated to form $NaAlO_2$; the escape of H_2O and CO_2 provides the necessary porosity.

The major problem which forced abandonment of the process was the significant decrease in granule strength over an extended period of cycling. The change in crystal structure in the cycle of absorption to regeneration and back again was apparently the main cause of the disintegration that also occurred in pilot-plant tests. Mechanical attrition probably also contributed, both in the fluidized bed tests made by CEGB and in the work with an entrained reactor by the Bureau of Mines. It is not clear whether or not granule deterioration can be avoided or minimized to an acceptable level so as to rejuvenate the process.

Scrubbing Processes

Except for the fact that the cation is not volatile, scrubbing processes based on use of a sodium salt as the absorbent are quite similar to those using ammonia. The same general type of regeneration methods are used, and high bisulfite/sulfite ratio is usually a process requirement. Absorption kinetics and rates of undesirable side product formation also appear to be similar.

Thermal Stripping: The main example of thermal stripping as applied to sodium system scrubbing is the Wellman-Lord process, which has been applied commercially on a 700-tpd sulfuric acid plant operated by the Olin Corp. at Paulsboro, New Jersey; two Claus plants (in Japan and the U.S.); and two oil-fired boilers in Japan (75-Mw unit at Japan Synthetic Rubber Company and 220-Mw oil-fired boiler at the Chubu power station). Additional units will be installed in the U.S. on the 125-Mw coal-fired boiler at the Mitchell station of Northern Indiana Public Service Company and two 330-Mw units at the San Juan station of the Public Service of New Mexico. Several others are planned in Japan.

The Wellman-Lord flowsheet has become well-known. A typical arrangement from the Japan Synthetic Rubber Plant at Chiba is shown in Figure 7.20. The large Chubu unit differs in only a few respects (Table 7.4 gives a breakdown of the test facility). Flue gas containing 1,800 ppm SO_2 is washed by the prescrubber installed in the lower part of the three-stage sieve-tray absorption tower.

The greater part of solids and SO_3 formed in the boiler is removed with the wash liquor. The wash liquor is filtered, blended with the Na_2SO_4 purge stream (discussed later), and neutralized, prior to release from the system. The partially cleaned gas rises in the tower while contacting a countercurrent flow of concentrated sodium sulfite solution which removes more than 90% of the inlet SO_2. The following scrubber operating parameters are typical.

Gas velocity (superficial), fps		5.4
pH, inlet		7.3 - 7.4
outlet		5.5
Solution concentration, (percent by weight)		
Na_2SO_3,	inlet	20
	outlet	7
$NaHSO_3$,	inlet	2
	outlet	20
Na_2SO_4,	inlet	5
	outlet	>5

Mist droplets are removed by the eliminator and demister combination. The eliminator consists of a two-stage vertical section housing; the first stage is a conventional chevron-type and the second a Tellerette packing. The demister is a horizontal section also using Tellerette packing. The gas is discharged at about 280°F after reheating in an oil-fired after-burner, requiring approximately 5% of the total energy consumption. The spent absorber, primarily $NaHSO_3$, is discharged from the absorption tower and is stored in a surge tank before it is pumped to a two-stage evaporator.

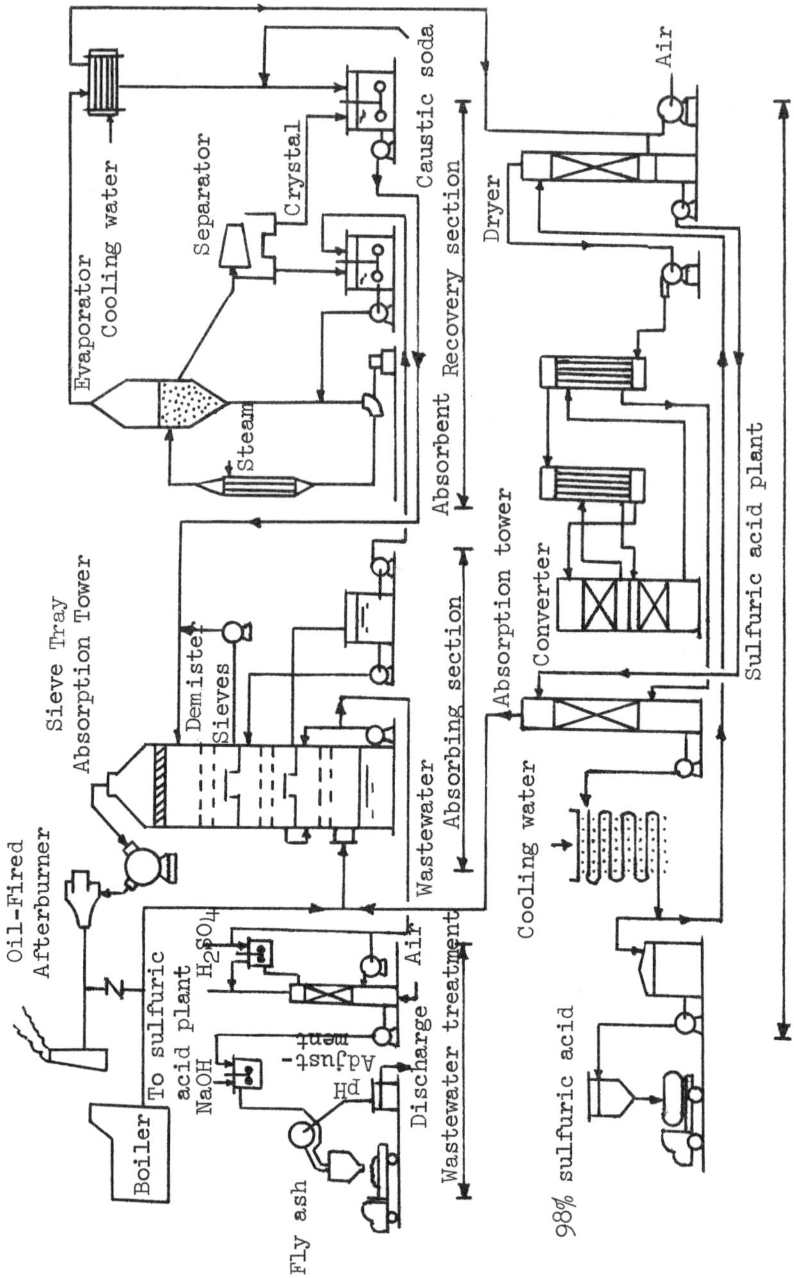

FIGURE 7.20: WELLMAN-LORD PROCESS

In the evaporator, the sodium bisulfite solution is heated to about 100°C by steam and decomposed into SO_2 gas and sodium sulfite.

$$2NaHSO_3 \rightarrow Na_2SO_3 + SO_2\uparrow + H_2O\uparrow$$

The SO_2 concentration of the gas leaving the evaporator is 90% after the water vapor is condensed out in a cooler. The concentrated SO_2 gas stream is sent to an acid plant. Sodium sulfite is crystallized out in the evaporator, separated by centrifuging, redissolved in the condensate from the cooler, and recycled to the absorber.

TABLE 7.4

Power station	Nishi-Nagoya Power Station, No. 1 plant (220 Mw)
Kind of plant	Wet system desulfurization plant using Wellman-Lord process
Gas rate for treatment	620,000 Nm^3/hr
Desulfurization rate	90% (before reheating)
Fuel	Fuel oil of 3% sulfur content
Raw material	Sodium hydroxide; 7 tpd makeup
By-product	98% sulfuric acid; 80 tpd maximum
Reheater	Oil-fired burner
Site size	4,500 m^2
Cost	$6,250,000
Operating schedule	Startup, November 1972; trial operation, May 1973; formal operation, September 1973
Manufacturer	Mitsubishi Kokoki Kaisha, Ltd.

To reduce buildup of Na_2SO_4 in the system, a portion of the mother liquor from the centrifuge is purged. This purge liquor is added to the purge liquor from the cooler. The bleed taken off in the cooler is required to remove contaminants which would build up in the system. The purge streams are subjected to wastewater treatment. The final wastewater is an essentially clear concentrated solution of sodium sulfate. This effluent eventually is sent to the bay.

A major feature of the Chubu unit is the extensive and highly automated control system. One manual control starts or stops the unit, all others are automatic. The main control is on the amount of liquor fed to the scrubber, which is varied to keep a uniform SO_2 concentration in the scrubber exit gas. The boiler load is used as an advance signal for changes. Only two operators are required.

The high degree of automation is required because of the widely fluctuating load. The capacity factor is only 42% and the boiler is shut down frequently. It does not operate over weekends at all. This wide variation has made other control measures necessary in addition to the automation.

The surge volume after the scrubber was made large enough so that the evaporator and acid plant could continue to operate at a minimum load during nonoperation of the boiler and scrubber.

The liquor is distributed over the sieve trays by a spray system in order to get good gas-liquid contact at low gas flow. The liquor weeps through the trays at low load and thus would not give good contact without the sprays.

The bypass damper is designed to open within 5 sec in the event the ID fan after the scrubber fails.

In the event of scrubber failure, the boiler is shifted to low-sulfur oil.

Another feature is that heat developed in the acid plant is used to preheat scrubber solution before the evaporator. Since the acid plant runs when the boiler and the main scrubber are down, a small auxiliary scrubber has been installed to scrub the acid plant tail gas during such period. Stored top stage liquor is used as the absorbent.

The mechanical equipment used at Chubu Electric is similar to that at the Chiba plant of Japan Synthetic Rubber. The entire scrubber and regeneration section is made of a high-chromium stainless steel alloy, reportedly slightly different in chromium content than that at Chiba.

Reliability has not been a problem at the Chubu Electric nor has it been at Japan Synthetic Rubber. The Chubu unit began operation September 1973 and Japan Synthetic Rubber started up July 1971. SO_2 removal efficiencies are in the order of 90%. Both plants utilize oil as a fuel. No data are available for coal-fired boilers.

The major drawback of the process is the requirement for a bleed to remove contaminants, primarily Na_2SO_4. Present data indicate that about 7 to 10% of the total incoming sulfur is lost as soluble Na_2SO_4, which requires 14 to 20% NaOH makeup. This is undesirable from an environmental viewpoint since future federal regulations for waste streams will probably prohibit such discharges. Sodium makeup costs are significant and will add appreciably to the overall operating cost. Attempts are being made to decrease the oxidation rate by means of introducing small quantities of oxidation inhibitors.

Acidulation: Regeneration of scrubber solution by acidulation is not as applicable in the sodium system as in the ammonium system because the sodium salts of strong acids either have little value (Na_2SO_4, NaCl) or are not in great demand (Na_3PO_4, $NaNO_3$). However, acidulation with bisulfate is usable because there is no coproduct sodium salt requiring disposal.

A system based on regeneration with bisulfate has been developed by Ionics-Stone and Webster and tested in a small pilot plant at the Gannon station of Tampa Electric in 1967-1968, and more recently in a 2,000-acfm pilot plant at the Valley station of Wisconsin Electric Power Company. A flowsheet of the system is shown in Figure 7.21. In contrast to thermal conversion of sulfate to bisulfate in the similar process in the ammonium system, the Ionics-Stone and Webster process uses electrolytic decomposition.

Entering flue gas is cooled from 290° ± 20°F to 120°F by direct water quench in the bottom of the absorber, after which the gas is contacted with aqueous caustic containing sodium sulfate and about 8% sodium hydroxide. Flue gas is returned to the stack from the top of the absorber. The caustic is converted to an aqueous mixture of sodium bisulfite, sulfite and sulfate containing the SO_2 and SO_3 removed from the flue gas.

FIGURE 7.21: UTILIZATION OF SOLUTION FROM SODIUM SYSTEM SCRUBBING OF STACK GAS BY DECOMPOSITION WITH $NaHSO_4$

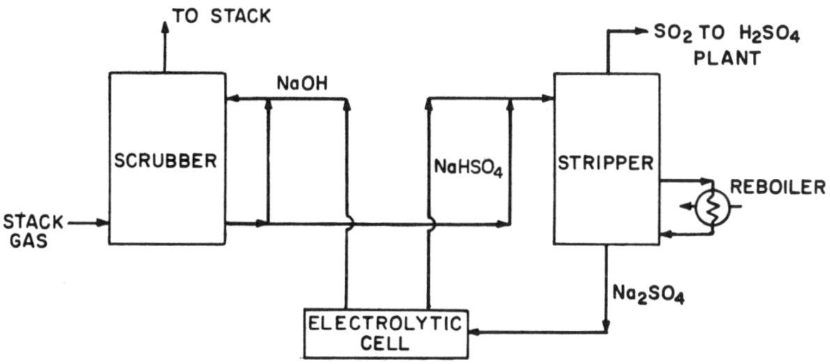

Process developed by Ionics-Stone and Webster

Above the quench section there are three packed sections each 10' high that contain 2" Tellerettes. For most of the test program the uppermost stage was used as a demister. Absorbent is recirculated around each absorbing stage. Prior to entering the stripper, the net effluent from the absorber is reacted with the sulfuric acid-sodium sulfate mixture recycled from the electrolyzers. The following reactions occur:

$$Na_2SO_3 + H_2SO_4 \rightarrow H_2O + SO_2 + Na_2SO_4$$

$$2NaHSO_3 + H_2SO_4 \rightarrow 2H_2O + 2SO_2 + Na_2SO_4$$

The operating pressure is 10 to 15 psig. The reboiler temperature is 240° to 250°F. Recovered SO_2 in the stripper overhead is at a purity of 98% or greater. The stripper bottoms stream is controlled to a pH of about 2.5 and contains 50 to 100 ppm of dissolved SO_2.

Stripped sodium sulfate solution from the stripper is cooled, pH adjusted to 8.5, and several parts per million of hydrogen peroxide added to precipitate ferric hydroxide. The latter and other heavy metal oxides and hydroxides are removed by filtration. The sodium sulfate is fed to the central compartment of the electrolyzers at a concentration of about 20%. A schematic diagram of the three-compartment (Type A) electrolyzer is shown in Figure 7.22.

The main components are an anode, a microporous diaphragm, a cation-selective membrane and a cathode. These components are separated from each other by flow directing spacers which also provide required gasketing. Sodium ions migrate through the cation-selective membrane toward the cathode under the influence of a direct current voltage impressed across the electrolyzer. At the cathode, water is electrolyzed to hydrogen gas and hydroxide anion:

$$H_2O + e^- \rightarrow \tfrac{1}{2}H_2 + OH^-$$

The hydroxide anions are electrically balanced by sodium cations entering the cathode compartment through the cation selective membrane. The effluent catholyte consists of NaOH plus Na_2SO_4, which after disengaging hydrogen is sent to a surge tank and then to the scrubber.

FIGURE 7.22: SCHEMATIC DIAGRAM OF A AND B ELECTROLYTIC CELLS

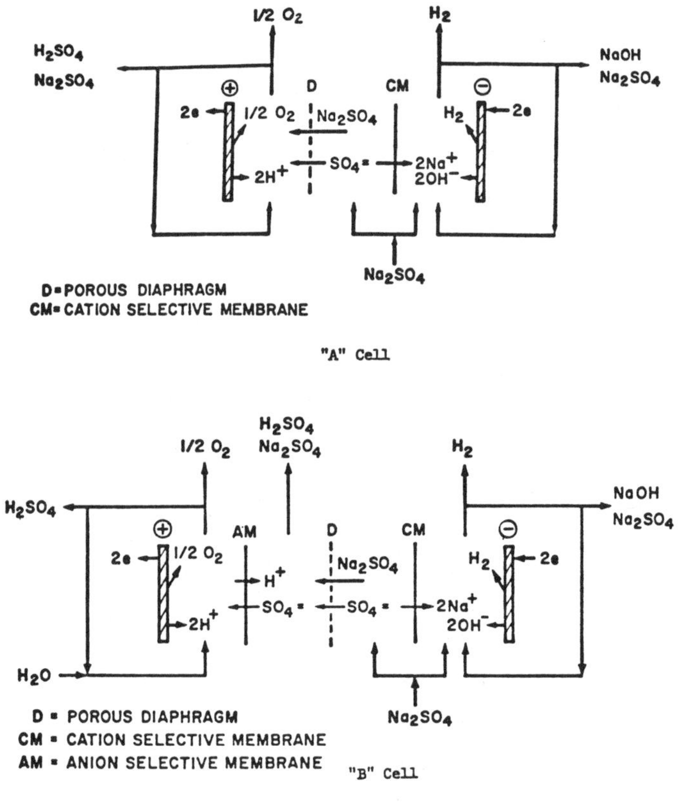

The function of the cation-selective membrane is to prevent physical mixing of the catholyte and center compartment feed streams. Essentially, only sodium cations from the center compartment pass through the membrane to combine with hydroxide anions produced at the cathode. Water may be fed to the catholyte compartment in sufficient quantity to produce 6 to 20% caustic.

However, the process is an essentially closed system. Therefore, to maintain water balance in the system it is necessary to feed recycled sodium sulfate solution to the catholyte compartment instead. This procedure does not introduce any problems and is the source of the sodium sulfate in the scrubber feed. The center compartment feed passes through the microporous diaphragm, into the anode compartment. At the anode, water is electrolyzed to hydrogen cations and oxygen:

$$H_2O \longrightarrow \tfrac{1}{2}O_2\uparrow + 2H^+ + 2e^-$$

The hydrogen cations combine with sulfate anions to form sulfuric acid. The diaphragm flow is designed to prevent hydrogen ions from migrating across the cation-selective membrane. Such flow must give enough linear velocity through the diaphragm to sweep hydrogen ions back into the anolyte compartment. The diaphragm should have a hydraulic resistance adequate to insure that such flow is substantially uniform over the entire surface.

A flow which is sufficient to sweep back most of the hydrogen ions carries with it about half of the sodium in the center compartment. Thus, at the anode, only part of the sodium sulfate in the feed stream is electrolyzed to sulfuric acid. The anode product is therefore a mixed solution containing both sulfuric acid and sodium sulfate. This product is the dilute acid solution that is recycled to release SO_2 in the stripper.

In the four-compartment (Type B) electrolyzers, a schematic diagram of which is presented in Figure 7.22, the reactions and operation of the catholyte compartment are exactly those described for three-compartment electrolyzers. Catholyte effluents from both electrolyzer types are combined before being sent to the absorber. The center compartment feed flows through the porous diaphragm, into the mid-anolyte compartment, from which it is discharged from the electrolyzer. This stream contains some sulfuric acid due to the inability of the anion selective membrane completely to exclude hydrogen ion.

The mid-anolyte effluent is combined with the anolyte stream from the three-compartment electrolyzers. The unique reactions of the four-compartment electrolyzer occur at the anode, where water is electrolyzed to hydrogen cations and oxygen gas. These hydrogen cations combined with sulfate anions entering the anode compartment to form sulfuric acid. The function of the anion-selective membrane is to exclude sodium from the anode compartment while allowing sulfate to enter. It also, to a considerable extent, prevents hydrogen cations from leaving the anode compartment. Thus, the anode reactions permit removing sulfate without losing sodium.

Sulfur absorbed as the sulfate will not be released by acidification with sulfuric acid and will result in a net sulfate ion increase in the system. The amount of oxidation will set the number of B cells required for removal of excess sulfate ion. B cells require slightly more capital investment than A cells and consume slightly more power. This provides further incentive for minimizing oxidation.

Under the test program operating conditions, oxidation levels in the absorber varied from about 2.6 to 4.0 lb/hr, with a mean of 3.4 lb/hr. The mean amount of SO_2 absorbed in the liquor was 28.1 lb/hr. The mean oxidation, therefore, represented about 12% of the SO_2 absorbed.

In general, oxidation was found to be affected by: the amount of liquid recirculated to each stage, higher rates increased oxidation; the S/C of the effluent liquor, high S/C's favored lower oxidation; and the concentration of SO_2 in the inlet gas, higher SO_2 partial pressure favored lower oxidation. Varying the concentration of sodium sulfate did not appear to affect the amount of oxidation. An obvious drawback to the process is the large amount of electric power required. A sizeable percentage of the total power output is needed for the electrolytic cells alone, which would affect availability of power as well as being costly if the power for the cells were required during periods of high customer demand.

Provision is made in the process, however, to store the scrubber solution during the hours of high power demand and to operate the cells only at night or other times when excess power production capacity is available. On this basis the developers assume that the cost of power is only the cost of fuel plus additional maintenance, as all other costs are the same whether the plant is operated at full or part capacity.

Other drawbacks to the process include the small size of the cell modules, cost of converting alternating to direct current, relatively high cost of the four-compartment cells required to purge excess sulfate, and the high cost for the solution storage required for the intermittent cell operation. Offsetting advantages are the high concentration of the SO_2 stream produced (suitable for conversion to sulfur), the relatively simple operation, and relatively low cost of the energy when used only in off-peak periods.

In comparison to the ammonia-scrubbing bisulfate process, the main question is how the electrolytic cell investment and operating cost compare with that of the thermal decomposition unit in the ammonia system. No conclusive data are available to answer the question.

Reduction: Autoclaving of Na_2SO_3 and $NaHSO_3$, similar to the Simon-Carves and Cominco processes in the ammonia system, is not very appropriate because of the large amount of almost useless Na_2SO_4 formed. However, direct treatment of sulfite-bisulfite solution with a reducing agent has received more attention than in the ammonium system. The major work has been done by the U.S. Bureau of Mines (Salt Lake City Metallurgy Research Center). These studies illustrate the very complex nature of sulfur chemistry, particularly when reduction of sulfites is attempted. Typical reactions are reported to be as follows.

$$H_2S + 3SO_2 \rightleftharpoons H_2S_4O_6$$

$$2NaHSO_3 + S \rightleftharpoons Na_2S_2O_3 + H_2O + SO_2$$

$$6Na_2S_2O_3 + 9SO_2 \rightleftharpoons 4Na_2S_3O_6 + Na_2S_4O_6 + Na_2S_5O_6$$

$$3Na_2S_3O_6 \rightleftharpoons 2Na_2SO_4 + 2SO_2 + Na_2S_5O_6$$

$$H_2S_4O_6 + 5H_2S \rightleftharpoons 9S + 6H_2O$$

Experience has shown also that lower polythionates ($S_2O_6^=$, $S_3O_6^=$) react with SO_2 to form the higher polythionates ($S_4O_6^=$, $S_5O_6^=$), and that sulfur is produced when thiosulfate reacts with polythionic acids. The system is highly complex and not yet fully understood, but it has been found that the side reactions can be controlled, and sulfur produced in a closed loop without building up a troublesome concentration of undesirable side products, by buffering the absorption solution at pH of 3.0 to 3.7. Organic acids are used for the purpose, and most of the work has been done with citric acid. Among the reactions believed to take place are the following (Ct = citrate radical):

$$(NaO)_3Ct + 3SO_2 + 3H_2O \rightleftharpoons 3NaHSO_3 + Ct(OH)_3$$

$$3NaHSO_3 + 6H_2S + Ct(OH)_3 \rightleftharpoons 9S + (NaO)_3Ct + 9H_2O$$

$$3NaS_2O_3 + Ct(OH)_3 \rightleftharpoons (NaO)_3Ct + 3S + 3SO_2 + 3H_2O$$

Thus the citrate suppresses side reactions and gives sulfur as the main product.

A general flowsheet for the process is shown in Figure 7.23. The gas is first cleaned and cooled and then scrubbed with the sodium citrate solution. The resulting sodium sulfite is reduced with H_2S and two-thirds of the product sulfur converted to H_2S for recycling.

FIGURE 7.23: SULFUR DIOXIDE RECOVERY BY SCRUBBING WITH SODIUM CITRATE SOLUTION FOLLOWED BY REDUCTION WITH H_2S

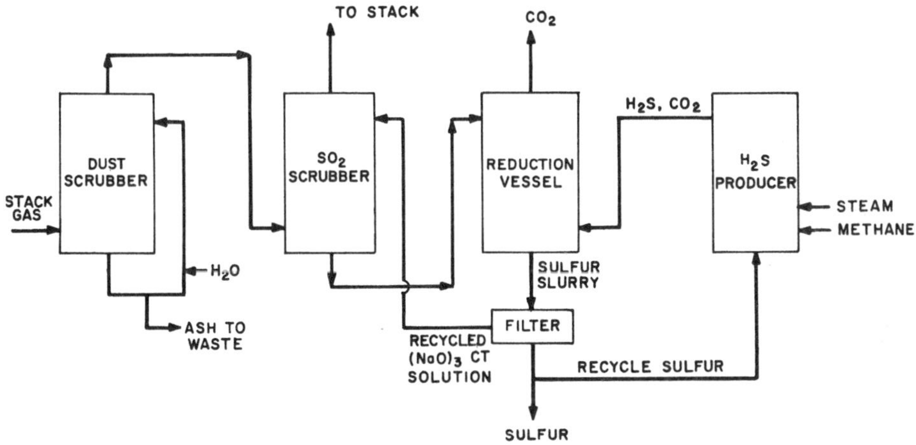

Process developed by U.S. Bureau of Mines (Salt Lake City Metallurgy Research Center).

There are three pilot plants in operation using either citric acid or an equivalent material: the Bureau of Mines' 1,000-scfm unit at Kellogg, Idaho (Bunker Hill smelter); Pfizer-Peabody-McKee's 2,500-scfm unit at Terre Haute, Indiana; and Stauffer's 100-scfm unit at Northeast Utilities' Norwalk Harbor station in Connecticut. The Stauffer method involves use of phosphate rather than citrate as the buffering agent but the operation otherwise is nearly identical.

The citrate-phosphate method has the outstanding advantage of being the only one in which elemental sulfur is precipitated directly from scrubber solution. On casual inspection it appears quite simple and straightforward; the scrubber effluent is merely treated with hydrogen sulfide (H_2S) and elemental sulfur precipitates in the solid form. A close study, however, shows several aspects that add complications and cost.

The fly ash from coal combustion must be separated at some stage to avoid contamination of the product. Pfizer and Stauffer are currently studying the problem but the Bureau will not get dust removal data until Phase III of the Bunker Hill project, in which dusty gas from the smelter will be treated. One of the main drawbacks to the citrate process is the relatively low pH in the scrubber and the resulting low efficiency when high bisulfite formation is attempted, which seems to be necessary. To get adequate SO_2 removal with high bisulfite formation, the scrubber effluent concentration must be limited to a low level, on the order of 2 to 10 g SO_2 per liter. In contrast, Wellman-Lord operates with 65 g/l or more regenerable SO_2 and ammonia processes up to 400 g/l. As a result of the low concentration, the regeneration system must be quite large for citrate and the cost correspondingly high.

Although the presence of thiosulfate (at least 0.3 molar) inhibits oxidation in the citrate system, some still occurs. The amount is not yet clear but has run as high as 1.3% in some tests. Although this is low as compared to Wellman-Lord (5 to 15%) it is still significant and will require evaporation and crystallization of a sidestream. For a 1,000-Mw plant on high-sulfur coal the sodium sulfate production could run as high as 15 to 20 tpd at full load. Disposition of this, as in the case of Wellman-Lord, could be difficult. In contrast, processes such as IFP, TSK, and Atomics International have no sulfate problem.

The Bureau of Mines' 0.5-Mw pilot plant started up early in 1975. The plant produces 1.2 to 1.5 tpd of sulfur. The work is divided into three phases: In Phase 1, cleaned gas going to the sulfuric acid plant (diluted 10 to 1 with air) will be scrubbed, to avoid the dust problem during initial operation, and H_2S will be purchased. The sulfur will be separated by flotation followed by melting at 130°C and 35 psi. In Phase II, the H_2S will be made from the product sulfur. In Phase III, the actual tail gas, dilute and dirty, will be treated.

A packed scrubber will be used, at about 4 fps and 20 ft or so of packing depth. Under these conditions very good absorption, over 90%, is obtained. An excess of H_2S is used to ensure complete regeneration since it is considered that otherwise good absorption would not be feasible. The excess H_2S is burned to SO_2 and recycled to the scrubber. Regeneration is accomplished in only one vessel (stirred) rather than the three used in earlier tests.

A kerosene-motor oil mixture is used for flotation. The regenerated liquor is cooled from 70° (the optimum regeneration temperature) to 50°C before recycling to the scrubber. Sulfur is converted to H_2S by melting, mixing with methane and steam, and passing through a bauxite catalyst at 700°C. The product gas contains about 77% H_2S and the remainder CO_2.

The Pfizer-Peabody-McKee 1-Mw pilot plant at Terre Haute was started up in June 1973. An early problem was in separating the sulfur from the product slurry. A proprietary separation method was developed. The Bureau method may be superior because it gives a relatively high sulfur concentration in the final separation step. Neither method, however, appears to be a completely satisfactory separating procedure. Another problem has been freezing of sulfur in lines; the Bureau found that full jacketing is essential. The Pfizer-Peabody-McKee group has recently proposed a 35-Mw demonstration at an unidentified utility.

The Stauffer Aqua Claus process was recently started up (late 1974) on a 100-cfm unit at Northeast Utilities' Norwalk station. Very little data are available on the process, which is purportedly quite similar to the citrate type except that a phosphate is used. Stauffer considers phosphate superior to citrate because it decomposes less and because it is cheaper in the first place. Like Pfizer, the sulfur is separated by a method (again proprietary) that does not require the use of kerosene. Stauffer claims that the kerosene would be an undesirable contaminant in marketing sulfur. The Chemical Construction Corporation has joined with Stauffer in promoting the process.

Another potential drawback to the citrate-type process is the requirement for methane or naphtha at some point in the regeneration operation. Universal Oil Products has developed a patented process called Sulfoxel which uses a sodium sulfite/hydroxide absorbent and produces elemental sulfur. The process was installed in 1973 on a 25-Mw prototype unit at Commonwealth Edison's State Line station. Very few data are available on this process even though it has been patented. Hydrogen sulfide is used to reduce sodium sulfite to thiosulfate which is then reduced with hydrogen or synthesis gas. Plant operation has been very sporadic since startup due to many problems. Several design changes have been made including replacement of the catalyst.

Atomics International has been developing for the last 4½ years a process (Aqueous Carbonate Process, ACP) that utilizes sodium carbonate solution as absorbent for removal of SO_2 from flue gases. The solution is atomized in a modified spray dryer, a device which has been in use for over 50 years, providing very effective contact efficiency. The SO_2 reacts with the sodium carbonate solution to make sodium sulfite and sulfate, which in turn are reduced and regenerated to carbonate in an aqueous regenerative subsystem with the attendant production of elemental sulfur. Inherent in the process is elimination of the need for reheat, since the flue gases are not saturated during contact with the scrubbing solution. The following reactions take place in the ACP process.

$$Na_2CO_3\ (aq) + SO_2 \rightarrow Na_2SO_3\ (s) + CO_2$$

$$Na_2SO_3\ (s) + \tfrac{3}{2}C \rightarrow Na_2S\ (s) + \tfrac{3}{2}CO_2$$

$$Na_2SO_4\ (s) + 2C \rightarrow Na_2S\ (s) + 2CO_2$$

$$Na_2S\ (s) \rightarrow Na_2S\ (aq)$$

$$Na_2S\ (aq) + H_2O + CO_2 \rightarrow Na_2CO_3\ (aq) + H_2S$$

$$Na_2CO_3\ (aq) + H_2O + CO_2 \rightarrow 2NaHCO_3\ (s)$$

$$2NaHCO_3\ (s) \rightarrow Na_2CO_3\ (aq) + H_2O + CO_2$$

$$H_2S + \tfrac{1}{2}O_2 \rightarrow H_2O + S\ (l)$$

The key component of the scrubbing system is a modified spray dryer which serves as a reaction chamber for the removal of SO_2. In the spray dryer, sodium carbonate solution is atomized by one or more high-speed centrifugal atomizers and mixed with the hot gas entering the dryer through vane-rings which surround the atomizers. The fine mist of solution droplets absorbs SO_2 while at the same time the thermal energy of the waste gas vaporizes the water without saturating or excessively cooling the gas. Thus the output of the spray dryer is a gas low in SO_2 but containing dry particles of sodium carbonate, sulfite, and sulfate, the reaction products from the contactor.

This powder is subsequently separated from the gas and collected for processing and regeneration. After product collection, the treated waste gas remains above its dew point and can be vented directly through a stack. Figure 7.24 shows a diagram of the key subsystems of the ACP regenerative system. The chemicals needed for the regeneration include makeup Na_2CO_3 (soda ash), a carbon source for the reduction step, and water. The outputs are ash, which comes mainly from the flue gas, and elemental sulfur. The product salt is conveyed, along with coke or any other source of carbon, to the molten salt reducer.

FIGURE 7.24: ATOMICS INTERNATIONAL REGENERATIVE PROCESS

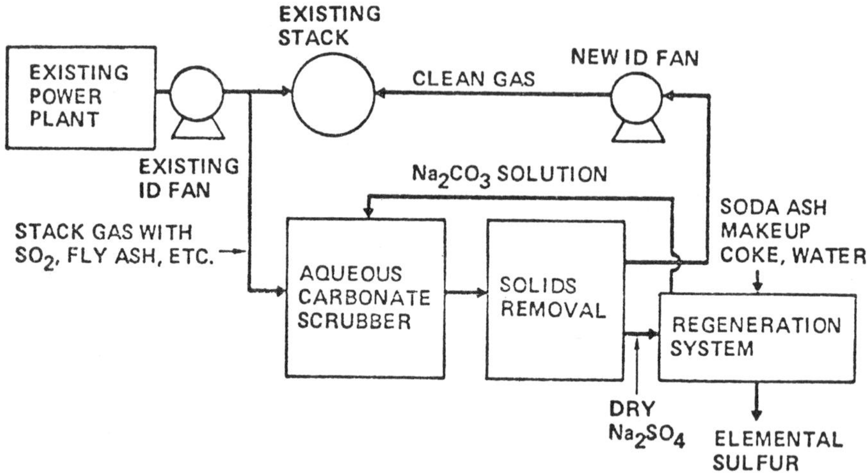

In the reducer, the salt is heated up, melted, and reduced by the addition of air and coke or coal. The air causes some reoxidation of sulfide for generating sensible heat, while the coke acts directly as a reductant to reduce the sulfite and sulfate to sulfide. The molten mixture flows from the reducer into a quench tank, where it is dissolved and thereafter processed as a low-temperature (below the boiling point of water) aqueous solution. The reducer off-gas is used as the source of CO_2 for subsequent carbonation steps and as a source of process heat.

The aqueous solution is cooled and filtered to remove any carbon or fly ash that is present. After filtration, the solution is precarbonated with pure CO_2 recovered in the decomposer. Final carbonation occurs in the bicarbonator-crystallizer with the reducer off-gas providing the necessary CO_2. Gases evolved from both the precarbonator and the bicarbonator-crystallizer are rich in hydrogen sulfide, and they are combined for subsequent recovery of elemental sulfur in the Claus plant. Tail gas from the Claus plant is returned to the scrubber for final cleanup. The product from the bicarbonator-crystallizer is a Na_2CO_3-$NaHCO_3$ slurry, which is decomposed to produce a Na_2CO_3 solution for return to the scrubber and to provide pure CO_2 for the precarbonator.

Development efforts to date have been concentrated on pilot demonstration of the scrubbing system and the reducer. Extensive test data have been generated with 5' and 7'-diameter spray-dryer scrubbers. Proprietary test results are also available from a 4'-diameter reducer.

The first pilot scrubber tests were conducted using simulated flue gas to establish the feasibility of reacting SO_2 with Na_2CO_3 solutions and slurries in a spray dryer. Subsequent tests were conducted at the Mohave station (Southern California Edison) where a 5'-diameter modified spray dryer was used to test SO_2 removal from a side stream of flue gas. It was used in over 100 tests at Mohave without operational problems. Most of the Mohave test data were obtained with flue gas containing 400 ppm or less SO_2, since this is characteristic for a power plant burning low-sulfur western coal.

A few tests were run at other SO_2 concentrations up to 1,500 ppm, but most of the high SO_2 concentrations were obtained using simulated flue gas in a 7'-diameter spray-dryer scrubber. After establishing that the SO_2 removal data obtained at Mohave were identical to those obtained when using simulated flue gas, an extensive range of tests was run with the 7' unit at SO_2 concentrations ranging from 200 to 8,000 ppm.

Figure 7.25 is a plot of some of the data taken during the Mohave test program. It illustrates the low liquid-to-gas ratio (L/G) characteristic of a spray-dryer scrubber. Only about 0.3 gal of a 5.5 weight percent Na_2CO_3 solution was needed per 1,000 scf of flue gas in order to obtain greater than 90% removal of the 400 parts per million inlet SO_2. Subsequent tests have confirmed that this same L/G can be used to remove greater than 90% of the SO_2 from gases containing from 200 to 4,000 ppm SO_2.

The concentration of Na_2CO_3 in solution is adjusted in proportion to the SO_2 concentration in the gas to provide sufficient alkalinity to neutralize the absorbed SO_2, but the L/G itself remains at about 0.3 gal/1,000 scf over this wide range of SO_2 concentrations. Above 4,000 ppm SO_2, it is necessary to increase the L/G to provide sufficient dissolved Na_2CO_3 to react with the SO_2 absorbed. However, even at 8,000 ppm SO_2, the required L/G is only about 0.6 gal/1,000 standard cubic feet as compared to most other scrubbers which would require 10 to 100 gal/1,000 scf for such an application.

FIGURE 7.25: SO_2 REMOVAL vs ABSORBENT FLOW IN ATOMICS INTERNATIONAL SPRAY DRYER ABSORBER

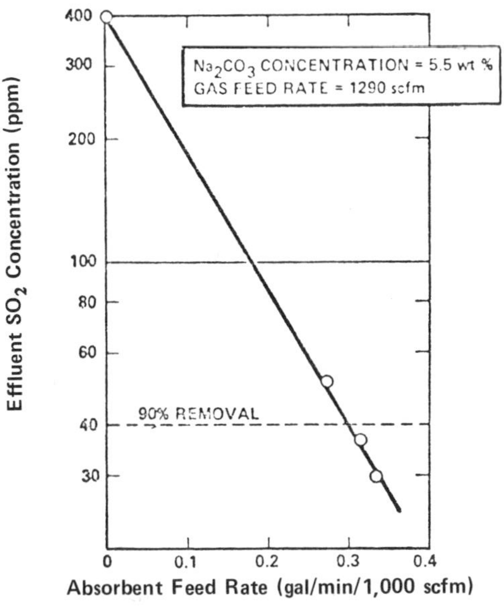

The proper operation of a spray-dryer scrubber also requires that a dry product be formed and subsequently removed from the gas stream. Pilot tests have shown that the product salts will be dry and collectable if the gas temperature at the dryer outlet is maintained about 20°F above its dew point. This also tends to minimize plume formation. The cyclone collectors used in the pilot tests removed from 89 to 99% of the product. Although this was excellent performance for mechanical collectors, particulate emission standards will require either additional collection devices (an electrostatic precipitator or bag house) in place of or in series with the cyclones.

Bench and pilot-scale tests of the various steps in ACP regeneration have been conducted at Atomics International laboratories and at their nearby field test facility. Reduction, quenching, filtration, precarbonation, carbonation, decomposition, and H_2S scrubbing have all been tested. The reduction technique is unique and makes the process significantly different from others. If the further development work is successful it could be a major improvement in sodium scrubbing recovery technology. In a parallel effort, the Office of Coal Research is sponsoring a demonstration of the Atomics International gasification system which employs a similar reducer as a combined coal gasifier-sulfur absorption unit.

Drawbacks include (1) large size of the spray dryer; (2) need for a second solids collection system; (3) relatively high heat requirement in the reducer (sensible heat is not recovered); and (4) difficulty in completely stripping H_2S out of the Na_2S-CO_2-H_2O solution. Further data are needed to evaluate these in comparison to other processes.

A somewhat similar sodium-reduction regeneration system (also proprietary) is being developed by Tsukishima Kikai Engineering Company (TSK) in Japan. In connection with work in the paper industry, TSK has licensed the Billerud process from SCA in Sweden and has built three plants in Japan. The process reduces sodium sulfite in waste liquors by injecting the liquor into an oil burner operating under partial oxidation conditions. The temperature is kept below the melting point of the salts so as to produce dry sodium carbonate and H_2S. Plants in operation or under construction include one in Sweden, one in Spain, one in the United States and four in Japan.

TSK has recently done pilot-plant work on adapting the Billerud method to reduction of Na_2SO_3 obtained in stack gas scrubbing. The process, called Sulfix is shown in Figure 7.26. Sodium sulfite solution from the scrubber is reduced in the reactor and the resulting Na_2SO_3 collected in an electrostatic precipitator. The collected material is dissolved and then oxidized to convert the small amount of NaHS formed to thiosulfate; otherwise the sulfide would decompose in the scrubber and evolve H_2S. Since a substantial amount of carbon is formed, the solution is filtered after oxidation and the carbon recycled to the reactor.

One of the main drawbacks to the process is that the reactor effluent gas contains only 4 to 7% H_2S, too low for good operation in a standard Claus plant. For this reason, a wet system probably would be used for converting the H_2S to sulfur. This part of the process is relatively undeveloped; the Takahax method is currently favored since it can accept the carbon not collected in the precipitator. The H_2S not converted to sulfur is burned and introduced into the main gas stream ahead of the scrubber. From 50 to 60% of the Na_2SO_4 formed in the scrubber is reduced each pass. The concentration reaches a steady state level.

FIGURE 7.26: TSK SULFIX PROCESS

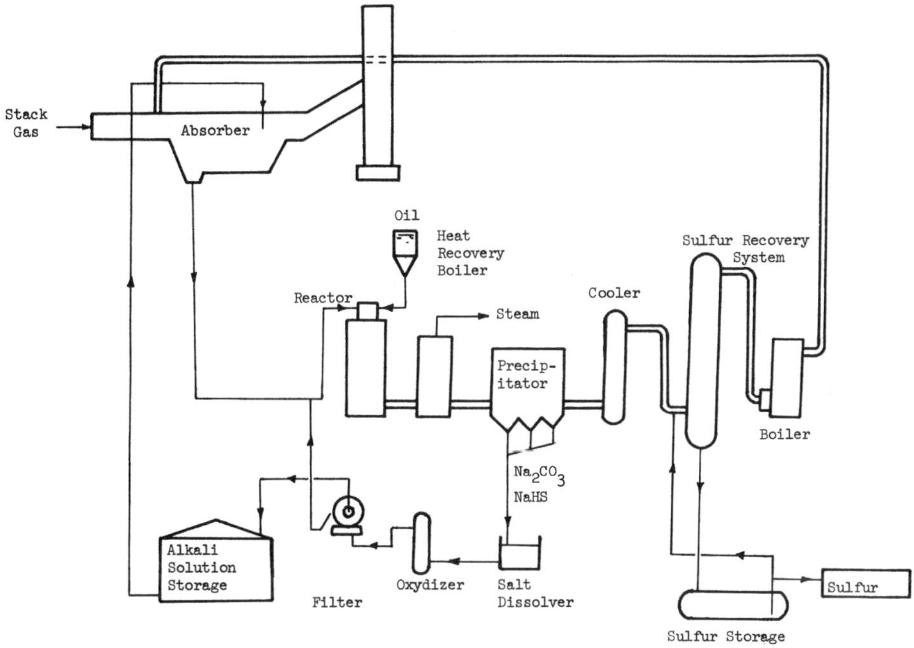

Other approaches, described only in patents, include: (1) reaction of sulfite with BaS to give Na_2S and $BaSO_3$, treating the Na_2S with H_2O and CO_2 to give H_2S, and reducing the $BaSO_3$ back to BaS; (2) separation of solid sulfite followed by reduction; and (3) reduction of sulfite solution with H_2S without benefit of citric acid.

Precipitation: Difficulty in regenerating sodium sulfites and sulfate has led some investigators to study precipitation methods, in which another cation is used to precipitate sulfites that are hopefully easier to regenerate. The main example of this is use of ZnO to precipitate $ZnSO_3 \cdot 2.5H_2O$, a process developed by Johnstone and coworkers at the University of Illinois in the late 1930s. The zinc sulfite is regenerated thermally, with the advantage that very little disproportionation occurs, whereas sodium sulfites disproportionate extensively on heating. The main reactions are as follows.

$$ZnO + 2NaHSO_3 + 1.5H_2O \rightarrow ZnSO_3 \cdot 2.5H_2O + Na_2SO_3$$

$$ZnSO_3 \cdot 2.5H_2O \xrightarrow{\Delta} ZnO + SO_2 + 2.5H_2O$$

Because there is some formation in the system of sulfate, which would not be decomposed in the calciner, a somewhat complicated side-stream treatment is included in the flowsheet to purge the system of sulfate as calcium sulfate (Figure 7.27). The SO_2 is absorbed by Na_2SO_3 solution; the effluent solution (S/C of about 0.86) reacted with ZnO in the main reactor; the $ZnSO_3 \cdot 2.5H_2O$ separated, dried, and calcined; and the ZnO recycled. A side stream of scrubber effluent is treated with lime to precipitate $CaSO_3$. The resulting slurry is treated with SO_2 to convert $CaSO_3$ to soluble $CaHSO_3$; the Ca^{++} thus supplied precipitates sulfate as $CaSO_4$, which is filtered off; thus the loss of regenerable sulfite is avoided and use of CaO is minimized.

FIGURE 7.27: RECOVERY OF SO_2 BY JOHNSTONE PROCESS

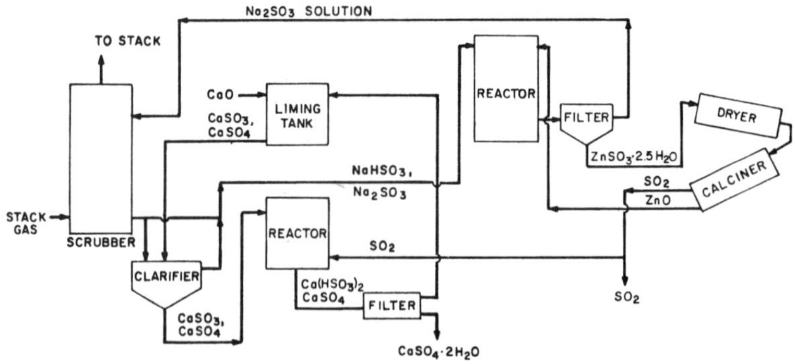

SO_2 is recovered by scrubbing stack gas with Na_2SO_3 solution, precipitating $ZnSO_3 \cdot 2.5H_2O$ by addition of ZnO, and decomposing with heat to SO_2 and ZnO. Process developed by Johnstone and coworkers at the University of Illinois.

Good results were obtained in small-scale pilot plant tests of the method. By keeping the Na concentration below 3 mols/100 mols water, formation of undesirable Na-Zn double salts was avoided. The $ZnSO_3 \cdot 2.5H_2O$ crystals settled well and could be dewatered easily. The crystals were decomposed after drying, in a falling-solids type of calciner, at a temperature of 675° to 740°C. Relatively little interest has been shown in the Johnstone process since it was first developed. However, Aerojet General did evaluate the method for EPA and concluded that it compares well with other recovery processes.

POTASSIUM COMPOUNDS

Much less work has been done on use of potassium salts as SO_2 absorbents than on ammonium and sodium. The cost of potassium carbonate, the usual form in which makeup is supplied, is higher than for Na_2CO_3 or NH_3, and there are few if any offsetting advantages.

Thermal Regeneration

Both Wellman-Lord and TVA have worked on thermal decomposition of $KHSO_3$. The general approach has been to separate solid potassium pyrosulfite ($K_2S_2O_5$), the anhydride of $KHSO_3$, and decompose it thermally, either by heating the solid in a kiln or by heating a solution or slurry in a stripper.

Separation of the solid pyrosulfite, which gives the maximum S/C of 1.0 can be accomplished in the potassium system because pyrosulfite is much less soluble than sulfite (Figure 7.28). As noted earlier, the reverse is true in the ammonium and sodium systems, that is, sulfite is much less soluble than bisulfite (the anhydride, pyrosulfite, does not form at operating temperatures in these systems).

FIGURE 7.28: THE SYSTEM K_2SO_3-$K_2S_2O_5$-H_2O

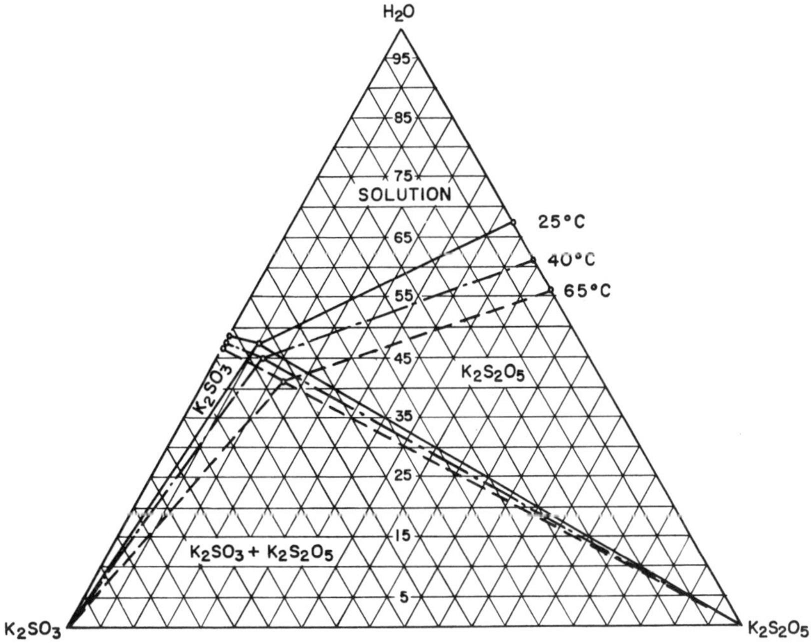

In the Wellman-Lord work, the pyrosulfite was separated by operating the absorber near pyrosulfite saturation and then cooling and crystallizing in a crystallizer. In the TVA variation, a solution of potassium pyrophosphate ($K_4P_2O_7$) was used as the absorbent. In this situation the pyrosulfite forms directly in the scrubber and the crystallizer therefore can be eliminated.

Thermal stripping has not been promising, even though the solid feed to the stripper has a higher S/C than can be attained in the liquid feed from ammonium or sodium scrubbing processes. Small-scale tests have shown that development of sulfite during stripping, by the reaction:

$$K_2S_2O_5 \xrightarrow{\Delta} K_2SO_3 + SO_2$$

drops the SO_2 vapor pressure so fast that the H_2O/SO_2 vapor pressure ratio, which controls the amount of heat required, soon becomes quite high. TVA work indicated an SO_2 vapor pressure of 23.6 mm Hg at an S/C of 0.987 (70°C and 4.44 mols S/100 mols H_2O) and 3.7 mm at an S/C of 0.966.

Thus in a practical situation there does not seem to be any good way of maintaining the high SO_2 vapor pressure of the pure pyrosulfite and avoiding the relatively low vapor pressure, and consequent high heat requirement, that develops as soon as a little of the bisulfite decomposes. Wellman-Lord built and operated a 50,000-scfm prototype unit at Baltimore Gas and Electric's Crane station (near Baltimore) in 1969. The unit was shut down after nine months because of the energy requirement for stripping, plus difficulties from plugging of the absorber by fly ash. The effort was shifted to the sodium system of absorption and stripping as described earlier.

The TVA stripping work (bench-scale) was also unsuccessful so the interest was shifted to thermal decomposition of the solid. Disproportionation to sulfate and sulfide was severe, as expected, but it was possible to evolve over a third of the sulfur as SO_2 at 400°F before formation of sulfate and sulfide stopped the decomposition. Thus if the solid residue were reduced to form H_2S, there would be the proper SO_2/H_2S ratio for the Claus process. The work being done by Atomics International and TSK on reducing sulfate to sulfide may be applicable to this system. No work in this area is currently being done.

Reduction

A unique potassium-based proprietary process has been proposed and tested by Consolidation Coal Co. The gas is scrubbed with a solution containing various potassium salts and the resulting solution reduced with CO at elevated temperature and pressure to evolve H_2S, which is converted to elemental sulfur by standard procedures. The process has been described in patents but no data have been published.

Tests of the scrubbing step have been carried out at Philadelphia Electric's Cromby station and the regeneration process has been studied in Consolidation Coal's laboratories. The process was selected by EPA (along with the Atomics International ACP and Shell copper oxide methods) as a candidate for the large demonstration project that EPA plans to cosponsor with a selected utility.

A slurry of K_2CO_3 in molten potassium thiocyanate (about 180°C) was used as the absorbent in a process announced by the Garrett Research and Development Co. The resulting K_2SO_3, insoluble in the system, is separated and roasted with coal to K_2S, which is converted to H_2S by reaction with CO_2 and H_2O. The K_2S is said to be converted to $KHCO_3$, which is separated by crystallization and calcined to give K_2CO_3 for recycling. The process does not cool the gas but it has several obvious drawbacks. Only laboratory tests were made. The process is no longer being developed.

MIXED CARBONATES

There have been several efforts aimed at reducing scrubbing equipment size and eliminating the need for reheat by scrubbing with a molten salt. The major effort has been by Atomics International, using a eutectric mixture of alkali carbonates (K_2CO_3, 35%; Na_2CO_3, 32%) at a scrubbing temperature of about 800°F. A simplified flowsheet of the process is given in Figure 7.29. Dust is removed as thoroughly as possible in a hot electrostatic precipitator. The gas is then scrubbed with the molten salt mixture at about 800°F.

FIGURE 7.29: MOLTEN SALT PROCESS FOR RECOVERY OF SO_2 FROM STACK GAS

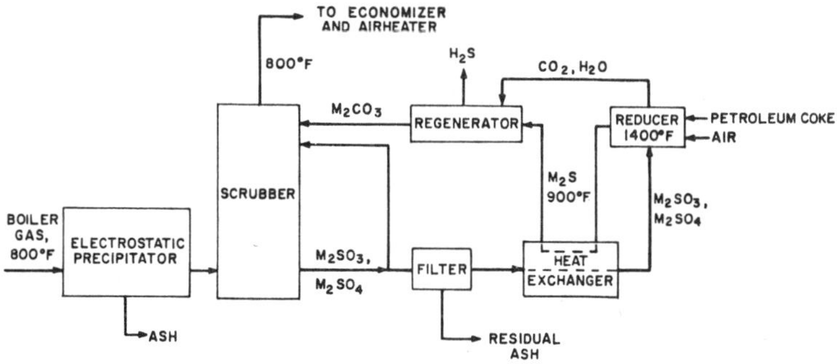

Process developed by Atomics International

Absorption rate is extremely rapid, allowing an L/G of about 1 gal/Mcf and a gas velocity of up to 25 fps. At the high temperature involved, however, equilibrium considerations limit carbonate/sulfite conversion to a maximum of about 40%. Something less than this would be expected in practice, making it necessary to recirculate a fairly large proportion of unreacted M_2CO_3 (M = alkali cation) through the system. Some oxidation of sulfite occurs, depending on several factors; about 10% seems indicated.

The effluent melt is filtered to remove residual ash (from which adhering melt must be recovered) and then preheated by heat exchange before the reduction step. It has been found that a temperature on the order of 1100° to 1400°F is necessary to get an adequate rate of reduction. At these temperatures disproportionation of sulfite to sulfate and sulfide occurs rapidly so that the actual material reduced is sulfate. Various reducing agents can be used; the use of petroleum coke is shown in Figure 7.29. The process is somewhat like sulfite/sulfate reduction in the kraft paper-pulping process, an operation that is well developed.

The mixture of alkali carbonates (eutectic composition) melts at 747°F to give a clear liquid with a viscosity of about 12 cp at 800°F. The specific heat is 0.4 and the density 2.0. The regeneration reaction is quite rapid and reaches equilibrium. The product gas contains about 30% H_2S and 35% each of CO_2 and H_2O. Most of the fly ash constituents are insoluble in the melt (except the alkali metal oxides). Iron and calcium compounds are slightly soluble but are desirable because they catalyze the reduction reaction. A cartridge-type filter with sintered stainless steel elements is favored for removing the ash.

The process was tested in a 10-Mw pilot plant at the Arthur Kill station of Consolidated Edison. Due to many problems, technical, financial, political, operational (manpower), the project has been shut down. The technical problem centered around severe stress corrosion purportedly due to the combination of four variables:

(1) absorption at 950°F,
(2) 2 to 20% sulfate in molten liquid,

(3) O_2 in flue gas
(4) stress in excess of 20,000 psi.

The combination of these four variables accelerated corrosion such that few meaningful operating data were obtained. It is felt now that by lowering the temperature down to 850°F and/or by reducing stress, the corrosion problem can be overcome. Early mist eliminator problems were also encountered but reportedly overcome. The reducer was never operated.

These technical problems were compounded by Consolidated Edison's financial difficulties, various political problems in the State of New York, and by Atomics International's manpower shortage. It is doubtful whether the project will ever be restarted.

ALKALINE EARTH ABSORBENTS

As described in the previous chapter, utilization or regeneration of alkali sulfite is usually carried out by a chemical reaction with an added material such as oxygen, an acid, or a reducing agent. Thermal regeneration is used only with the bisulfite, which is more difficult to form in the scrubber. Heating of sulfite causes so much disproportionation that it is not usually attempted. The situation is different for alkaline earth sulfites, some of which can be decomposed at elevated temperature (on the order of 500° to 1,200°C) to give SO_2 and regenerate the oxide with only a minor amount of disproportionation. The two that have received the most attention are calcium and magnesium. (Magnesium is not strictly an alkaline earth but is closely related.)

None of the alkaline earth methods have been used commercially but several companies have done pilot or prototype studies. Alkaline earths have also been studied in processes not involving thermal regeneration, including reduction to sulfur and producing a salable grade of calcium sulfate.

MAGNESIUM

Most of the alkaline earth processes use a magnesium compound as the absorbent and, with one exception, the leading magnesia processes involve scrubbing with a slurry of MgO or $Mg(OH)_2$ to produce $MgSO_3 \cdot 6H_2O$, which is heated to evolve a rich stream of SO_2 and produce MgO for recycling. Differences in processes include type of scrubber, other materials added, type of regenerator, operating pH, and method for cleaning the SO_2 stream.

Analogy to Magnesium-Based Wood Pulping

The magnesium system for removing SO_2 from stack gases is quite similar in many respects to magnesium-based pulping in the paper industry, which has been adopted by some companies as a means of reducing or eliminating air and water pollution. Of the four sulfite pulping systems—calcium, magnesium, ammonium, and sodium—magnesium is the one generally considered to be the most amenable to regeneration of the pulping chemical, which in sulfite pulping is bisulfite. The first plant using the system, generally called Magnefite, was built in the U.S. in the 1940s; several are now in operation. A generalized flowsheet is given in Figure 8.1. Wood chips are digested with $Mg(HSO_3)_2$ solution to dissolve out the lignin, the cellulose fibers (pulp) are separated, and the waste liquor concentrated to 55 to 70%. The dissolved material is made up of various organic constituents, including sugars and lignosulfonic acid, plus the magnesium combined in various compounds.

FIGURE 8.1: RECOVERY OF SO_2 AND MgO IN MAGNESIUM-BASED SULFITE PULPING PROCESS

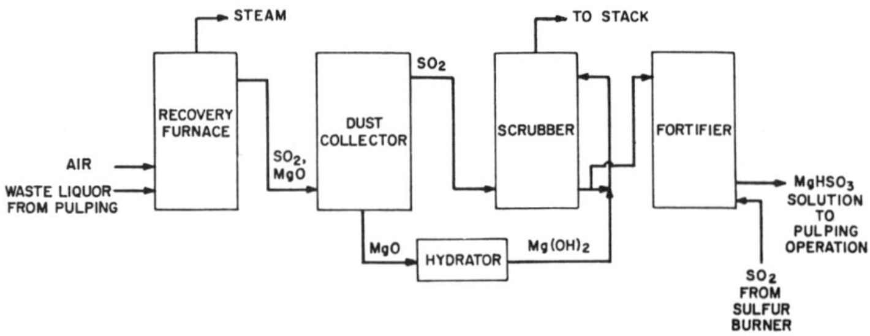

The concentrated liquor is burned in a recovery furnace. An oxidizing atmosphere is maintained by introducing a closely controlled amount of air slightly in excess of that required for combustion. Under these conditions the carbon burns out of the various organic compounds leaving MgO and SO_2. The combustion gas passes over boiler tubes to produce steam and then into a dust collector to remove the MgO, which is then hydrated and pumped to the scrubber. Before hydration, the MgO is washed to remove Na, Ca, and K sulfates that would otherwise accumulate in the system. Hydration requires an extended retention time and heating of the slurry to 180° to 200°F. Both packed scrubbers and venturis have been used. The preference seems to be for a series of venturis. The SO_2 can be removed typically to a level of less than 250 ppm (inlet of 1 to 1.5%). Precooling to 110° to 150°F is practiced. Fortification of the scrubber effluent liquor with SO_2, to convert any sulfite to bisulfite, completes the cycle, the liquor going to the chip digester.

Thus there exists a commercial process for scrubbing a waste gas containing SO_2 with $Mg(OH)_2$-$MgSO_3$ slurry and reducing the SO_2 content to less than 250 ppm. Applying the method to stack gas scrubbing would require several process changes, of course, but the basic principles appear to be the same. The main differences are that (1) the pulp mill gas contains more SO_2 (1.0% or more vs 0.1 to 0.4%), (2) the object in pulping is to make soluble $Mg(HSO_3)_2$ whereas in stack gas scrubbing $MgSO_3$ is desired, and (3) the calcining step in the pulping operation involves decomposition of complex metal organic compounds rather than simple $MgSO_3$.

It is not clear to what extent side reactions such as disproportionation take place in the recovery furnace; published data cover overall Mg and S loss without designating the part lost by sulfate formation in the furnace. It is reported that 20 to 30% of the waste liquor sulfur content is in the form of sulfates, much of which is formed in the digestion (wood pulping) step. It is claimed that this is reduced to sulfite in the furnace. Since the MgO ash is washed to remove soluble salts, $MgSO_4$ should leave the system with the wash water, as well as any other soluble products of disproportionation or other side reactions, but no data on the amounts lost in this way appear to be available. If the furnace temperature is high enough, of course, the $MgSO_4$ is decomposed thermally to MgO; this is probably what happens in most plants.

Scrubber Design and Performance

Data on vapor pressure for the system $MgO\text{-}SO_2\text{-}H_2O$, in the range of conditions applicable to stack gas scrubbing, are not very plentiful. The data available were obtained mainly for the magnesium-based pulping process and therefore are out of range of stack gas SO_2 recovery, for which partial pressure of SO_2 in the gas is much lower and sulfite content of the scrubber solution is considerably higher. Other basic data useful in scrubber design are also scarce. Therefore until basic studies are made it appears that design must be based on empirical considerations. Various scrubber types have been used, including venturis, mobile bed, and packed bed.

Opinions as to construction material appear to differ. In the pulping industry, where the need to make bisulfite requires relatively low pH in the scrubber, brick-lined towers have been used. For the higher pH involved in stack gas scrubbing, where the objective is to produce $MgSO_3$ crystals, some have recommended use of mild steel, although a lining of rubber or other erosion-resistant material may well be required to avoid erosion by the slurry. Other investigators have used lead or other corrosion resistant materials.

Absorption efficiency for SO_2 in the system seems to be very good, although $Mg(OH)_2$, the form in which the Mg is fed to the scrubber system in most processes, is extremely insoluble. However, since the scrubber is operated like alkali scrubbing systems, with direct recirculation from scrubber outlet (in respect to liquor flow) to inlet and with dissolved sulfite in the recirculating liquor, the $SO_3^{=}$ buffers the solution and provides a supply of OH^- ion:

$$SO_3^{=} + H_2O \rightleftharpoons OH^- + HSO_3^-$$

$$SO_2 + OH^- \longrightarrow HSO_3^-$$

In pilot plant tests, removal efficiency of over 95% has been obtained at reasonable liquor recirculation and MgO feed rates. About 10 gal/mcf has been adequate in a mobile bed scrubber when the stoichiometry was such to give the sulfite; higher MgO/SO_2 does not increase absorption significantly. Sulfate concentration also does not appear to have an adverse effect unless the pH is low. Hydration of the MgO before introduction into the scrubber, while practiced in some processes, is probably not necessary unless the MgO is below average in reactivity. Scaling in the scrubber system, as in lime/limestone scrubbing, can be a serious problem, particularly in the recirculation system. The mechanism appears different from that in lime/limestone scrubbing, but some of the factors that decrease scaling are common to both processes. Conversion of $MgSO_3$ to $Mg(HSO_3)_2$ in the scrubber should not develop scaling because $Mg(HSO_3)_2$ is quite soluble. Thus conversion back to sulfite by reaction with MgO or $Mg(OH)_2$ is the likely cause of scaling, wherever it occurs in the system.

In contrast, scrubber scaling occurs in the calcium system because of generally lower Ca salt solubility and consequent tendency for crystallization in the scrubber. For this reason high liquor rate is not as helpful for avoiding scaling in the magnesium system; high liquor velocity, however, has been found to be helpful. Selection of construction materials that have relatively low scaling tendency, plus providing an adequate number of seed crystals in the recirculation tank, should also be beneficial.

Oxidation in the scrubber is a problem as in most scrubbing processes. However, the usual practice is to operate the calciner under reducing conditions in order to minimize oxidation during calcination; hence any sulfate in the sulfite fed to the calciner would be reduced and would only increase the amount of reducing agent required rather than make a separate desulfating step necessary as in many other processes. The degree of oxidation in the scrubber depends on several factors, as will be discussed under the specific processes. This automatic purge of sulfate is an important advantage of Mg system scrubbing.

Drying and Calcination

In most processes the $MgSO_3$ is dried before calcining, the main objective being to remove the large amount of hydration water associated with $MgSO_3$ crystallized from solution. The type of hydrate formed in the scrubber is therefore important. Literature data generally indicate that $MgSO_3{\cdot}6H_2O$ and $MgSO_3{\cdot}3H_2O$ are the stable forms, respectively, below and above about 40°C. Since the scrubber temperature is likely to be about 50°C, formation of $MgSO_3{\cdot}3H_2O$ is indicated. Plant tests, however, with one or two exceptions, have shown that hexahydrate is the species formed, and recent laboratory work indicates that the hexahydrate is stable to a temperature higher than that reported by some investigators.

Some investigators have reported that $MgSO_3{\cdot}6H_2O$ loses water of hydration in steps, and others that the hexahydrate can be converted to the trihydrate by heating. Recent tests indicate, however, that solid hexahydrate can not be converted to the trihydrate by heating. A slurry can be converted, however, by heating to a temperature (about 175°F) at which hexahydrate dissolves and recrystallizes as trihydrate. Various types of dryers have been tested, including the fluidized type. The dryer temperature has ranged from 400°F in fluidized dryers to 700°F in the rotary type.

The main questions in calcination are the temperature level required and the extent of side reactions. Since many investigators have studied the thermal decomposition of $MgSO_3$, it should be possible to evaluate the situation with some confidence; the main difficulty, however, is that in a practical process the decomposition temperature is likely to be attained by direct contact with combustion gas, whereas much of the information reported is based on tests in which indirect heating or heating by a stream of hot nitrogen was used. If direct heating is used, the composition of the heating gas becomes important.

For heating by indirect means or with an inert gas, published data indicate that the nature of the decomposition products depends on the temperature. At lower temperatures (300° to 500°C), sulfate, thiosulfate, SO_2, and elemental sulfur are formed, with sulfate predominating. At higher temperature levels thiosulfate decreases, sulfate and sulfur increase, and evolved SO_2 becomes predominant. Above 900°C all side products are unstable and are replaced by SO_2 (Figure 8.2).

FIGURE 8.2: HIGH TEMPERATURE EQUILIBRIUM DIAGRAM FOR $Mg\text{-}O_2\text{-}H_2\text{-}S$ COMPOUNDS

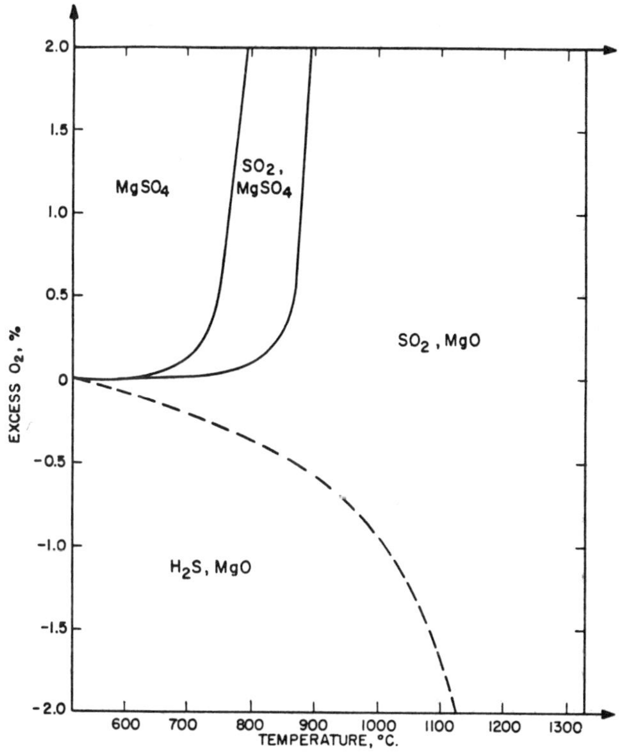

Adapted from Sillen and Anderson, Svensk Papperstidning, årg. 55, Sept. 4, 1952, pp. 622 to 631

The unique feature of this behavior is that sulfide, a usual major product of disproportionation, does not occur. Various mechanisms have been proposed for this, mainly associated with the relatively low decomposition temperature and the formation and stability of thiosulfate at the lower temperatures. The equations shown on the following page will illustrate this.

$$MgSO_3 \longrightarrow MgO + SO_2$$

$$4MgSO_3 \longrightarrow 2MgSO_4 + MgS_2O_3 + MgO$$

$$MgS_2O_3 \longrightarrow MgSO_3 + S$$

The absence of sulfide formation is fortunate as it could cause evolution of H_2S on recycle to the scrubber (or to a hydrator before the scrubber). Laboratory tests, in which the CO_2, H_2O, and O_2 contents of the heating gas (in contact with the $MgSO_3$) were varied, indicate that $MgSO_3$ is resistant to CO_2 and H_2O in the gas (at 500° to 600°C) but that a reducing atmosphere should be used to minimize reaction with oxygen. Even under mildly reducing conditions, however, some oxidation occurs and there is minor conversion to reduced forms of sulfur, mainly thiosulfate.

The pilot plant and full-scale work has been at considerably higher temperatures, on the order of 850° to 1,200°F, and with reducing agent present (usually coal or coke). Very little information has been made available on degree of oxidation and formation of reduction products. It appears, however, that the high temperature plus the reducing agent is effective in minimizing formation of side products, either oxidized or reduced.

A major consideration in any solids recycle process is whether the heating step involved in regenerating the solid affects absorption efficiency. There appears to be some difference of opinion on this point in regard to MgO regeneration. The high calcination temperature could sinter the MgO and make it too unreactive for reaction with $Mg(HSO_3)_2$ to produce $MgSO_3$. However, since the MgO is hydrated to $Mg(OH)_2$ in the scrubber circuit, and since the hydration requires recrystallization, it would appear that even if the MgO were somewhat dead-burned, the only difficulty would be a longer hydration time required, and that the hydrated material would be as reactive as if no sintering had occurred.

Various types of calciners have been tested or proposed. Most of the work has been with Herreschoff furnaces or rotary kilns but fluidized calciners are generally considered to be more promising. The situation is similar to that in other industries—such as limestone calcination, phosphate ore calcination, and ore roasting—where the uniform temperature and lower cost of fluidized reactors have been quite advantageous.

The gas stream evolved in the calciner contains SO_2 in an amount that varies with the particular process; concentrations ranging from 5 to 15% SO_2 have been reported. Factors affecting the SO_2 concentration include degree to which the $MgSO_3$ is dried (since any drying done in the calciner requires more combustion gas), calcining efficiency (degree of heat loss and consequent amount of combustion gas), and amount of $MgSO_4$ in the feed (sulfate requires more gas volume because of the need for reducing gas and for more heat input to the reduction reaction). If a dryer is used there is no particular difficulty in getting an SO_2 concentration high enough for feed to a sulfuric acid plant; the concentration appears too low, however, for economical conversion to sulfur.

The gas contains entrained solids, mainly MgO. Various ways of removing the dust from the gas have been proposed, including venturi scrubbers and bag filters. If any fly ash from the main plant (power plant or smelter) accumulates in the recycle stream, there is the problem of separating MgO (for recycle) from the ash. Various means for doing this have been proposed.

Process Variations

The magnesium-based system for sulfur oxide scrubbing has been studied by several research organizations, in the USSR, the Federal Republic of Germany, the United States, and Japan. Although the processes are all based on the general flowsheet described above, each has distinctive variations.

NIIOGAZ: A major effort has been carried out in the USSR on developing the magnesia process. A 6,000 scfm pilot plant has been operated in an ore sintering plant since 1957 and a 9,000 scfm unit has been tested on stack gas from a power plant. A flowsheet is given in Figure 8.3.

FIGURE 8.3: NIIOGAZ (USSR) MAGNESIA PROCESS FOR REMOVING SULFUR OXIDES FROM STACK GAS

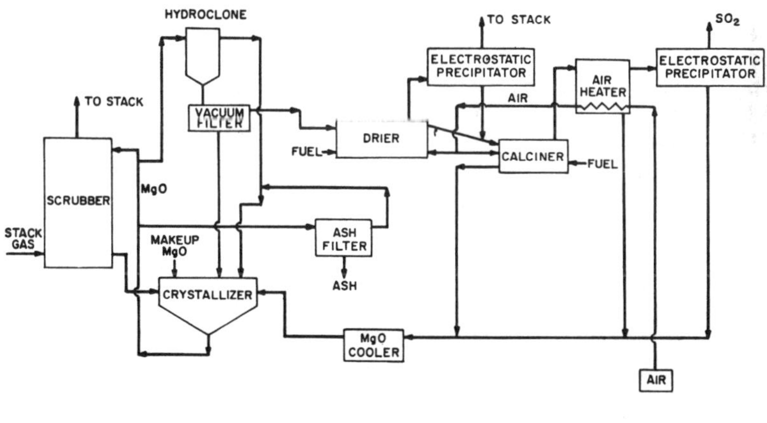

The power plant stack gas contained 0.25 to 0.3% SO_2 and 2 to 3 g/scm (0.9 to 1.3 g/scf) of dust. The ash was removed by filtering a side stream of scrubber solution. The scrubber, packed with egg crate wood packing, performed well; SO_2 absorption was 95% at a pH of 6.1 to 6.3, gas velocity of 1 to 2 m/sec (3.3 to 6.6 ft/sec), liquid flow of 10 m^3/m^2/hr of scrubber cross section (about 23 gal/mcf), and liquor retention time in the crystallizer of 15 minutes. The recirculating slurry contained 8% $MgSO_3 \cdot 6H_2O$ crystals, 2% $MgSO_3$ in solution, 10 to 15% $MgSO_4$, and a maximum of 4% ash. The product side stream of slurry was concentrated to 33% $MgSO_3 \cdot 6H_2O$ in the hydroclone and then filtered.

After drying to 4 to 10% H_2O, the crystals were calcinated at 800° to 900°C. Both a rotary kiln and a fluidized calciner were tested; the rotary kiln gave a higher SO_2 concentration in the off gas but required more retention time. At the 10 to 15% $MgSO_4$ level in the solution, it was reported that there was no oxidation. No scaling was reported; possibly the high pump erosion reported indicated an erosive situation throughout the system (because of ash and sulfite crystals) that prevented scaling.

In the tests on ore sintering gas, a scrubbing efficiency of 92% was obtained in a single-stage venturi scrubber. A very large scrubber assembly (21 scrubbers, 1,770,000 scfm of gas) has been constructed and is being operated on lime/limestone slurry while development of the magnesia process is being completed. Installation of a 300 Mw system at a power station is also being considered.

Grillo: The Grillo-Werke AG, a zinc producer in West Germany (Duisburg-Hamborn), carried out an active program for several years (beginning in 1964) on development of magnesia scrubbing. The process and technique developed is generally similar to that described above except that some MnO_2 is carried along with the MgO in the circuit; the mol ratio is 3 to 6 mols of MgO per mol of MnO_2.

A generalized version of the process flowsheet is shown in Figure 8.4. The absorption is carried out in a cocurrent scrubber and a side stream of slurry is dried in a spray dryer. The dried material is mixed with coal and calcined in a fluid bed furnace at 850° to 900°C. The offgases are cleaned in standard dust removal equipment and then sent to a sulfuric acid plant.

FIGURE 8.4: GRILLO MAGNESIA SCRUBBING PROCESS

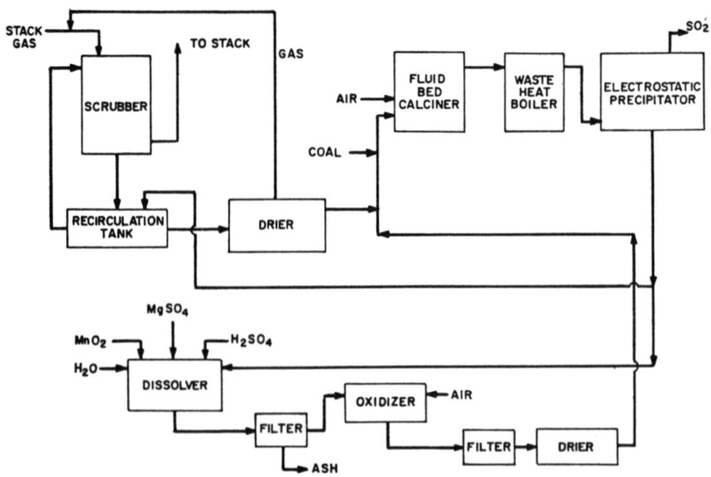

The regenerated absorbent is recycled, with a side stream treated to separate ash; since the ash tends to react with manganese in the furnace and form insoluble manganese silicate, a chemical treatment with H_2SO_4 is used to leach out the manganese and avoid loss in the filtration step where the ash is separated from the soluble Mg and Mn compounds. An oxidizing step can be included to precipitate and remove iron compounds dissolved from the ash that would otherwise accumulate in the system. The main source of the ash is the coal used in the calcining step. The process has been tested mainly on oil-burning power plants so far, so that little ash comes into the scrubber. If it were applied on a coal burning plant, the heavy load of ash could be removed before the scrubber as it is in other processes. Typical data on the process are as follows.

Scrubber volume	1.4×10^{-4} m^3 gas/hr
Absorbent required	1.5 g/g SO_2
Absorbent loss	<3% per cycle
Pressure drop	200 mm H_2O

In operation, the pH of the fresh slurry is 8.0 to 8.5, dropping to about 4.0 in the scrubber. An absorption efficiency of 90 to 95% is attainable without difficulty. The solids drawn off contain 17 to 21% sulfur as sulfite, bisulfite, and sulfate; about half the sulfur is present as sulfate, presumably because the MnO_2 promotes oxidation.

The scrubber exit gas temperature is reported to be about 70°C. The very short retention time in the scrubber is said to prevent full humidification of the gas. The Grillo process is unique in regard to carrying MnO_2 in the circuit. It is claimed that the MnO_2 increases absorption rate and also prevents loss of MgO activity by sintering. An optional feature of the process, particularly for oil-burning plants, is dry absorption of SO_3 with part of the absorbent at 300° to 450°C. By this means the air heater can be operated at lower temperature, to recover more heat, without corrosion by SO_3 condensing as H_2SO_4 on the heater surfaces.

A major concept in the Grillo work has been separation of the scrubbing and regeneration steps as completely independent operations, with loaded absorbent collected from one or more power plants (or other SO_2 producing plants) and regenerated at a central installation. The original intent was to use existing pyrites-burning H_2SO_4 plants for regeneration, since the coal-absorbent mixture can be burned like pyrites. The coal and $MgSO_3$ were to be shipped already mixed so that the mixture could be used in the H_2SO_4 plant with a minimum of change in plant design and operation. Certain difficulties have developed in applying the central regeneration concept to existing pyrites plants. Building a new regeneration facility may be better, in which case the coal probably would not be mixed in at the power plant.

The process has been tested in a 12,000 scfm pilot plant on sulfuric acid plant tail gas and in a 16,000 scfm unit at a power plant (in Wesseling, Germany). The test periods extended for about 1.5 years each. Grillo considers the process developed and has offered it to the public. The process appears to have lost ground, however, to those being tested on a full scale in the U.S. and Japan.

Chemico-Basic: The method developed by the Chemico-Basic Corp., called the Magnesium-Base SO_2 Recovery Process, is shown in Figure 8.5. Particulate matter is first removed in a venturi and the SO_2 is then scrubbed with $MgSO_3$-$Mg(OH)_2$

slurry in a second venturi. Crystals of $MgSO_3 \cdot 6H_2O$ are separated from a side stream by centrifuging (plus any $MgSO_4 \cdot 7H_2O$ formed in the scrubber circuit) and dried to the anhydrous form. Carbon is added to reduce sulfate and the mixture calcined to evolve SO_2 and regenerate MgO. The calciner exit gas contains 15 to 16% SO_2.

FIGURE 8.5: CHEMICO-BASIC MgO SYSTEM FOR SCRUBBING STACK GAS CONTAINING DUST AND SO_2

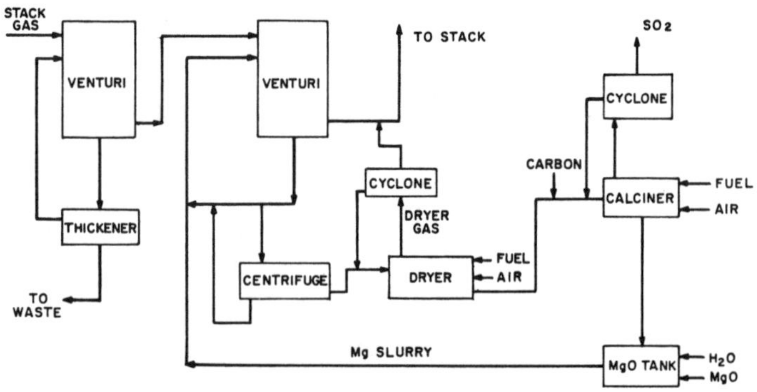

Dryer offgas is introduced into the main power plant gas stream (after the scrubber) to supply reheat (about 15°F). The SO_2 from the calciner is converted to H_2SO_4. Dust is first removed by a cyclone and wet scrubber in series, after which the saturated gas is cooled to 100°F in a direct contact cooler. The cleaned, cool gas then enters a standard sulfuric acid plant of the wet type. In another version of the process, dust and SO_2 are removed in the same scrubber (Figure 8.6). To separate the ash from the sulfite formed by SO_2 absorption requires that the sulfite be in solution in the scrubber effluent liquor, which makes it necessary to operate at relatively low pH to produce soluble bisulfite and to have another step in which sulfite is precipitated. In the flowsheet shown, it appears that this is done mainly in the reaction tank. Addition of MgO or $Mg(OH)_2$ at this point should convert bisulfite to sulfite and cause $MgSO_3 \cdot 6H_2O$ to crystallize from the clear liquor obtained after ash removal.

The flowsheet in Figure 8.6 illustrates the central regeneration system that Chemico-Basic has promoted. The dried $MgSO_3$ produced at several points would be shipped to a single central process plant for regeneration. The principal activity has been the large prototype test facility at a Boston Electric Company plant in Everett, Mass. (Mystic Station No. 6).

The gas from a 155 Mw, oil-fired boiler was scrubbed in a venturi scrubber, the resulting $MgSO_3$ separated from the scrubber bleed stream by centrifuging, and the solid product dried in a countercurrent drier. The dried magnesium sulfite was shipped to Essex Chemical Corporation (near Providence, Rhode Island) where it was calcined in an oil-fired kiln and the SO_2 converted to H_2SO_4 in a standard sulfuric acid plant. The test system was started up in April 1972 and closed down in June 1974.

FIGURE 8.6: CHEMICO-BASIC SYSTEM FOR MgO SCRUBBING USING SINGLE SCRUBBER FOR DUST AND SO_2

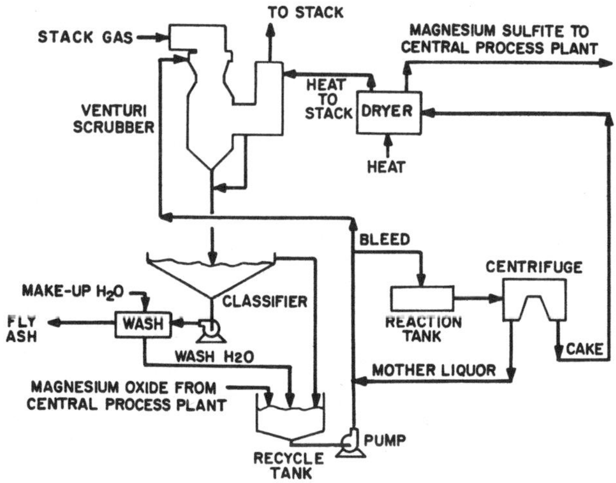

Many problems, particularly of the mechanical type, were encountered in the test program. The final assessment by Boston Edison was that although operation was not very good a second unit could be built with a high level of confidence that it would work reliably. Changes that probably would be made include (1) use of a cocurrent rather than countercurrent drier, (2) rubber coatings in equipment to reduce erosion, and (3) various changes in the solids handling system. In early 1975 Boston Edison was still considering whether or not to install the process on other boilers.

Chemico-Basic has also installed an MgO system for the Potomac Electric Power Company at the coal-fired Dickerson station; the unit handles half the gas from a 190-Mw boiler. For this installation there are two venturis in series, one for particulate and the other for SO_2. The drier is a cocurrent type and the $MgSO_3$ was shipped to Essex Chemical for regeneration (in the same kiln used for the Boston Edison product). The system was started up in September 1973 and the test project was to be completed in early 1975. Good SO_2 removal was obtained but there were many mechanical problems. Another Chemico-Basic system (162-Mw) has been installed at an Idemitsu Kosan plant in Japan.

United Engineers: United Engineers and Constructors has constructed a 120-Mw MgO scrubbing system at Philadelphia Electric's Eddystone station. One of the unique features is that a fluidized drier is used. The unit was in the preliminary startup phases in early 1975. Philadelphia Electric has agreed with EPA to install the process on other boilers at the Eddystone and Cromby stations if the tests of the initial unit at Eddystone are successful.

Other: In the U.S., B & W has been active in developing magnesia scrubbing in the paper pulp industry. In addition, B & W carried out a test program for EPA in which the scrubbing step of a magnesia process for power plant SO_2 recovery was tested.

In Japan, the Tsukishima Kikai Company (TSK) has developed an MgO process generally similar to those in the U.S. A full-scale unit has been installed at the Onahama Smelting and Refining Company to recover SO_2 from copper smelter gas.

CALCIUM

Calcium compounds (oxide and carbonate) are used mainly for throwaway processes, but there has been some effort to find acceptable ways for recovering the sulfur value from the product ($CaSO_3$ or $CaSO_4$). These involve both wet and dry processes, most of which have not been developed as far as the magnesia-based methods.

Chemistry of Absorption Product Regeneration

The products obtained in scrubbing or dry sorption with lime or limestone can be regenerated in the same way as in the magnesia system, by thermal decomposition coupled with use of reducing agent to assist in decomposition of sulfate. The main difference is that higher temperature is required in the calcium system and that the chemistry is somewhat different. The main difference in the chemistry is that undesirable formation of sulfide occurs in the calcium system whereas little or none is developed in regenerating $MgSO_3$. The main reactions in $CaSO_3$ decomposition have been reported to be as follows.

$$CaSO_3 \longrightarrow CaO + SO_2$$

$$4CaSO_3 \longrightarrow 3CaSO_4 + CaS$$

$$2CaSO_3 + SO_2 \longrightarrow 2CaSO_4 + S$$

$$2CaO + 3S \longrightarrow 2CaS + SO_2$$

The sulfide predominates at higher vapor pressures of SO_2 and under reducing conditions, but formation is reduced by increase in temperature. Sulfate formation is also inhibited by higher temperature; at very high temperatures, CaO is the thermodynamically stable phase (Figure 8.7). The $CaSO_3$ begins to decompose at fairly low temperatures (600° to 700°C) to give CaO and SO_2, but to minimize formation of $CaSO_4$ and CaS higher temperature is required. The distribution of sulfur compounds in the solid residue remaining after heating calcium sulfite (Table 8.1) shows that a temperature of 1000° to 1100°C practically eliminates sulfide and gives low sulfate.

The tests in which these data were obtained were made in an atmosphere of nitrogen. In TVA tests, in which the atmosphere contained CO_2 and H_2O plus either oxygen or methane (to give a reducing condition), similar results were obtained except that sulfate was somewhat higher under conditions that gave low sulfide. Since sulfate is present because it is formed both during decomposition and during the original sorption step, the chemistry of sulfate decomposition is also important. The following reactions are involved in reduction with reformed gas.

$$(1)\quad CaSO_4 + CO \longrightarrow CaO + SO_2 + CO_2$$

$$(2)\quad CaSO_4 + H_2 \longrightarrow CaO + SO_2 + H_2O$$

$$(3)\quad CaSO_4 + 4CO \longrightarrow CaS + 4CO_2$$

$$(4)\quad CaSO_4 + 4H_2 \longrightarrow CaS + 4H_2O$$

$$(5)\quad 3CaSO_4 + CaS \longrightarrow 4CaO + 4SO_2$$

Reactions (1), (2), and (5) are favored by high temperature, and (1) and (2) by high CO/CO_2 and H_2/H_2O ratio. However, (3) and (4) are promoted far more than are (1) and (2) by high reducing gas ratio (to the fourth rather than the first power); at a ratio of 1:2 or higher, sulfide begins to be formed at most temperatures. Calculations show that CaS is the main product at 1880°F, 95% CaO and 5% CaS at 2060°F, and 94% CaO and 6% CaS at 2240°F, all at 1:2 CO/CO_2 and H_2/H_2O ratio. Straight thermal decomposition of $CaSO_4$, without reducing agent, requires such a high temperature that it does not appear practicable.

TABLE 8.1: THERMAL DECOMPOSITION OF CALCIUM SULFITE*

	Composition of Residue			
	Percent Sulfur in Residue as			Percent Sulfur as
Temp., °C.	Sulfite	Sulfide	Sulfate	SO2 Evolved
650	79.2	4.6	15.2	0.38
800	3.8	23.2	71.3	3.9
900	4.0	15.3	59.8	22.0
1000	-	0.36	6.1	91.7
1100	-	0.12	4.2	93.3

*Heated in current of nitrogen for 2 minutes.

FIGURE 8.7: HIGH TEMPERATURE EQUILIBRIUM DIAGRAM FOR $Ca\text{-}O_2\text{-}S$ COMPOUNDS

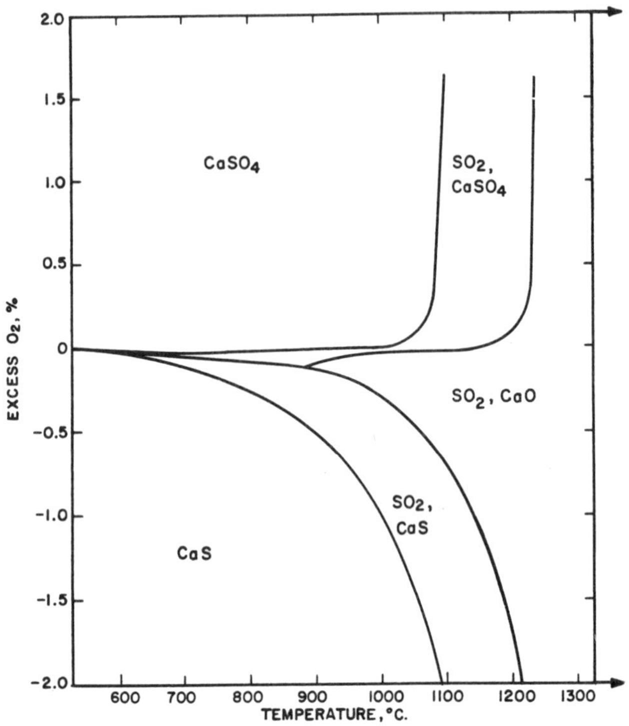

Adapted from Sillen and Anderson, Svensk Papperstidning, årg. 55, Sept. 4, 1952, pp. 622 to 631

Analogy to Gypsum Reduction

The presence of sulfate makes the situation analogous to that in processes for decomposition of naturally-occurring calcium sulfate or of by-product sulfate from other industries. A large amount of work has been done, going back for 50 years or more, on decomposition of gypsum ($CaSO_4 \cdot 2H_2O$) to give either sulfur or sulfuric acid. The objective has been either to use natural gypsum, a low-cost ore, as raw material or to convert by-product gypsum from phosphoric acid production to sulfuric acid, thereby recycling the acid.

Reaction with Silicates: Possibly the oldest process for decomposing gypsum is heating with silicates and reducing agent to evolve SO_2 and produce a clinker suitable for cement production. The earliest method was the Muller-Kühne process, developed in the 1910 to 1920 decade; gypsum was ground with clay, iron ore, and coal or coke, followed by roasting in a rotary kiln. The iron ore and clay were added in the proportions required to give the composition of standard Portland cement in the product clinker. The kiln gas, containing 7 to 8% SO_2, was cleaned and converted to sulfuric acid. About 5 to 10% of the kiln feed was evolved as dust.

Since the Muller-Kühne method was first developed, various improvements have been made and several large plants have been built, mainly in Europe. A simplified flowsheet of the Chemico OKC process is shown in Figure 8.8.

FIGURE 8.8: CHEMICO-OKC PROCESS FOR MAKING SULFURIC ACID AND CEMENT FROM GYPSUM

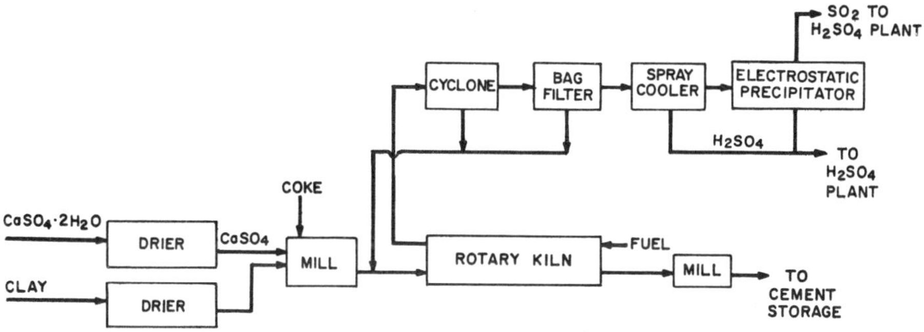

The clay and gypsum are dried, removing the hydrate water in the gypsum, and ground with coke (plus iron ore if needed). The mixture is calcined at 2600°F (temperature in the combustion zone), cooled, and ground. The gas, which leaves the kiln at about 1100°F, is cooled to 600°F, passed through cyclones and a bag filter to remove dust, cooled and further cleaned in spray coolers (which also remove SO_3 formed in the kiln), and passed through an electrostatic mist collector to remove H_2SO_4 mist (also from the SO_3 formed in the kiln). The clean, cool gas is then converted to H_2SO_4 in a standard wet-type plant.

In other processes silica is added and carbon omitted, as the affinity of the SiO_2 for CaO is sufficient to give desulfurization at a satisfactory rate. In tests at TVA, about 72% desulfurization was obtained (in 120 minutes at 1250°C) at an SiO_2:S mol ratio of 0.7 and 95% at a ratio of 1:1. Iron oxide added to produce cement clinker also promoted the reaction. In most of the commercial plants, however, some coke is used to further accelerate the decomposition.

The H_2SO_4/cement process should be applicable to utilization of the $CaSO_3/CaSO_4$ produced in lime/limestone processes for treating stack gas. It is reported that the method is under study by the National Coal Board in England. The main drawback is the very large amount of cement produced and the difficulty in marketing it.

Reduction to CaO and SO_2: Reduction of gypsum directly to SO_2 and CaO, so that the CaO can be recycled, is somewhat difficult because of the side reactions that occur in the absence of a reactant such as silica. Formation of sulfide is the main consideration because of the odor problem that would result if the sulfide reacted with H_2O and CO_2 on return of the regenerated solid to the scrubber. The major work on reduction to CaO apparently is that done in the past few years at the University of Iowa. The process involves reduction of gypsum by hydrogen or carbon monoxide rather than carbon, at about 2200°F (1205°C). Close control of temperature and gas composition is required in order to get good decomposition and negligible formation of sulfide, a requirement that is best met in a fluidized bed decomposer. Furthermore, the fluidized bed is suitable for carrying out both the reduction and the production of the reducing gas in the same piece of equipment.

In pilot plant work to date, the process has appeared stable and easy to control. The refractory furnace lining has resisted furnace conditions well, and dust evolution from the granular charge has not been excessive. In some runs over 90% desulfurization has been attained with no detectable sulfide formation and with an SO_2 content in the product gas of 5 to 10%. Further work is required to identify the optimum conditions for minimizing sulfide; it is expected that some sulfate will have to be left undecomposed to accomplish this.

Application of the method to $CaSO_3/CaSO_4$ from an SO_2 recovery operation should be feasible. The amount of reducing agent needed presumably could be lower than in gypsum reduction because of direct SO_2 evolution from most of the sulfite.

Reduction to CaS: Difficulty from sulfide formation can be avoided by purposely reducing the gypsum to sulfide and then decomposing it by reaction with CO_2 and H_2O to give H_2S, which would be converted to sulfur. This is an old process, used in Germany during World War I. Waste calcium sulfide was converted to sulfur even earlier, by Chance and Claus (England) in the late 1800s. In recent times, Dorr-Oliver and the U.S. Bureau of Mines (Salt Lake City Metallurgy Research Center) have worked on the process.

In the Dorr-Oliver studies, by-product gypsum from the phosphoric acid industry was roasted in fluidized beds in the system shown in Figure 8.9. In the first reactor the gypsum is dried and dehydrated and the temperature raised to a level at which residual hydrocarbon from the other reactor is burned. In the second reactor reducing gas formed by incomplete combustion of hydrocarbon fuel reduces the $CaSO_4$ to CaS. Reduction as high as 98% has been attained. A major drawback, for the particular raw material used, was that about 25% was so fine that it could not be retained in the fluidized bed and had to be removed before the reduction step.

In the Bureau of Mines work, reduction to CaS has been accomplished with coal in small-scale rotary kiln tests and with reducing gases in a fluidized bed. With bituminous coal, 94 to 98% reduction was attained in the kiln; the reduction rate was comparable in the fluidized bed but dusting was a problem. Both the Chance-Claus method of sulfide decomposition and another one that gives Na_2CO_3 as an end product have been studied.

The Na_2CO_3 process involves the reactions shown on the following page.

$$CaS + H_2S \xrightarrow{H_2O} Ca(HS)_2$$

$$Ca(HS)_2 + 2NaR\ (resin) \longrightarrow CaR_2 + 2NaHS$$

$$2NaHS + 2CO_2 + 2H_2O \longrightarrow 2NaHCO_3 + 2H_2S$$

$$2NaHCO_3 \xrightarrow{\Delta} Na_2CO_3 + CO_2 + H_2O$$

$$CaR_2 + 2NaCl \longrightarrow 2NaR + CaCl_2$$

Thus the H_2S on its way to the Claus process is used with CaS to convert NaCl to Na_2CO_3. Reduction to CaS has been attempted on a plant scale in a plant in Texas, using natural gypsum as the raw material. Details on the operation are not available, but it is understood that the plant was operated for a short time and then shut down because of decline in the price of sulfur. Reduction to CaS should also be applicable to $CaSO_3$ but no data on the reaction seem to be available. Avoiding the sulfide problem, plus production of sulfur rather than H_2SO_4, may make the CaS route the choice over reduction to CaO.

FIGURE 8.9: DORR-OLIVER PROCESS FOR REDUCING GYPSUM TO CALCIUM SULFIDE

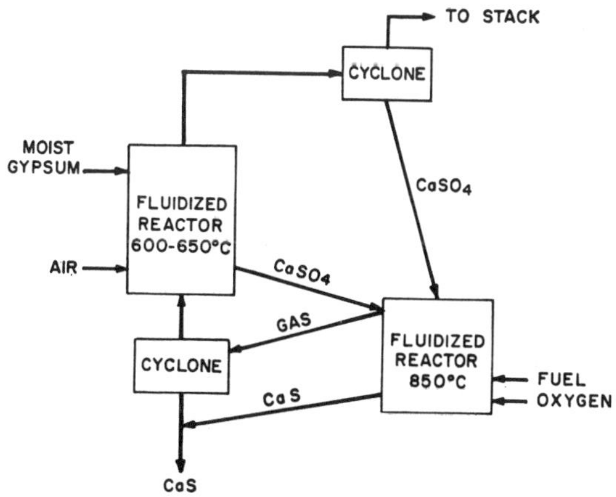

The sulfide is converted to elemental sulfur by standard methods.

Process Variations in Regeneration

Various organizations have studied regeneration of the $CaSO_3/CaSO_4$ obtained in removing SO_2 from stack gas. Most of the work has been done in dry systems, particularly in conjunction with fluidized combustion (see Chapter 2)—both complete combustion and partial (to give a fuel gas). The work on sorption has been described in Chapter 2 under throwaway processes; only the investigations of regeneration methods will be discussed here.

Esso Research and Engineering (U.S.): In a project carried out for EPA, Esso Research and Engineering has studied regeneration of $CaSO_3/CaSO_4$ produced in a fluidized SO_2 sorption vessel (small scale). At a 1:2 CO/CO_2 or H_2/H_2O mol ratio, little CaS was formed (1 to 2% of total sulfur) but with H_2 alone sulfide formation was high. The SO_2 concentration in the effluent gas had a first order dependence on the inlet reducing gas concentration. Methane was unsatisfactory as a reducing agent.

The regenerated CaO retained its activity through at least 8 cycles but declined thereafter. It was postulated that a total exposure time to regenerator temperature of 15 hours should be feasible before significant loss of activity. There was indication that some constituent of the fly ash contributed to the decline in reactivity.

Esso Research Centre (England): Two types of fluidized lime sorption/regeneration methods have been studied by the Esso Research Centre in England; both have been operated only on oil. The first method involves the usual procedure (see Chapter 2) of burning the oil in a fluidized bed of limestone, thereby sorbing the SO_2 as $CaSO_4$. In the batch-type experimental arrangement used, regeneration was carried out in the same bed by increasing the fuel rate until 55% of the theoretical air was being fed; with more air than this the SO_2 content of the effluent gas dropped off.

There was some loss of sorption efficiency over several regeneration cycles—from 95 to 65% in 15 cycles. A fairly high removal efficiency could be maintained by removing fines and adding fresh makeup; the amount corresponded to about 14% of the theoretical feed rate required for complete utilization on a once-through basis. In other words, the process is partly throwaway and partly regenerative, at a 1:7 ratio.

The second method, considered to be more promising, involves operating the combustor with a deficiency of air to produce a fuel gas and fix the sulfur on the lime particles as a layer of CaS. This would allow installing the process on an existing plant because the fuel gas could be burned in the existing power plant burners. A flowsheet of the method is shown in Figure 8.10. About 30% of the stoichiometric air for full combustion is fed and no steam is used, which gives a fuel gas rich in hydrocarbons (cracked products) rather than carbon oxides. Good sorption efficiency was obtained at 800° to 900°C but dropped off at 950°C; sorption was complete at the lower temperatures until the sulfur content of the bed reached 6%, after which a slow decline occurred.

The CaS is moved to the regenerator through a defluidized transfer line. Solids are carried into the line by splash and then flow to the regenerator by gravity. Pulses of air reverse the flow occasionally to move regenerated CaO back into the gasifier-sorber. The regenerator off-gas contains about 15% SO_2. There is some attrition in the process and also some loss in sorption efficiency in extended cycling. The fines from both vessels are caught in cyclones and returned; however, a portion is discarded both to prevent choking the system with fines and also to allow makeup to keep sorption efficiency up. For a given degree of discarding and makeup, the system levels out at some SO_2 removal efficiency; at a makeup rate corresponding to about 55% of stoichiometric the steady state removal is 80%, whereas at 100% stoichiometric it is practically 100%. It has been suggested that the makeup be gypsum rather than lime, thereby increasing production and in effect making H_2SO_4 (or S) at low cost.

FIGURE 8.10: LIME ABSORPTION/REGENERATION PROCESS DEVELOPED BY ESSO RESEARCH CENTRE

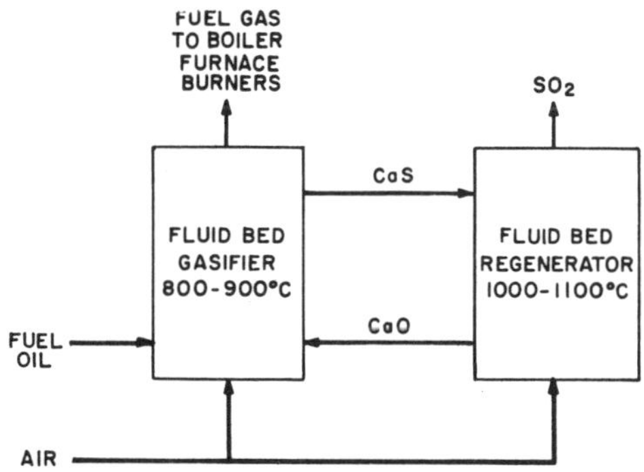

A continuous pilot plant to demonstrate the flowsheet in Figure 8.10 has been constructed under EPA sponsorship. The process appears to have considerable promise because of its relative simplicity and because sorbing the SO_2 as a thin film in a surface reaction on the lime granules (about one-eighth inch in diameter) gives very fast sorption and high removal efficiency. The resulting need to recycle large volumes of sorbent is no great handicap because the fluid transfer from sorber to regenerator is easy and inexpensive as compared to the conveyor type of transport in other processes. One drawback is that the process may be difficult to adapt to coal firing because of the different conditions required in gasifying coal and the effect of the coal ash on the operation.

Consolidation Coal: The use of dolomite in a fluidized sorber has been studied by the Consolidation Coal Co. Sorption efficiency was 79% at a stoichiometric (Ca/S) ratio of 0.95 and 88.5% at a ratio of 1.89 (1800°F, once-through feed of dolomite, and 8.6 to 12.6% S accumulated in dolomite). Regeneration was carried out at 1950°F by partial combustion of CO. Up to 96% conversion of $CaSO_4$ to CaO was obtained, but the CaO lost efficiency fairly rapidly with continued cycling.

Dolomite was used because of its wide availability, although the fact that the Mg portion did not enter into the reaction made a large amount of sorbent necessary. It was estimated that makeup for a process involving regeneration would be only about 20% of the sorbent amount required for once-through, throwaway operation.

TVA: Dry sorption followed by regeneration has the drawbacks, as discussed in previous chapters, that sorbent loses efficiency in cycling and that $CaSO_4$ is produced rather than $CaSO_3$—which makes regeneration more difficult and more costly. By analogy to the MgO system, wet scrubbing should be better in regard to these problems. Nevertheless, although lime/limestone wet scrubbing is relatively well developed, there has been little emphasis on regenerating the scrubber effluent solids in processes of the type used in the MgO system.

Small-scale work has been done at TVA on the flowsheet shown in Figure 8.11. Dust is removed in a prescrubber to avoid carrying it through the regeneration system. The SO_2 scrubbing can be operated on either a recovery or throwaway basis; for recovery the absorbent is mainly regenerated CaO and for throwaway limestone is fed to the delay tank. In the recovery mode of operation, oxidation in the scrubber is minimized as much as possible by the methods described earlier.

The $CaSO_3/CaSO_4$ crystals are removed in a centrifuge or filter and fed directly to the calciner. Here no special attempt is made to reduce sulfate; the calciner is operated with main emphasis on avoiding sulfide formation, which probably requires the presence or formation of some sulfate (or possibly introduction of air near the end of the calcination to burn CaS to CaO). The calcined product is then slaked and subjected to a minerals beneficiation procedure to separate a fraction concentrated in sulfate, which is discarded as a purge. The enriched $Ca(OH)_2$ is fed to the delay tank along with makeup limestone.

FIGURE 8.11: LIME SCRUBBING WITH REGENERATION OF EFFLUENT SOLIDS

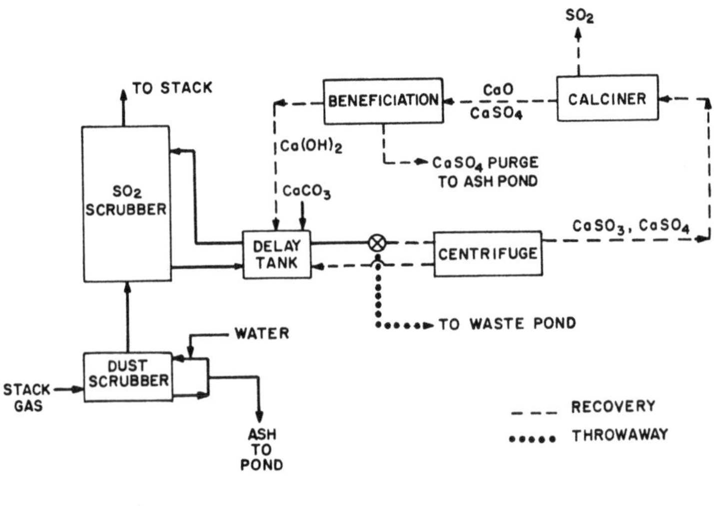

The system can be operated on a throwaway basis when necessary. For throwaway operation, the scrubber effluent is diverted before the centrifuge and sent directly to the waste pond. The solids can be separated if desired, of course, and transported away from the plant in solid form. In the throwaway mode, all the calcium is supplied as limestone. The process has several potential advantages over MgO scrubbing and regeneration.

(1) It should be possible to omit the dryer since the hydrate is $CaSO_3 \cdot \frac{1}{2}H_2O$ rather than $MgSO_3 \cdot 6H_2O$ and therefore much less dilution of SO_2 with combustion gas would occur in the calciner from supplying heat to remove water.

(2) The scrubber section could be installed as a first step and recovery added later if the situation so dictated. This is an important consideration because industry is faced with the imminent necessity for removing SO_2 from stack gases but there are no fully proven processes available. It would be better to keep investment at a minimum, by installing a throwaway process, until more experience is gained.

(3) The full process is versatile; in the event of market failure or other product distribution problems, operation could be changed to a throwaway basis on short notice. The necessity for producing acid at all times to avoid pollution would be avoided, whereas if the acid tanks became full for some reason or other in the MgO process, the entire operation would have to be shut down.

(4) Calcium sulfite is relatively insoluble as compared to magnesium sulfite and therefore should be easier to store as a buffer between the scrubber and regeneration sections.

(5) The purge after the calciner serves the purpose of avoiding the cost of reducing sulfate and of removing impurities such as residual ash and soluble salts that build up in the system. The purge is feasible only because limestone is cheap—which also makes normal process losses less expensive.

The main disadvantages, in comparison to MgO, are the tendency to sulfide formation and the generally higher temperature required for calcium sulfite decomposition. However, if sulfate reduction is not attempted the temperature may be little higher than for the magnesium system, where the necessity for sulfate reduction requires a temperature higher than for simple sulfite decomposition. As for sulfide formation, pilot plant tests may be necessary to resolve the question; further work on the Iowa State gypsum reduction method should help in this respect, since it seems likely that if sulfide formation is not the problem in gypsum reduction it should not be in calcium sulfite decomposition. If the problem cannot be solved, the route of reducing all the way to CaS can be followed and sulfur produced rather than sulfuric acid.

All parts of the process are fairly well tested except the beneficiation step, which would be similar to that used in the phosphate industry for removing limestone from phosphate ore. Slaking of the calcined ore gives very fine particles of $Ca(OH)_2$ that can be separated on a size basis from the larger phosphate particles. The calcined calcium sulfite/sulfate would be expected to behave similarly as tests have shown that the sulfate can be grown to fairly large crystal size.

Other: There have been several other studies on lime/limestone sorption followed by regeneration but little information on them has been reported. Pope, Evans, and Robbins, in a project carried out for the Office of Coal Research, have studied fluidized combustion of coal. In the course of this work, limestone was used in the bed to absorb SO_2 and it was found possible to periodically remove the sulfur by running the bed under reducing conditions. No sulfide formation was reported but this probably would have burned off during the sorption portion of the cycle. The Still process in Germany was reported to include regeneration of the sulfur-laden product. No data on the regeneration have been made available, however, and the poor performance in the sorption step appears to have caused loss of interest in the method. Several other organizations, including the M.W. Kellogg Company, Combustion Engineering, and Ontario have done work on $CaSO_3$-$CaSO_4$ regeneration.

Nonregenerative Methods

Some of the work on utilization of the absorption product has not involved regeneration of either CaO or SO_2. Instead the product is used as is with perhaps some treatment to improve its properties for some end use, mainly as a construction material. In Japan, practically all the $CaSO_3$ produced in the several lime/limestone and double alkali installations now operating is oxidized to gypsum and used in making cement or wallboard. The slurry from the scrubber is first treated with H_2SO_4 or SO_2 to reduce the pH, thereby getting sulfite into solution, and then blown with air to oxidize the sulfate.

METAL OXIDES

In addition to alkali and alkaline earth metals and their compounds, metal oxides such as those of Mn, Fe, Cu, and Zn have been investigated widely as sorbents for sulfur oxides. One of these, CuO, is being tested on a large scale in Japan and on a pilot scale in the U.S. Another, involving MnO_2 as the sorbent, was recently abandoned after full scale testing in Japan. In most of the processes that have been proposed, including the CuO and MnO_2 systems, the metal oxide has been used in solid form in a dry system; work on slurry scrubbing has been somewhat limited.

Like most of the other absorbents, metal oxides were used first in other fields for sorbing SO_2. For example, SO_2 has been used in recovering manganese from low-grade ores. A gas stream containing SO_2 is scrubbed with a slurry of the ore, giving soluble $MnSO_4$. Formation of various undesirable products, including sulfites, dithionates, and thiosulfates, is inhibited by operating at low pH. The $MnSO_4$ is decomposed at 1100° to 1200°C to give Mn_3O_4 and SO_2.

Much of the metal oxide development for SO_2 removal from stack gases has been aimed at avoiding cooling the gas below normal stack temperature. For power plant use, such a dry-sorption system ideally should (1) be operable at air heater exit temperature (so that it can be added to existing plants), (2) produce no side products, (3) maintain sorbent strength and efficiency over extended cycling, and (4) preferably be amenable to sorption and regeneration at the same temperature and in the same length of time.

CHEMISTRY OF METAL OXIDE SORPTION

A great deal of effort has gone into what might be called the simplest approach to SO_2 removal from stack gas, that is, sorption at one temperature and desorption at a higher one. As discussed in previous chapters, this is most easily done if bisulfite can be produced in the sorption step, because the bisulfite decomposes easily at a temperature not much higher than that at which sorption was carried out. The drawback is that the bisulfite can be formed only by wet scrubbing and under certain conditions that increase cost.

Sulfites are easier to make and can also be decomposed by heating although a higher temperature is required than for bisulfite. Some sulfites can be decomposed in the 400° to 800°F range, however, which is a practicable operating level. For the alkali and alkaline earth metals considered up to this point, the main problem with sulfites is the disproportionation that occurs in the heating steps. Magnesium is an exception, and therefore sorption by $Mg(OH)_2$ followed by heating the resulting magnesium sulfite is a leading process; the wet scrubbing system used to form the sulfite, however, has the usual problems of wet processes.

Because of these problems, attention has been given to finding a metal oxide that can absorb SO_2 in the dry state with a minimum loss in free energy, the objective being to avoid need for supplying a large amount of energy in regenerating the sorbent back to the oxide form. A survey of the pertinent thermodynamic data on 48 metal oxides was made by TRACOR for EPA. The primary data sought were heat of formation, absolute entropy, heat capacity, and phase transition heat and temperature.

The first consideration was that the sorption reaction be feasible—that is, have a negative free energy change—in the range of 100° to 750°C, since, for power plants, this covers the practical range of boiler temperature at which the sorption step can be operated. This criterion alone eliminated many oxides, including those of the alkalis and alkaline earths, and indicated that the oxides of the transition metals are the most promising.

The next criterion was whether or not the sorption product could be decomposed thermally to give a stream of SO_2 sufficiently concentrated for economical conversion to sulfuric acid or sulfur, and at an acceptable energy input. Only 16 oxides survived this test. The sorption kinetics were considered next. Of the 16 remaining oxides, only 6 were found by experiment to give an adequate sorption rate, copper, chromium, iron, nickel, cobalt and cerium. In earlier work, the U.S. Bureau of Mines found that the oxides of Mn, Co, and Cu were the most effective sorbents of a large number tested. The same results were obtained by workers in the Tokyo Institute of Technology.

For some of the oxides, for example, iron and chromium, adsorption is the principal mechanism when the sorption is carried out at low temperature (below about 200°C) whereas chemical reaction takes place over a higher temperature. The degree of adsorption, however, does not appear to be high enough for a practical process.

In a study by the AVCO Corporation, the best sorbents were found to be the oxides of iron, copper, and manganese, each carried on an alumina base. Each sorption product could be regenerated thermally at about 650°C. Russian work has indicated the following sequence of increasing sorption activity: ZnO, Fe_2O_3, CuO, Co_2O_3, Mn_2O_3.

Most of the work mentioned thus far has involved simple heating as a means of regeneration, on the basis that a chemical regeneration such as reduction of sulfite or sulfate to metal sulfide or H_2S would be unduly complicated and that cost of reducing agent would be a major drawback. Moreover, since sorption in these studies was carried out in the dry state to avoid cooling the gas, oxidation to sulfate was promoted, in most cases to complete oxidation.

Although thermal decomposition and dry sorption simplify matters, they are not necessarily superior. As has been seen in the earlier discussion of alkali and alkaline earth sorbents, chemical reaction can be a promising way of regenerating a sorbent and wet scrubbing gives sulfite, which is easier to regenerate than sulfate. Even when thermal decomposition of sulfite is involved, a small amount of reducing agent may be included to reduce any sulfate present and to inhibit disproportionation.

The metal oxides can be regenerated in these ways also and some of the processes involve regeneration by chemical reaction. The main advantage of the metal oxides is that the sulfates of some of them can be decomposed thermally at a temperature low enough for practical operation, whereas this is not generally true for the alkalis and alkaline earths. Thus the metal oxides offer another option for sorbent regeneration.

PROCESS VARIATIONS

Manganese Oxides

The great affinity of manganese oxides for SO_2, even in the dry state, has led to a large amount of research on manganese-based sorbents. Studies have indicated that at a temperature on the order of 350°C, MnO_2 takes up SO_2 faster than any other sorbent.

Early process work on manganese (in the 1950s) was carried out by the U.S. Bureau of Mines and by TVA, on dry and wet sorption, respectively. In the Bureau tests, (1) sorption in the dry state was favored by elevated temperature [23 to 31 grams SO_2 sorbed per 100 grams MnO_2 during period of 90% SO_2 removal at 130°C (265°F); 53 to 61 grams at 330°C (625°F)], (2) $MnO_{1.88}$ was the best ratio of Mn to O_2, (3) precipitated $MnO_{1.88}$ performed better than an electrolysis product, (4) manganese oxide was superior to the oxides of other metals, and (5) the product of sorption by $MnO_{1.88}$ was the sulfate.

The $MnSO_4$ was regenerated by reaction with NaOH to give Na_2SO_4 which was then subjected to electrolysis, producing NaOH and weak sulfuric acid. Thus the process was somewhat similar to alkali scrubbing methods (see Chapter 7); the main function of the manganese was to allow sorption without cooling the gas. A major disadvantage of the method is the low product acid concentration and the high cost of electrolyzing all the way to NaOH and H_2SO_4 as compared to going only to $NaHSO_4$ as in the Ionic-Stone and Webster process.

In the wet scrubbing method studied by TVA, the gas was scrubbed with a slurry of MnO_2; manganese dithionate $[Mn_2(SO_3)_3]$ and H_2SO_4 were formed and good SO_2 removal was obtained. The acid was neutralized by adding more MnO_2 to the solution after the scrubber, thus forming $MnSO_4$.

Various ways of regenerating the dithionate-sulfate solution have been proposed, most of them involving separation of solid $MnSO_4$ and heating to give MnO_2 and SO_2. Methods for separating the $MnSO_4$ include (1) oxidation of dithionate to sulfate followed by vacuum evaporation, and (2) heating under pressure to decompose dithionate and crystallize sulfate (because of the inverse temperature-solubility relationship of $MnSO_4$). Heating $MnSO_4$ at about 900°C (with 5 to 10% carbon added) in a gas low in oxygen content is reported to have given essentially complete decomposition and a gas containing over 99% of the sulfur as SO_2. Without the carbon, about 10% was present as SO_3.

The use of ores to supply Mn has been studied by some because return from Mn production added to that from SO_2 recovery would help process economics. The absorption efficiency of ores varies widely, however, and impurities retard the MnO_2-SO_2 reaction. Other drawbacks are (1) the manganese market cannot absorb the large quantities that could be generated by power plants and smelters and (2) the ore is not located well for most power plants.

Undersea ore as well as the usual types has been considered. Loose nodules of ore are found at 1,000 to 20,000 ft under water in some ocean areas. Use of such ore has been studied by Kennecott Copper Corp., by a process in which gas containing SO_2 is passed through towers packed with the nodules. The towers are flushed periodically with water to remove the soluble $MnSO_4$. Work on a similar process has been done in Czechoslovakia. These ore-type processes, although originally considered promising, are no longer being developed.

Another possibility is to react MnO_2 with SO_2 in a dry sorption system and then decompose the resulting $MnSO_4$ by heating. Although this simplifies the process, it brings in the problem of activity loss resulting from repeated cycles of sorption and regeneration. The dry sorption-wet regeneration and wet absorption-dry regeneration methods avoid this because the absorbent goes through a dissolved stage in each cycle. Thus gradual sintering and loss of porosity are not a problem.

The major effort on use of manganese oxide has been that of Mitsubishi Heavy Industries in Japan. The process is of the dry sorption-wet regeneration type, similar to that studied by the Bureau of Mines except that ammonia is used to regenerate the manganese oxide rather than NaOH. A flowsheet is shown in Figure 9.1.

Finely-divided activated manganese oxide ($MnO_{1.5\text{-}1.8}$) is fed into an entrained sorption vessel operated at 100° to 180°C, caught in a cyclone, and the greater part recycled (about 7:1 recycle). The concentration of sorbent in the gas is kept at about 150 to 200 g/m³; particle size averages about 40μ. The gas leaving the cyclone passes through an electrostatic precipitator where the dust content of the gas is reduced to less than 10 mg/m³, of which less than 15% is manganese. The loaded sorbent is treated with ammonia in a tower through which air is blown. The reactions are considered to be as follows, where x = 1.5 to 1.8 and i = 0.1 to 1.0:

$$MnSO_4 + 2NH_4OH \longrightarrow Mn(OH)_2 + (NH_4)_2SO_4$$
$$Mn(OH)_2 + \tfrac{1}{2}(x-1)O_2 + (i-1)H_2O \longrightarrow MnO_x{\cdot}iH_2O$$

The process was tested in 1967 on a prototype plant treating 150,000 Nm³/h flue gas (approximately 55 Mw) from an oil-fired boiler, and then tested more recently on a commercial size 150 Mw oil-fired unit of Chubu Electric Power. Both of these test units were on oil-burning power plants unequipped with dust removal equipment. Therefore, although the amount of dust (soot) in the gas was small, it accumulated in the closed loop sorption-regeneration system. To avoid this, a sidestream of the liquor from the reactor was treated to remove the soot by adding a small amount of kerosene to serve as a flotation agent.

The $(NH_4)_2SO_4$ solution separated in the filter was crystallized and finished by standard methods. A good quality of sulfate was made. The web sorbent cake was mixed in with the dry recycle stream and returned to the entrained reactor. The process has the advantages of (1) good SO_2 sorption (over 90% removal from gas containing only 0.11% SO_2), (2) low pressure drop in the sorption system, (3) no cooling of the gas, (4) maintenance of sorbent activity because of the dissolution step, and (5) applicability to existing plants (since sorption is carried out at 100° to 180°C).

The two main problems that eventually caused the termination of the development project at Chubu Electric were the troubles encountered with slurries in the regeneration section and the difficulties of selling ammonium sulfate. If the ammonium sulfate were used as an intermediate material, by any one of several methods described in Chapter 7, there might be a limited market. Not enough data is available to fully evaluate the slurry regeneration problems.

FIGURE 9.1: MITSUBISHI MANGANESE OXIDE PROCESS

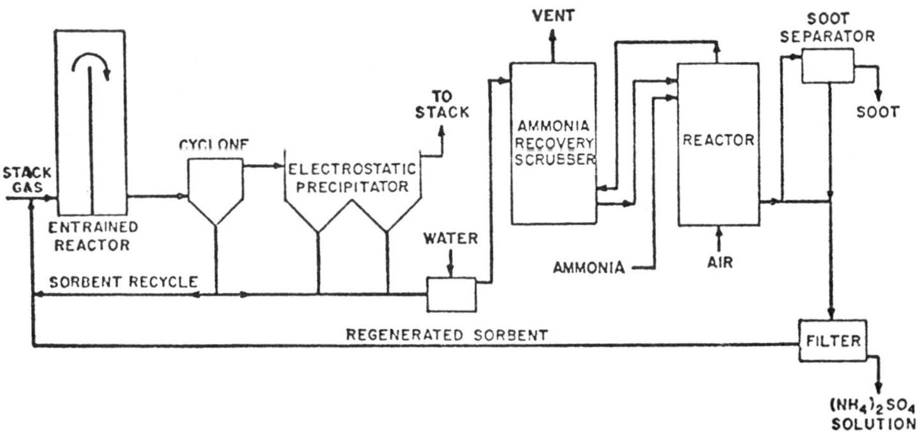

Even if the above problems were alleviated, the resistance by some to the use of manganese oxide, on the basis that escape of the material to the atmosphere would be dangerous to health, would be a serious hurdle to overcome. The amount escaping actually should be quite small, as the low resistivitiy of the oxide makes collection in the precipitaator very efficient, claimed to be on the order of 99.98%. Nevertheless, there has been opposition to installation of even test facilities. The Japanese Institute of Public Health has reported that the toxicity of manganese dioxide is low but that the manganous form produced by reduction during the SO_2 sorption step is quite harmful.

Copper Oxide

A considerable amount of attention has been given to copper oxide as an SO_2 sorbent. In the dry state CuO sorbs SO_2 well at 300° to 550°C and can be regenerated thermally at 700°C and higher. Most of the work has involved regeneration by reduction, however, which has the advantage that the reduction temperature is low enough to allow sorption and regeneration at the same temperature.

The use of CuO was studied in early work by the U.S. Bureau of Mines. The major recent studies have been by the Bureau and by Shell in the Netherlands. Much of the recent work has been concerned with developing a sorbent with a physical structure capable of giving both good sorption and long life.

Since preparation of sorbent granules from precipitates or powders was generally unsuccessful, attention has been centered on impregnation of the copper into a porous, strong support. Alumina and macroporous silica seem to have given the best results as supports. Various impregnating salts have been tested, including $CuSO_4$, $CuCl_2$, $CuBr_2$ and $Cu(NO_3)_2$; $CuSO_4$ seems satisfactory. The optimum Cu content of the impregnated granules appears to be 4 to 6%. The sorption reaction is:

$$CuO + SO_2 + 0.5O_2 \longrightarrow CuSO_4$$

Since the $CuSO_4$ is reduced to the metal during regeneration, oxidation to CuO also takes place during the sorption step:

$$Cu + 0.5O_2 \longrightarrow CuO$$

The two reactions together give a ΔH of -112.5 Kcal/mol (25°C). The Cu oxidation is the fastest so the SO_2 appears to have an adequate amount of CuO to react with except during a very short period at the beginning of the sorption step when presumably only Cu is present, which does not react rapidly with SO_2. The SO_2 absorption is first order in regard to SO_2 concentration and amount of copper in the sorbent.

Rate of sorption is affected by temperature and, of course, by the amount of SO_2 already sorbed. Diffusion within the granules seems to be the controlling kinetic mechanism, although some of the work indicates that chemical reaction rate may be at least partially controlling.

The CuO also sorbs SO_3 under normal operation. Under some conditions, however, the exit SO_3 concentration can be higher than at the inlet. The CuO apparently catalyzes oxidation of SO_2 to SO_3, so that if some malfunction interferes with SO_2 sorption the SO_3 content in the outlet gas can increase to an unacceptable level.

Thermal regeneration would require a temperature somewhat above 700°C to get an acceptable rate. The main problem is loss of absorption efficiency and of sorbent granule strength at the high temperature. Regeneration by reduction can be carried out at much lower temperatures; about 400°C seems to be the optimum for at this level fast absorption can be obtained with a minimum of adverse physical and chemical phenomena.

The reducing agent can be H_2, CO, or a light hydrocarbon although H_2 appears to be the best. Various reduction reactions can be postulated.

$$CuSO_4 + 2H_2 \longrightarrow Cu + SO_2 + 2H_2O$$
$$CuSO_4 + 4H_2 \longrightarrow CuS + 4H_2O$$
$$CuS + H_2 \longrightarrow Cu + H_2S$$
$$Al_2(SO_4)_3 + 12H_2 \longrightarrow Al_2S_3 + 12H_2O$$

The sulfide formation seems to be promoted by low temperature and by use of hydrogen as the reducing agent. Higher temperature promotes sulfide reduction and even higher levels produce $Al_2(SO_4)_3$ reduction; a small amount of $Al_2(SO_4)_3$, on the order of 1 to 2%, forms in the support from reaction with SO_3 during the sorption step. Thus use of hydrogen requires close temperature control in regeneration both to avoid returning unreactive copper to the sorption step and also to reduce hydrogen consumption; the data indicate that the temperature should be quite close to 400°C.

Use of methane reduces sulfide formation; the CH_4 consumption is approximately that indicated by reaction:

$$CuSO_4 + 0.5CH_4 \longrightarrow Cu + SO_2 + 0.5CO_2 + H_2O$$

The regeneration rate, however, is slower than for hydrogen; about 500°C is required for rapid regeneration although 450°C may be adequate with excess methane. The regeneration reduces the sorbent loading from about 5% to 1% S.

Physical deterioration of the sorbent, a major drawback to the alkalized alumina method, does not appear to be a serious problem in CuO sorption. There are several probable reasons for this, including (1) less thermal stress because of regenerating at about the same temperature as in the sorption step and (2) much less reaction of the Al_2O_3 with metal or SO_2, thereby reducing strain due to cyclic change in crystal structure. The problem is not entirely eliminated, however, as the exothermic sorption reactions can cause a sharp increase in sorbent temperature during a short period just after return of the sorbent from regeneration. Such temperature increase must be kept to a minimum because the thermal stress can cause sorbent granule attrition.

The main potential causes for loss in sorbent efficiency appear to be (1) plugging of pores by dust from the gas, (2) maldistribution of Cu in the pores as a result of continual recrystallization (which is aggravated by high temperature), (3) buildup of $Al_2(SO_4)_3$ in the granules, and (4) gradual recrystallization of Al_2O_3. These problems were originally investigated by Shell in the Pernis pilot plant (500 scfm) in the Netherlands operating on a sidestream from an oil-burning furnace.

The unit was of the swing-type with twin sorbent beds; slide-type dampers were used to shift the gas stream from one bed to the other. The cycle time was 1 to 2 hours. Sorbents were tested for long periods, 3,000 to 7,500 hours and 2,000 to 5,000 cycles. Over 20,000 hours of operation were logged at this pilot plant.

Fines formation was not serious but appeared to be worse for the sorbents containing the most copper (overall range was 3 to 9.3%). Some sodium and vanadium deposited from the gas but no serious adverse effects were noted. No Cl^- was absorbed and there was no detectable volatilization of copper as $CuCl_2$. The problem of dust deposition in pores was avoided by using the Shell parallel-bed reactor (see later description).

The Bureau of Mines work has been on a small scale but a conceptual process design has been evolved (Figure 9.2). Sorption is carried out in a fluidized bed and regeneration in a moving bed, with heat exchangers between them since sorption is to be at 300° to 330°C and regeneration at 400° to 440°C. Dust is removed before the sorber by an electrostatic precipitator. Pickup of sulfur in the sorber is from 1% at the inlet to 2.5% at the outlet. In the regenerator 80% of the methane reacts to give a gas containing 38% SO_2. The regenerated sorbent is cooled slightly by preheating the reducing agent (natural gas) and then is heated incidentally to 465°C because of Cu oxidation by the carrier air.

Moving the sorbent back and forth aggravates the attrition problem as compared to fixed-bed operation. The Bureau of Mines estimates a sorbent loss of 0.025% per cycle. For a one-hour cycle time, this gives a sorbent life of about 6 months. Sorbent makeup is estimated to account for about 15% of the total operating cost. The Bureau of Mines design is aimed at making sulfuric acid. The gas from the regenerator, which contains unreacted methane, is mixed with enough additional methane to support combustion, burned, and cooled to condense moisture. It is then reheated and sent to the sulfuric acid plant.

In the U.S., UOP (licensee of the Shell process in this country) has installed a 0.6-Mw pilot plant at Tampa Electric's Big Bend Station. The purpose is to test the system on a coal-fired boiler; previous tests with a dummy reactor in the Netherlands has indicated that the reactor can accept fly ash without fouling.

FIGURE 9.2: BUREAU OF MINES CONCEPTUAL DESIGN FOR SO_2 SORPTION-REGENERATION SYSTEM USING CuO AS THE SORBENT

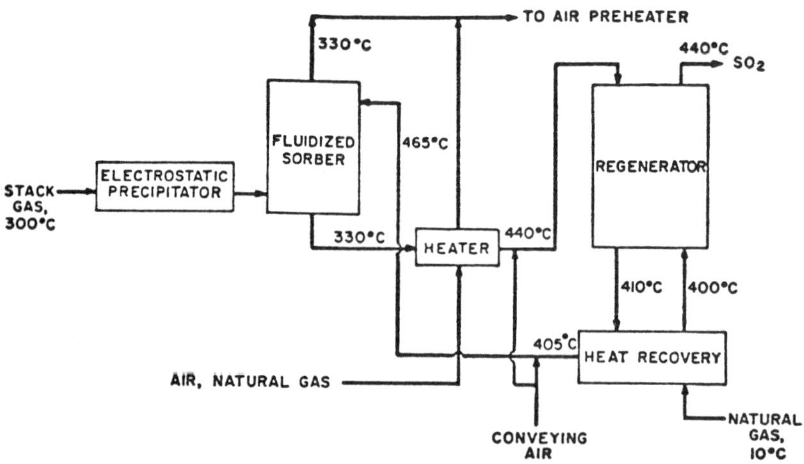

FIGURE 9.3: SIMPLIFIED PROCESS FLOWSHEET OF SHELL UNIT AT SHOWA YOKKAICHI SEKIYU REFINERY

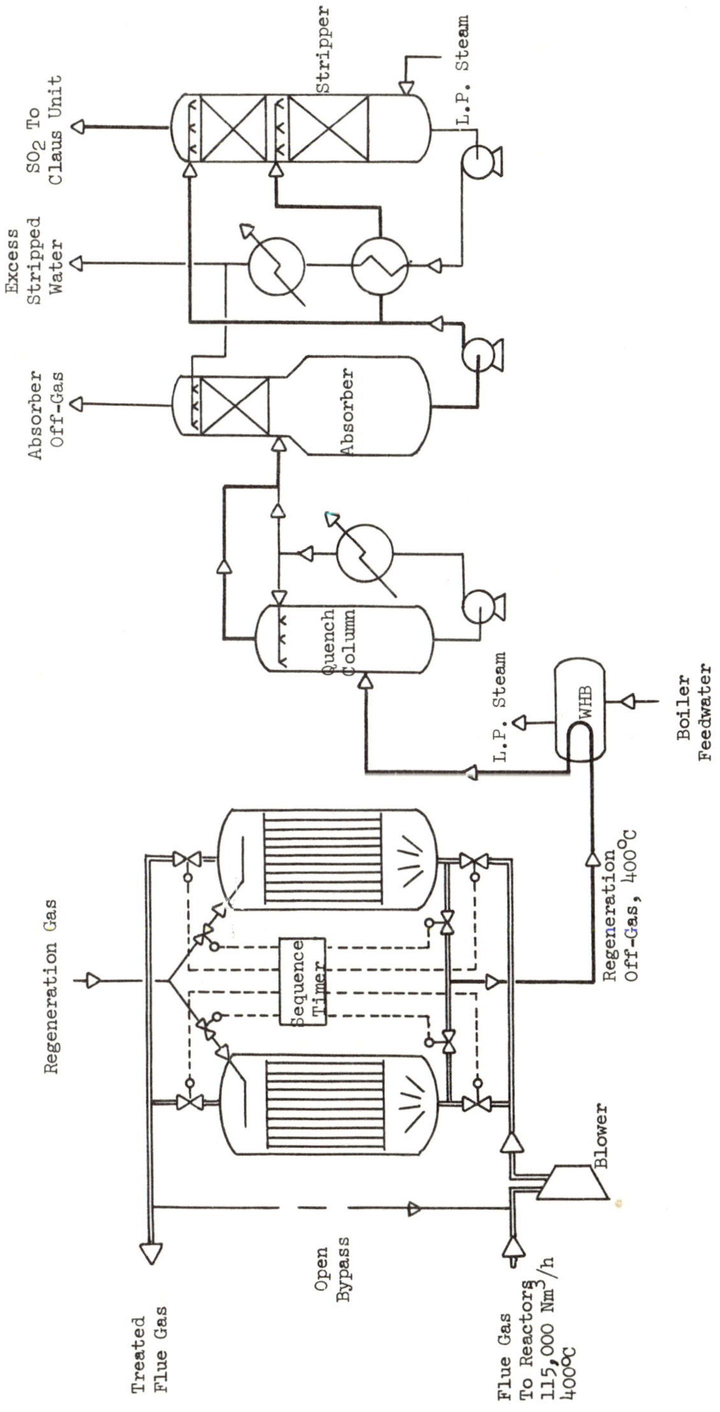

The first full-scale Shell copper oxide process on an oil-fired boiler commenced operation at the Showa Yokkaichi Sekiyu (SYS) refinery in Japan in August 1973. A flowsheet of the SYS unit at Yokkaichi is given in Figure 9.3. Gas at 400°C (750°F) flows through one of two parallel absorbers, where the SO_2 is absorbed by CuO impregnated into alumina granules. After about 2 hours of acceptance flow, dampers are operated to shift the gas flow to the other reactor. A reducing gas (hydrogen at SYS) also at 400°C, is then passed through the loaded reactor, producing a discontinuous off-gas stream containing SO_2, water vapor, and inerts.

The hot SO_2-laden gas evolved during regeneration is passed through a waste heat boiler and then into a water absorber-stripper system. The purpose of the latter is to prevent surges in the gas flow to the Claus unit. By designing adequate liquid surge volume into the absorber vessel, the stripper can be operated continuously. Excess water vapor is also removed in the absorber-stripped system.

At SYS, H_2S for the Claus unit is available from the refinery. For a power plant it would be necessary to convert part of the product sulfur to H_2S or to use a process such as that developed by Allied Chemical which can reduce SO_2 by reaction with methane to the final product-elemental sulfur.

The main design feature of the process is the parallel passage reactor (Figure 9.4) which allows absorption at low pressure drop and without absorbent blockage by particulate brought in with the gas. The absorbent granules are held between stainless steel gauze layers to form very thin sections, into which the SO_2 diffuses at an adequate rate from the gas flowing through the passages between the layers.

The degree of absorption varies over the absorption period, high in the beginning and lower at the end. The average is on the order of 90%. The absorbent granules are quite small in order to get as full a utilization as possible during the absorption period. The absorbent section is also made quite thin for the same purpose.

The absorbent is an extruded type, loaded into the sections by a very careful vibration procedure developed by Shell. The sections are modular; 68 sections are required to make up each 40-Mw reactor at SYS. This has the advantage that sections can be removed for cleaning or maintenance without major interference with operation. The reducing gas is introduced at the top of the reactor. One benefit of this is that buoyancy of the hydrogen promotes plug flow at the low velocity involved.

Methane is not considered by Shell to be a good reducing agent because it gives a reduction rate too slow to allow reduction in the same or less time than required for sorption. Methane is cheaper but if some passes through the regenerator unreacted it complicates operation of the Claus reactors. Hydrocarbons higher than methane can also be used, but complicate operation by forming coke-type deposits on the sorbent; although this burns off during the sorption portion of the cycle, it adds to the undesirable development of high temperature at the beginning of the cycle.

The main mechanical requirement is a set of dampers that will close reliably and seal tight enough to prevent gas leakage. If the stack gas and the hydrogen should mix an explosive mixture might be formed. Tests have shown the gas compositions to be such, however, that actually no explosive mixture results from any proportion in mixtures. Nevertheless, the reactors are purged with steam between cycles and the dampers are carefully designed to prevent leaks.

The dampers are an eccentric butterfly type manufactured by Gebruder Adams in Bochum, Germany. They are circular and sit against a Stellite ring mounted inside the circular duct in a slanted position (top downstream of the lower part). On the damper periphery there is a seal ring made of thin stainless steel sheets and asbestos sheets laminated together with the edges bearing on the Stellite duct ring. There is a groove in the lamellar ring into which steam is introduced; there is always a positive flow of steam into the duct either up or downstream, thus preventing gas leakage since the steam is at considerably higher pressure than the gas.

FIGURE 9.4: PARALLEL PASSAGE REACTOR FOR SHELL CuO PROCESS

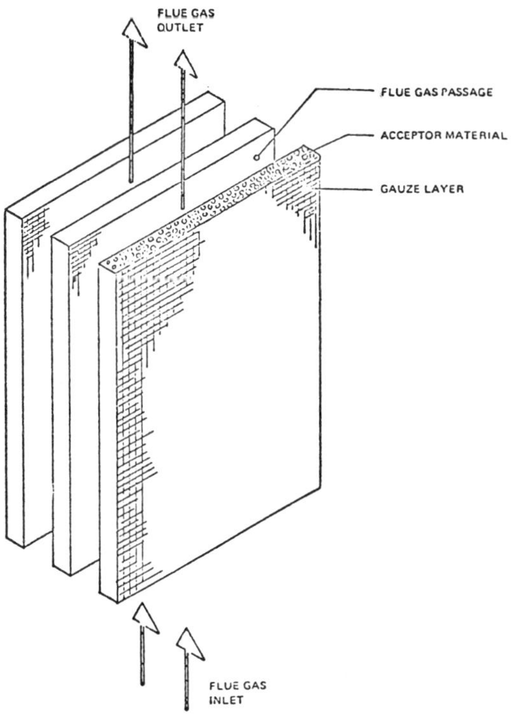

Performance at SYS has been generally good, with an average removal efficiency of 90% and a hydrogen consumption of about 0.2 lb H_2/lb S recovered. The acceptance test was made in September, 1973 during which SO_2 removal was 92% (1,800 ppm inlet).

There is little or no NO_X removal in normal operation. However, Shell is working on a combined SO_2-NO_X system and has had some promising results at Yokkaichi when injecting NH_3 on the upstream side of the absorber. There has been no loss of absorbent activity yet. It is expected that life will be about 3 years.

The primary advantages of the process are: (1) no flue gas reheating is required, (2) no slurry-related scaling and plugging problems, (3) sulfur is the end product, (4) the system has potential for combined SO_X-NO_X removal, and (5) overall plant thermal efficiency could be increased since the air heater could be operated at lower temperatures in SO_3-free flue gas.

However, the system has the following major intrinsic disadvantages: (1) large reactor volume, (2) extra reducing agent required for reducing all the way to copper, (3) inherent retrofit difficulties, (4) requirement for a wet system for SO_2 absorption and stripping. The cost of SO_2 recovery by the process has been estimated by UOP to be competitive with other regenerable processes.

Zinc Oxide

Testing of zinc oxide as a sorbent is currently limited to small-scale work, although pilot-plant studies have been made in the past. The early work by Johnstone and coworkers at the University of Illinois has been described in Chapter 7. Johnstone evidently considered slurry scrubbing to be infeasible, as he scrubbed with Na_2SO_3 solution and then precipitated $ZnSO_3$ by adding recycled ZnO to the scrubber effluent. In more recent work, particularly at TVA (1971), good results have been obtained by scrubbing with ZnO slurry to form $ZnSO_3$, which can be separated and decomposed thermally (as Johnstone did) or regenerated in other ways. The patent literature indicates that Metallgesellschaft in Germany and research institutes in Czechoslovakia have considered a similar approach.

A possible advantage of direct scrubbing is that for high Na^+ concentrations double salt formation ($Na_2SO_3 \cdot 3ZnSO_3 \cdot 4H_2O$) can be a problem in the Johnstone process. The material precipitates and is thus included in the material going to the calciner, where the Na_2SO_3 is either inactive or decomposes to sulfide and sulfate.

Whatever the method for producing the $ZnSO_3$, the problems in calcination are the same. One of these is disproportionation, which occurs when practically any sulfite is heated in the absence of a reducing agent. Johnstone reported 2 to 3% disproportionation which, although low, is troublesome because the disproportionation products build up in the system.

Aerojet General has also studied $ZnSO_3$ disproportionation. It was found that carrying out the decomposition at the relatively low temperature of 275°C, requiring about 2 hours for completion, essentially prevented disproportionation to sulfide and sulfate but that some oxidation to sulfate occurred if there was any oxygen present in the calciner atmosphere. The decomposition rate could be increased by inclusion of water vapor in the heating gas.

From the analogous work on use of magnesia as an SO_2 absorbent, it should also be possible to decompose $ZnSO_3$ without undesirable side reactions if a small amount of reducing agent were included in the calciner charge or in the heating gas. Even if oxidation can be prevented during calcining, there is still the problem of sulfate formation in the sorption step and accumulation in the closed system. One solution is to promote oxidation and decompose $ZnSO_3$, which makes the situation similar to that in use of Mn and Cu.

However, reduction and chemical regeneration, the methods generally used in the Cu and Mn systems, have not been applied to $ZnSO_4$; apparently only thermal decomposition has been considered. Thermal decomposition is attractive because it not only takes care of the oxidation problem but also should reduce capital costs, assuming that the decomposition produced SO_3 and the oxidation section of the sulfuric acid plant could therefore be eliminated. The main drawback to this is that SO_3 dissociates partially to SO_2 and O_2 at high temperature:

$$ZnSO_4 \longrightarrow ZnO + SO_3$$
$$SO_3 \longrightarrow SO_2 + 0.5O_2$$

In work by Aerojet General, $ZnSO_4$ was decomposed at 750° to 850°C but the reaction was slow. Dissociation of evolved SO_3 to SO_2 was negligible at 800°C but about 10% at 850°C. In work some years ago by Catalytic Construction Company and International Minerals and Chemical, a pilot plant study of $ZnSO_4$ decomposition by heating was carried out.

The objective was not SO_2 removal from stack gases but rather conversion of ammonium sulfate, a fertilizer material in oversupply, to sulfuric acid; the $(NH_4)_2SO_4$ was reacted with ZnO at high temperature and the resulting $ZnSO_4$ decomposed to SO_3. Complete decomposition of the $ZnSO_4$ could be obtained at 850°C in a moving bed reactor with a countercurrent flow of solids and with a sweep gas.

Loss of absorbent strength and activity did not appear to be problems in this work but the tests were not of long duration. For application to SO_2 recovery by absorption in ZnO slurry, where the reaction presumably takes place mainly in the liquid phase, loss of ZnO activity should not cause any difficulty.

In such a process, however, there are the problems of getting the soluble $ZnSO_4$ out of solution. Small-scale tests have shown that the sulfite can be oxidized fairly easily with air while in solution. The sulfate then must be crystallized out, perhaps by taking a side stream from the scrubber, oxidizing, concentrating, and cooling. Or the relatively insoluble sulfite could be separated, added to a saturated sulfate solution, oxidized, and the crystallized sulfate separated. Another method, as is currently being done in Japan, is to reduce the pH of the side stream; since the bisulfite is more soluble than the sulfite form of Zn more oxidizable material is available and hence more oxidation occurs.

This applies to complete sulfite oxidation, so that the oxidation of SO_2 in a sulfuric acid plant would be replaced by oxidation of sulfite in solution. The system could also be applied only to the sulfate formed unavoidably by oxidation in the boiler and scrubber, with the main operation being formation of $ZnSO_3$ followed by thermal decomposition. However, from an economic standpoint, it might be better if thermal decomposition of sulfates were included as a process step.

Oxidation of sulfite in solution is a basic process approach that may be applicable to metals other than zinc. In fact, the sulfates of Mn, Cu, and Al all are reported to decompose at a lower temperature than for $ZnSO_4$. The solution step eliminates the loss in activity over extended cycling that is the main problem when both sorption and regeneration are carried out in the solid state, and production of SO_3 rather than SO_2 should be a major advantage in making sulfuric acid. Some dissociation of SO_3 to SO_2 could be taken care of by scrubbing out the SO_3 in strong H_2SO_4 and recycling the SO_2 back to the main scrubber. Moreover, SO_2 could be recycled back to the $ZnSO_4$ decomposer to reduce conversion of SO_3 to SO_2, or it could be recycled back to the oxidizer to reduce pH and increase oxidation.

Operation of such a process should be carried out in an opposite manner to that of sulfate reduction (as in the Shell and Bureau of Mines CuO methods) or $MgSO_3$ thermal decomposition. Instead of adding reducing agent, excess oxygen should be present to inhibit SO_3 dissociation, and oxidation should be promoted throughout the process.

There have been some efforts toward using ZnO in a dry sorption system but the material is inactive except in the presence of a relatively large amount of water vapor at low temperature [analogous to the Still process with $Ca(OH)_2$]. Aerojet General has studied sorption under these conditions in a fluidized bed. The reaction is considered to be:

$$ZnO + SO_2 + 2.5H_2O \xrightarrow{H_2O} ZnSO_3{\cdot}2.5H_2O$$

The gas was first cooled and humidified in a prescrubber and then passed through the fluidized bed. Up to about 68% of the ZnO was converted to $ZnSO_3{\cdot}2.5H_2O$ in batch tests up to the breakthrough point of 90% removal.

Iron Oxides

Iron oxide has been included with other metal oxides in some of the broad studies on the subject. It appears to rank well with MnO_2 and CuO in affinity for SO_2 and the sulfate has an especially low decomposition temperature. In the AVCO study, Cu, Mn, and Fe were the favored oxides and the final six from the TRACOR screening included iron oxide. Little work has been done specifically on iron oxides, however. Siemens-Schucketwerke in Germany has done small-scale work with iron oxide as an SO_2 sorbent, dry sorption followed by thermal regeneration, but the development does not appear to have been carried very far. TVA made exploratory small scale tests with iron oxide slurry solutions but these were unpromising in that absorption efficiency was low.

Other

Various other metal oxides have been considered but little work appears to have been done on them. Of the six that survived the TRACOR screening, Ce, Cr, Fe, Co, Ni, and Cu, only Cu and Fe have received much attention. Of others that have been considered, Esso Research and Engineering has patented the use of vanadium oxides (V_2O_3 or V_2O_4). At 800°F, these are said to react with SO_2 to give $VOSO_4$. Regeneration is carried out by heating to 1100°F in an atmosphere of SO_2 (to prevent oxidation to inactive V_2O_5); the product is SO_3. The SO_3 is scrubbed out with 60% H_2SO_4 to produce 98% H_2SO_4 and the SO_2 recycled to the regenerator.

Work has been done by Brookhaven National Laboratory on use of uranium oxide as an SO_2 sorbent:

$$UO_2 + SO_2 + O_2 \longrightarrow UO_2SO_4$$

Good sorption was obtained at 300°C and the sorbent picked up about 80% of the stoichiometric amount of SO_2 before breakthrough. Oxygen is necessary for sorption below 500°C. The process proposed has the unique feature that the UO_2SO_4 is reduced with elemental sulfur.

$$S + UO_2SO_4 \longrightarrow UO_2 + 2SO_2$$

Thus the SO_2 would either have to be reduced to sulfur or sulfur purchased for the regeneration step. In the latter case, one ton of purchased sulfur would join with a ton of recovered sulfur to give two tons of sulfur as sulfuric acid.

ADSORPTION

The use of activated carbon or other adsorbents to remove sulfur oxides from waste gases would appear to be attractive because adsorbed compounds normally can be removed from the adsorbent by simple thermal stripping at relatively low temperature; if this could be accomplished in SO_2 removal, the troublesome side reactions encountered in most regeneration processes might be avoided. Unfortunately, the highly active adsorbent surface catalyzes oxidation and the resulting SO_3 reacts with moisture in the gas, forming H_2SO_4 in the pores of the adsorbent because of the high dew point of H_2SO_4 in the gas.

Thus one of the main problems in using adsorbents is getting the product acid out of the pores. Various methods for doing this have been used or proposed including: (1) washing out with water, (2) heating the loaded adsorbent to induce reaction between carbon and acid so as to give a rich SO_2 stream suitable for acid production, and (3) reaction with a reducing gas to give sulfur as the final product.

Several types of adsorbents have been studied, but the generally higher adsorptive capacity of activated carbon has made it the favorite. The principal current work on carbon processes is in Japan, the United States and West Germany. Some of the processes have been studied on a prototype scale and two are currently installed on full-scale, commercial boilers in Japan.

CHEMICAL AND PHYSICAL FACTORS

Sorption of SO_2 by activated carbon can be both chemical and physical. The degree of each is affected by the presence of oxygen in the gas; various studies have shown that an oxygen-free carbon surface has limited total SO_2 sorption capacity. The presence of water vapor is also important, presumably because it reacts with the SO_3 formed by reaction of sorbed SO_2 with O_2, thus increasing the degree of chemisorption at the expense of physicosorption. Fortunately, most stack gases contain O_2 and H_2O in amounts adequate for this to take place.

In the absence of O_2 and H_2O, chemisorption in the range from 50° to 300°C is quite small, on the order of 1% by weight. The chemisorption is independent of temperature; in contrast, physical adsorption decreases sharply with increase in temperature, from several percent at 60°C to none at 250°C.

With O_2 and H_2O in the gas, fairly good sorption rate and capacity have been attained, although metal oxides appear to be superior in this respect. The sorption capacity of carbon obviously depends on many factors, but a sulfur loading of 2 to 7% seems to be a practicable range. In contrast, a sorbent such as CuO or MnO_2 can be loaded to 20% sulfur or more. Thus, the inventory of sorbent and the size of the sorption vessels must be relatively large in carbon processes.

The sorption temperature should be as low as possible, of course, in order to get maximum sulfur pickup per cycle. Several of the efforts, however, have been aimed at retaining at least part of the stack gas buoyancy and therefore the sorption temperature used has been higher than that in wet scrubbing processes. Another consideration is that cooling the gas may cause corrosion from condensation of SO_3 already present in the gas, particularly if indirect cooling is used. This can be prevented by using a direct cooler to scrub out the SO_3 with water, which is practiced in some processes by passing the gas through a cooling tower where it is cooled by contact with product acid from the adsorption section, thereby concentrating the acid. In the large-scale test units and commercial installations, a fairly wide range of sorption temperature has been used.

Sorption rate can be increased by impregnating the adsorbent with a metal salt. Various materials have been used, mainly salts of metals such as copper, iron, nickel, cobalt, manganese, chromium and cerium, which are the metals whose oxides have been found to be good SO_2 sorbents. Presumably their function when used with carbon is to catalyze SO_2 oxidation and thereby clear the adsorptive surfaces so that more SO_2 can be adsorbed. It has also been claimed that impregnation with catalyst maintains sorption efficiency over extended cycling.

The type of carbon is an important consideration, both in regard to immediate sorption efficiency and to maintenance of efficiency in extended operation. Various types have been used, ranging from specially prepared, highly adsorptive carbon to untreated coal. Tests with coal have shown that the adsorptive capacity varies only slightly from one coal to another. The capacity, of course, is much lower than for activated carbon, 2.5 to 3.0 grams SO_2 per 1,000 grams coal versus 140 grams for activated carbon, but coal is so much cheaper that its use has been proposed. Most investigators, however, have used an activated carbon, although the degree of activation (development of porosity) before use has often not been as much as is usual for activated carbon. The regeneration method has in some cases helped in this respect by renewing surfaces and increasing porosity.

Although the principal work has been done with carbon, other adsorbents have also been studied. In tests with silica gel, only about one-third as much SO_2 was sorbed as with activated carbon, but it is reported that the silica gel is more easily activated. Bentonites have been tested, as well as molecular sieves containing synthetic zeolites; the latter sorbed up to 29 grams SO_2 per 100 grams adsorbent. It has been reported that adsorbents remove nitrogen oxides and gaseous chlorides from the gas as well as sulfur oxides; however recent data on commercial installations do not confirm the reports.

In addition to limited capacity, adsorption processes have the drawback that gas velocity through the sorption vessel must be relatively low in order to allow time for the SO_2 to diffuse into the adsorbent pores. Sorber gas velocities range from 1 to 4 feet per second. Another problem is susceptibility of the highly reactive carbon to oxidation by the oxygen in the gas. In one earlier case, this was so rapid as to produce a runaway increase in temperature and consequent burning of the carbon.

The carbon is also subject to the other drawbacks of solid sorbents, including attrition in transport between sorber and regenerator, stoppage of pores by dust in the gas, and loss of activity due to formation of unreactive compounds on pore surfaces over a long period of use. Nevertheless, adsorption by carbon has some important advantages, including relative simplicity and freedom from side reactions in regeneration.

PROCESS VARIATIONS

The major carbon processes employ various methods of handling solids in the sorber and the regenerator. The problem is to get good contact between gas and adsorbent in the sorber without developing hot spots, and to regenerate with a minimum of chemical and physical stress on the adsorbent. Although there are some differences in design of the adsorption equipment, the main variations have been in the regeneration step.

Regeneration by Heating

Since the sulfuric acid formed in the sorption step is so intimately associated with the carbon, a convenient method of regeneration is to merely heat the loaded carbon so as to induce reaction between the carbon and acid. At the temperatures normally used, the reduction is only to SO_2.

An inert gas is used in some of the processes to carry heat into the carbon granules and to carry the SO_2 out. The gas is usually made by burning a fuel without an excess of oxygen. Also, steam has been proposed in at least one patented process and in another the power plant stack gas is used (with pretreatment to remove oxygen). Hot sand is mixed with the loaded carbon in still another variation.

Reinluft: The first of the major carbon processes, the Reinluft (clean air), employs thermal regeneration. The method was developed in Germany, beginning in about 1957.

In the first version of the method, both sorption and regeneration were carried out in a single tower with the carbon moving downward continuously through sorption sections at the top and desorption sections at the bottom; these sections were made up of several parallel tubes or shafts. Gas at 250° to 300°F entered the first sorber section, flowed upward through the second, and out the top to the stack. The loaded carbon flowed down to the desorber where inert gas at about 750°F was passed through to strip out SO_2; at this temperature, the carbon reacted with H_2SO_4 deposited in the pores to give SO_2, CO_2 and H_2O.

$$2H_2SO_4 + C \longrightarrow 2SO_2 + 2H_2O + CO_2$$

Thus although carbon is consumed in the regeneration, the amount is only 10% of the weight of SO_2 removed from the gas.

Even this small consumption, however, is costly if an expensive grade of activated carbon is used. In the Reinluft work, it was found that cokes from low temperature carbonization of young fuels, such as peat and lignite, were satisfactory although not initially very active. In the cyclic operation the consumption of carbon in each cycle enlarged the pores and thus developed high activity after a few cycles.

Reinluft pilot plants were operated at the Volkswagenwerk plant in Germany in 1959 and at Duisburger Kupferhutte (near Essen) in 1962. These were followed by larger test units at Carbosulf (Cologne, 1965-1967) and at the STEAG power plant in Lunen (1966-1968); gas volumes were 32,400 and 19,400 scfm, respectively.

A major problem in the larger plants was self-ignition of the carbon in the sorption section, which made it necessary to shut down and remove the sorbent in order to quench it. Several things contributed to the ignition, including fine sulfur in some of the gases treated (gas from Claus furnace at Carbosulf) and irregular downward movement of the carbon in the feed hopper which left some of it in contact with air for too long a time.

Several means were tried in efforts to eliminate the problem. The final solution adopted was to divide the feed hopper, and use of flow controls to ensure uniform downward flow of carbon in all shafts.

Other changes have been proposed by Chemiebau, which has licensed the original Reinluft process. Chemiebau proposes five units for a 300 Mw power plant. The adsorber in each unit would have a cross sectional area of 992 ft^2.

The Chemiebau design is shown in Figure 10.1. The adsorbent carbon, between 0.1 and 0.3 inches in particle size, flows downward countercurrent to the stack gas. Undersize is screened out after the adsorber and used as fuel in the boiler. In the desorber, hot gas desorbs the SO_2 in an internal heat exchange arrangement by which the carbon is first heated and then cooled. A side stream of SO_2 rich gas from the circulating gas system is sent to a sulfuric acid plant.

The Reinluft method has several advantages, including low cost of sorbent, simplicity of regeneration, and commercial grade of product. Offsetting these is the relatively large production of fines, both from attrition and also because the consumption of carbon in each cycle enlarges pores and weakens the granule structure. No commercial units are currently planned, although Commonwealth Associates has licensed the process in the U.S.

In Figure 10.1 which illustrates the Chemiebau design for the Reinluft plant, both the adsorber and desorber contain a set of vertical tubes through which adsorbent carbon moves downwardly.

FIGURE 10.1: CHEMIEBAU DESIGN FOR REINLUFT PLANT

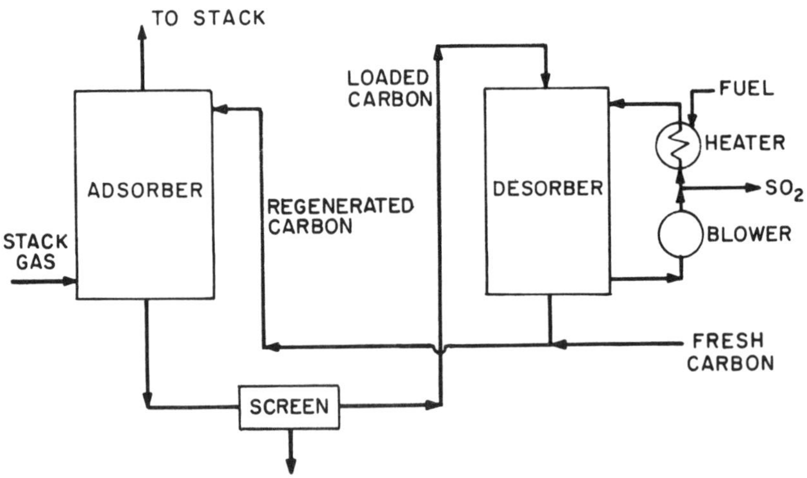

Bergbau-Forschung: A variation of sorption on carbon followed by thermal regeneration has been developed by the Bergbau-Forschung, a research institute in the hard coal industry located in Essen, Germany. The Bergbau improvement over the Reinluft is mainly in design of the regeneration system to avoid local overheating. In the Reinluft a hot gas was introduced whereas in the Bergbau hot sand is mixed in with the carbon.

FIGURE 10.2: BERGBAU-FORSCHUNG CARBON PROCESS WITH THERMAL REGENERATION

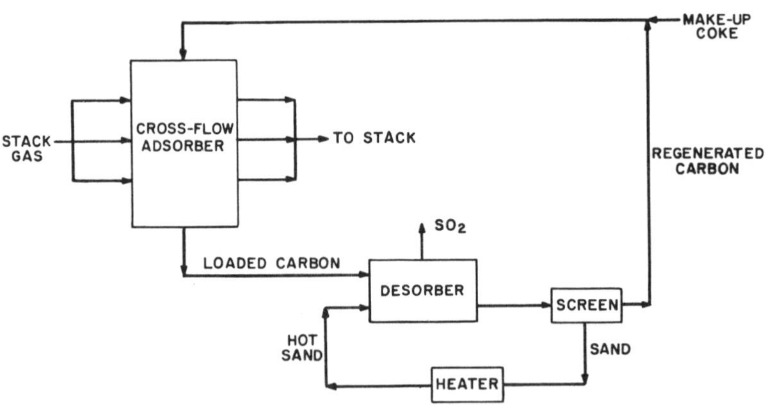

Bergbau did only small-scale and pilot-plant (2,000 m^3/hr) work on the process. The German federal government then became interested and decided to fund a demonstration system; STEAG agreed to installation of the unit on one of the Lunen boilers. Startup was initially scheduled for January 1974 but various problems caused delays. Some of the equipment was operated in mid 1974 but because of several mechanical problems including inadequate bucket elevator operation, failure of oil pump bearings, insufficient sand-char separation capacity, and damper leakage, start-up of the full system did not occur until December 1974.

The Luen unit is sized to handle 90,000 scfm, equivalent to about 45 Mw. In the adsorber the gas passes through a bed (1.75 meters) of carbon pellets (nine millimeters in diameter). The temperature of the bed is 250° to 300°F. The absorber is constructed in such a way that the gas passes down into the center of an annulus, formed by a downward moving annular column of carbon, and horizontally outward through the carbon. The carbon moves out of the adsorber, is elevated to the top of the regeneration unit, mixed with hot sand (about 1100°F), and passed down through a reactor in which the heat picked up from the sand causes the adsorbed H_2SO_4 in the pores to react with carbon and become reduced to SO_2. The acid loading on the incoming char is about 10% by weight and the SO_2 content of the evolved gas (mixture of SO_2, CO_2 and H_2O) from 20 to 30%. The sand is screened out and the carbon recycled.

The hot SO_2 stream passes to a unit designed by Bamag (a German engineering company) in which the SO_2 is reduced to sulfur over a catalyst by reaction with city gas (reported to be coke oven gas). It is expected that a sulfur yield of 93 to 94% can be obtained.

The adsorber dimensions are 7.5 meters overall diameter of the annular bed and about 10 meters in height. This illustrates one of the main drawbacks to carbon adsorption, the low gas velocity (approximately 1 foot per second) allowable and the resulting large reactor size.

The carbon is made from hard coal by a process developed by Bergbau. It is said to be relatively attrition resistant and to have a high ignition temperature, both of which make it superior to that used in the Reinluft process. The process for making it is reported to be critical but the type of coal used is not necessarily important. Careful temperature control is required during the coking step. Production cost is high but should be reduced if a large demand develops.

Gas for the unit is taken from the 250 Mw Lunen boiler. The SO_2 content is about 1,200 ppm and the design removal efficiency is 80%. Since one of the Bergbau advantages is that it also has dust removal capability, various inlet dust concentrations will be tested. The objective is 150 mg/Nm^3 in the outlet gas, which is the regulatory requirement. The main problem is increased pressure drop at high inlet dust loadings; the limits in this respect will be determined.

Two sand heating systems will be tried at Lunen, one in which the combustion gas (city gas), part of which is recycled, entrains the sand and carries it up through a large pipe back to the top of the regenerator. The sand is disengaged from the gas in a cyclone separator. The recycled combustion gas passes from the cyclone to a gas-to-gas heat exchanger (to preheat combustion air), through water sprays, to the carbon cooler, and then back to the sand-heating circuit. There is some gas leakage which bleeds off excess combustion gas from the system to a stack. The alternate heating system is a fluidized sand heater at the base of the pipe that carries the heated sand upwards.

In the earlier work reported above, the pilot plant was operated for over two years (6,000 hours actual running time). The carbon loss averaged 1% per cycle, for 30 hour cycle periods at 80% removal. At this rate, about three charges of char would be required per year. The presence of fly ash had no effect on char attrition, even at 4 to 5 g/Nm^3 inlet. Ash collection was good (150 mg/Nm^3 or lower in the outlet gas).

Adsorption of NO_X will also be tested at the Lunen unit. In a nine week earlier pilot-plant test 40 to 60% of the NO_X was removed and only nitrogen was found in the gas from the regenerator. Fundamental studies have indicated that the char probably reduces NO to N_2 at the particular temperature and pressure. At low temperature, below 100°C, this does not occur and the only effect is adsorption of NO_2.

The main questions on the Bergbau process are effect of continued operation on gas pressure drop, char attrition rate through all the handling steps involved, operability of the sand heating and hot sand transfer system, and turndown capability of the SO_2 reduction unit. In regard to cost, the very large adsorber size, the complicated char regeneration system, and the need for a gaseous reductant are major drawbacks. Moreover, the Lunen plant is designed for 80% removal of 1,200 ppm SO_2, whereas in the U.S. over 90% removal of 2,500 ppm or higher is being required in some cases; the Bergbau cost is likely to increase rapidly with higher removal efficiency. On the other hand, elimination of the reheat requirement and the ability to remove dust and NO_X from the gas are major advantages.

In a parallel effort in the United States, Foster Wheeler (a licensee for the Bergbau Forschung dry adsorption process) designed, engineered and is constructing a large scale demonstration unit (20 Mw) for startup in 1975 at Gulf Power's Scholz station. There are several design differences between the two demonstration units, including a two stage adsorber and a fluidized bed sand heater being used at Scholz; however, the main difference will be in the SO_2 reducer.

Foster Wheeler has developed a proprietory process (Resox) in which SO_2-rich gas is passed through a vessel containing crushed coal. Here, SO_2 is reduced to gaseous elemental sulfur and the liberated oxygen combines with a portion of the coal to form carbon dioxide. Reduction in the Resox reactor is carried out at 1200° to 1500°F. The gases leaving the reactor enter a sulfur condenser where gaseous sulfur is condensed and the tail gas recycled back to the adsorber to remove residual sulfur values.

Sumitomo-Kansai: In Japan the Sumitomo Shipbuilding and Machinery Co. and the Kansai Electric Power Co. have collaborated in developing a carbon process based on thermal regeneration. A pilot plant (10,000 m^3/hr) was installed at Kansai's Himeji Second power station in October, 1968, and tested through June, 1969. This led to a larger scale test on a 250 Mw boiler at Kansai's Sakai Port power station. Operation on a side stream equivalent to 62.5 Mw was begun in February, 1972, giving a sulfuric acid production of about 20 tons per day (80% removal; 2% sulfur in fuel oil). The demonstration unit is still in operation.

The process employs cross-flow adsorption as in the Bergbau-Forschung method and regeneration is accomplished by a hot inert gas as in the Reinluft. The flowsheet for the process is shown in Figure 10.3

Ash in the gas from an oil-fired boiler is removed in a mechanical collector ahead of the adsorber. The total flow of gas to the recovery system (150,000 acfm) is split between two parallel vessels containing moving beds of granulated carbon. The system was designed to remove 80% of the SO_2 when oil containing 2% sulfur is burned. Actually, oil with 0.75% sulfur is now being used to conform to air quality regulations and the inlet and outlet SO_2 concentrations are 400 and 20 to 40 respectively.

The carbon is introduced at the top of the adsorber and moves downward at a rate controlled by a rotary lock at the bottom while the gas flows across the bed; inlet gas temperature is 250°F. Low gas velocity is required (approximately 1.5 ft/sec) which makes the plant extraordinarily large for the equivalent 62 Mw gas flow. The capacity of the activated carbon for adsorption of SO_2 is best when the charge is new; the activity decreased to about 90% of the initial value and then leveled out. Attrition of the granulated carbon is excessive and is at least equivalent to annual replacement of the charge; Sumitomo is studying re-agglomeration of the fines.

The loaded adsorbent is transferred by an inclined conveyor (Redler type) to the top of a single desorption tower where once again the moving bed is contacted with gas in a crossflow configuration. For desorption, the inert sweep gas is formed by controlled combustion of butane and heated, when recycled, to about 660°F in a heat exchanger. The SO_2 concentration of the desorber exhaust gas is controlled in the range of 10 to 20%.

A sidestream from the recirculating sweep gas in the desorber is processed in a miniaturized sulfuric acid plant to produce 20 tons/day of 98% acid. The gas is passed through high-efficiency cyclones, diluted with air, and dried in a strong acid-scrubbing tower before it enters the converter. The acid plant is a conventional contact type with vanadium catalyst. The tail gas from the acid plant is returned to the SO_2 removal system.

There are apparently no plans for installation of the Sumitomo process on larger scale. Operation of the large prototype system at Kansai will provide a good basis for comparing the merits of this process with alternative methods. At least 6 months of additional operation will be necessary to establish carbon activity and makeup requirements.

The major advantage of this process is the production of concentrated sulfuric acid in a system that may be applied to existing power plants without appreciable reduction in gas temperature. Disadvantages include high investment because of the need for large gas contacting equipment in addition to a sulfuric acid plant, high operating cost because of attrition of carbon in the moving-bed system, and the need for precise atmosphere control in the desorber.

Regeneration by Washing

Adsorption accomplishes what many researchers have sought, namely, oxidation of SO_2 at a temperature low enough to allow adding the process on to an exist-

FIGURE 10.3: SUMITOMO PROCESS—SAKAI PORT

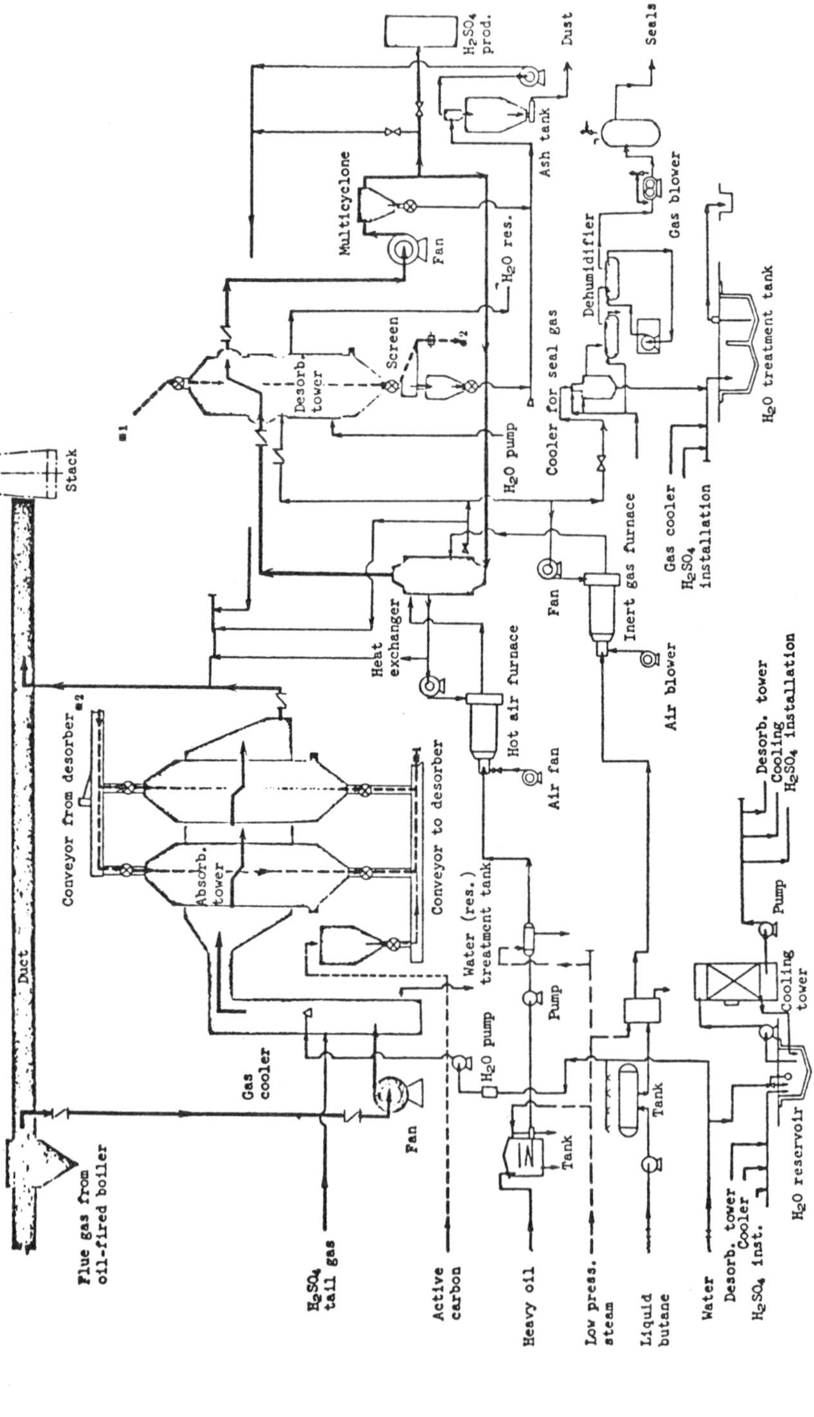

ing plant. Moreover, conversion of SO_3 to H_2SO_4 is accomplished in the same operation, thus concentrating the SO_2 to the necessary degree and producing acid from it all in one step. Each carbon granule, in effect, becomes a miniature plant that is equivalent to the complicated series of equipment assemblies required in more complex processes—sorption, desorption, sorbent recycle, SO_2 gas cleaning, SO_2 oxidation, and SO_3 absorption.

Such a process would seem to be ideal, and since the acid is already produced, reducing it back to SO_2 as in the process just described appears quite illogical. For this reason many research efforts have been aimed at getting the acid out of the carbon granules without decomposing it.

Washing with water has been the main approach but the results have been discouraging because of the large amount of water required and the consequent low concentration of the product acid. The acid formed in the granules is probably lower than commercial grade in the first place; although no measurements appear to have been made, it should be even lower in strength than the 80% obtained in the Monsanto catalytic oxidation process (see next chapter), because the condensation temperature is lower in the adsorption method. The additional wash water brings the concentration down to a very low level on the order of 5 to 15%.

The practicality of making and using such low strength acid is dubious. Concentration to an intermediate level may be feasible, but going to the commercial grade (92 to 98%) would be a very difficult operation. Probably the best approach is to concentrate to an intermediate level and use the acid in a chemical or fertilizer plant located adjacent to or near the plant emitting the SO_2. Phosphoric acid production, for example, can be carried out with sulfuric acid at as low as 60% concentration.

Various methods have been made to use the waste heat in stack gases for concentrating the acid. Thus in a power plant the weak acid from the recovery unit could be used in a venturi to remove dust and to cool and humidify the gas. The main drawbacks, as compared to removing dust in an electrostatic precipitator, are the loss in gas buoyancy resulting from the cooling, the possible difficulty in separating dust and acid, and the probable extraction of impurities from the ash by the acid. As an alternative, a precipitator could be used to remove the dust and an evaporation column to concentrate the acid. This eliminates the opportunity of using a wet scrubber for dust removal, which usually is economical if a wet system is involved anyway, but the overall arrangement may be the best.

An incidental advantage of concentrating the acid by heat recovery from the stack gas is the lower adsorption temperature, which should improve the adsorbent capacity. The situation is not clear, however, as to the net effect of adsorption temperature. There are some data to indicate that the rate of SO_2 oxidation may become limiting at low temperature and that use of a catalyst impregnated in the carbon pores may therefore by required. However, work by Hitachi in Japan (see later discussion) indicates that although the rate of SO_2 oxidation decreases with temperature, the decrease is more than offset by increased physical adsorption. The Hitachi data are shown in Figure 10.4.

FIGURE 10.4: ADSORPTION OF SO_2 IN TWO HOURS ON ACTIVATED CARBON IN THE PRESENCE OF O_2 AND H_2O

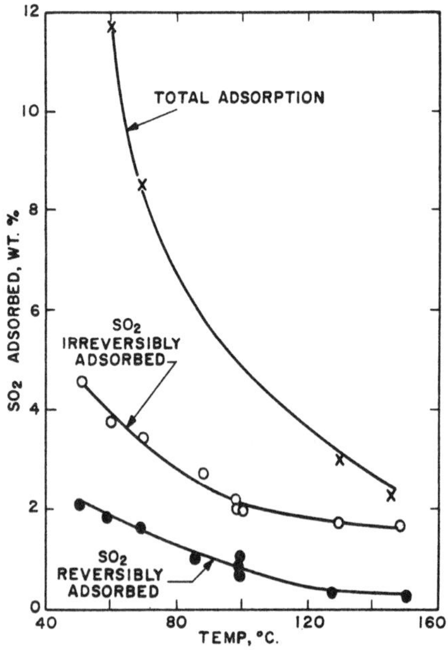

The reversibly adsorbed SO_2 is physically adsorbed and can be decreased by reducing the pressure. The irreversibly adsorbed SO_2 is mainly present as H_2SO_4, removable only by washing or heating.

It can also be argued that concentrating the acid by heat recovery from the gas is impractical because cooling the gas makes it necessary to reheat it, and the heat for reheating might as well be used in the concentrator to simplify the situation. The validity of this depends on several factors, including relative amounts of heat recovered from the gas and put back into it, size of the heat exchangers involved, the relative degrees of the mist elimination problem, means for applying supplemental heat beyond that recoverable from the gas, value of increased sorption capacity, and condensation problems in the adsorber. A complete analysis of the situation would require a considerable amount of further study.

Application of the washing method to sulfuric acid plants has the advantage that the weak acid from the washing step can be used without concentration, as diluting liquor in the SO_3 absorber. The use of catalysts seems particularly applicable when the regeneration method is washing.

Although catalyst impregnation of the carbon has been specified (in patents) in conjunction with thermal regeneration, the burnout of carbon and formation of unrecycled fines in thermal regeneration would appear to make catalysts inappropriate. In regeneration by washing, on the other hand, the longer life of the adsorbent and the generally lower temperature level makes catalysis more promising.

Corrosion is an obvious problem in wet regeneration processes because of the highly corrosive nature of weak sulfuric acid. The preferred materials are rubber lined or plastic coated steel.

Lurgi (Sulfacid): The Sulfacid process, developed by Lurgi Apparetebau Gesellschaft in Frankfurt, Germany, is somewhat unique in that adsorption and desorption are carried out simultaneously in the same adsorbent bed. The bed is washed continuously to remove H_2SO_4 as fast as it is formed. A flowsheet for the process is shown in Figure 10.5.

As shown in the flowsheet, the gas is first cooled, by scrubbing with weak acid from the adsorber, to a temperature of 60° to 70°C. The venturi shown is used when dust removal from the gas is required. The cocurrent flow in the venturi limits the concentration attainable to 25 to 30%; it is reported that if a countercurrent cooler-evaporator is used the concentration can be increased to about 70% with power plant stack gas.

FIGURE 10.5: SULFACID PROCESS FOR REMOVING SO_2 FROM STACK GAS

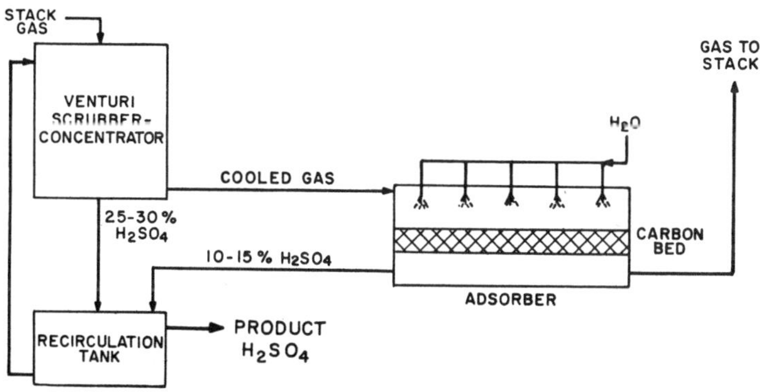

For sulfuric acid plants, where the incoming gas is at lower temperature, the scrubber is used mainly to saturate the gas and provide the required moisture. Waste steam can also be added for this purpose. The cooled and saturated gas passes into the adsorber where it flows through a shallow bed of carbon cocurrently with the wash water. The carbon is a high quality, activated type, with promoters impregnated in the pores. The constant washing of acid from the pores appears to maintain a high level of activity.

After several years of pilot plant operation there was no loss of adsorbent effectiveness; however, since this was on dust-free waste gases it is not known what effect dust from power plant or smelter operations might have. No gas reheat is shown, as the plants of the type to which the process has been applied (see below) commonly emit waste gas at a temperature little higher than ambient.

Since the prescrubber introduces waste gas impurities into the product acid, Lurgi proposes using the acid for fertilizer production, ore leaching, metal pickling, and other rough uses. It is proposed alternatively that the crude acid be cracked to SO_2 and steam at 800° to 900°C and the SO_2 reconverted to clean H_2SO_4.

Several small Sulfacid units are in commercial operation, on waste gases from industrial boilers and from plants making sulfonation products, sulfuric acid, and titanium dioxide. The problems in application to large, coal-fired boilers do not appear to have been worked out. The amount of carbon required for a 500 Mw boiler would be on the order of 3,500 tons (0.25% SO_2 in inlet gas; 95% removal) at a cost of probably over $500 per ton. Pressure drop in the adsorber is expected to be only about 8 inches H_2O. The space requirement would be on the order of 65 by 180 feet and the maximum height about 130 feet.

The Sulfacid method is a relatively simple one with several outstanding advantages. It is not as well suited to production of commercial grade acid as some of the other processes, perhaps, but for sulfuric acid plants and for making crude acid to be used in adjacent fertilizer or other plants that can use such acid, the process is especially appropriate.

Hitachi: One of the major carbon processes has been developed by Hitachi, Ltd. in Japan, in pilot and prototype test units located at the Goi station of the Tokyo Electric Co. near Tokyo. This led to the construction of a large scale (385,000 acfm) system to treat one-fourth of the gas from one 600 Mw oil-fired boiler at Tokyo Electric's Kashima station. The process employs fixed beds of carbon with the gas flow alternated from one to the other; use of the fixed beds eliminates the cost of moving adsorbent from sorption to regeneration and back again and avoids the resulting mechanical attrition and granule breakup.

In order to avoid deposition of particulates in the packed bed, gas to be treated by the Hitachi process should be relatively free of dust. At the Kashima plant, Lurgi electrostatic precipitators reduce the dust loading to less than 0.02 gr/scf.

The adsorption system consists of two parallel trains of three towers in series. Figure 10.6 depicts the size of the absorption towers. Each tower has three fixed beds of activated carbon (particle size about one-fourth inch) and operates at a gas velocity of about 1.5 ft/sec. The vessels are mild steel coated with a phenolic resin. The operation of each tower is cyclic with adsorption, washing, and drying occurring sequentially in each bed.

During the wash stage, gas flow is diverted by louvered dampers to another bed. The total cycle time is 63 hours: 7 hours wash, 20 hours dry (adsorption also takes place during this period), and 36 hours adsorb. The adsorbed SO_2 oxidizes in place to SO_3 and is collected as 20% H_2SO_4. Removal efficiency is reported to be about 80%.

In early prototype tests at the Goi plant, dilute acid was concentrated to 65% by submerged combustion. However, operating problems and poor economics indicated that production of gypsum would be a better approach for the larger plant at Kashima. Acid from a surge tank is metered into a stirred reaction vessel where the stoichiometric amount of pulverized limestone is added. The operating conditions are proprietary. The limestone is ground in a rubber-lined vertical attrition mill. Large gypsum particles are formed in the neutralization operation and batch centrifuges are used to separate the product. After the gypsum is removed, the liquor is discharged to the sea. Soot in the liquor would plug the carbon pores if the effluent were recycled for washing the beds.

FIGURE 10.6: HITACHI INSTALLATION AT KASHIMA

Without a concentration step the acid strength produced is only 20%, which is too dilute for commercial use. Concentration of dilute sulfuric acid is a messy and expensive operation. Use of limestone to produce gypsum may be practical in Japan where there currently is a market for gypsum, but in the United States it would be an expensive throwaway system. The fact that the Japanese have not planned further installations after demonstration on a large scale would indicate that the alternatives are more practical. Operation with effluent discharge after precipitation of gypsum is a further disadvantage.

Regeneration by Reduction

Since thermal regeneration comsumes expensive carbon and regeneration by washing gives a weak acid that is difficult to convert to a useful product, researchers have sought better methods for regeneration. One approach is to reduce the acid all the way to H_2S or to elemental sulfur, both of which are concentrated products that can be volatilized from the carbon without any dilution.

The main work in this area appears to be that of the Westvaco Corp., a major producer of activated carbon. The feasibility of using such an approach was confirmed in bench scale and small pilot equipment whereby H_2S in contact with the sulfuric acid on the carbon resulted in conversion to elemental sulfur which was then stripped off the carbon by heating. Part of the sulfur was reacted with an outside source of hydrogen to produce the needed H_2S. More recently a 20,000 cfh (0.15 Mw) integral pilot plant using flue gas from an oil-fired boiler was run for 300 hours.

A simplified flowsheet of the process is shown in Figure 10.8. SO_2 is adsorbed in a fluidized bed of carbon operated at a gas exit temperature of about 220°F. It is claimed that residual dust passes through the fluid bed without impairing the adsorption efficiency.

The loaded carbon is lifted to the top of a second fluidized bed reactor where reaction of H_2S with H_2SO_4 to produce sulfur takes place; temperatures near 300°F are required for the reaction.

$$H_2SO_4 + 3H_2S \xrightarrow[\text{carbon}]{\text{activated}} 4S + 4H_2O$$

FIGURE 10.7: WESTVACO CARBON ADSORPTION PROCESS

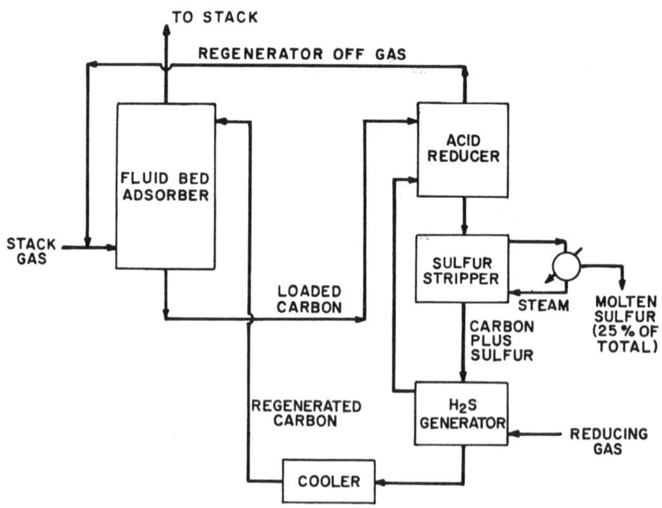

The sulfur-loaded carbon is then treated with hydrogen at elevated temperature in a third bed (fluidized or moving bed) to volatilize the sulfur and convert three-fourths of it to H_2S. The gas stream is cooled to condense the product sulfur and the H_2S passed on to the fluidized reducer.

The removal of SO_2 during the integral run over the 300 hour period averaged 94% from an inlet concentration of 2,000 ppm. Gas velocity was approximately 3.0 feet per second, which is higher than for other carbon processes. No NO_x removal was observed.

During the integral run the carbon was cycled through the system some 21 times based on a calculated carbon residence time of 15 hours. The attrition rate measured during this period showed an initial decrease and then stabilization at a rate of 0.26 pound per hour. The data indicated that nearly all of this attrition occurred in the fluidized beds of the SO_2 adsorber. Based on a 500 Mw unit operating 7,000 hours per year and activated carbon at $500 per ton, this would amount to an annual replacement cost of $1,500,000. Westvaco believes that with a combination of larger particle size and carbon with improved hardness this attrition rate can be reduced by about 90%.

It was shown important to keep careful control of the temperature in the acid reducer. If the temperature is too high, part of the acid is reduced to SO_2 and cycles back to the adsorber. Since a constant gas volume was employed, no data on turn-down capability was obtained. This is a problem whenever a fluid bed unit is used.

Work on regeneration by reduction apparently has been done also by Shell in the Netherlands. A 1965 patent covers reduction with methane at 300° to 500°C, with steam added to increase reaction rate. By adjusting the reduction temperature, it is claimed that an SO_2/H_2S mixture of proper ratio for the Claus process can be obtained.

CATALYTIC OXIDATION AND REDUCTION

As the preceding chapters have shown, most of the work on SO_2 removal from waste gases has been concerned with concentrating the SO_2, usually by sorption-desorption, as a first step in recovery. Many researchers, however, have worked toward avoiding the expense and complexity of sorption-desorption by converting the SO_2 to recoverable compounds in situ, that is, within the main gas stream.

Most of this effort has been aimed at either oxidizing the SO_2 to SO_3 or reducing it to sulfur. If sulfur is produced, it is recovered as is, but if the oxidation route is followed, SO_3 is either converted to H_2SO_4 or absorbed by some other material to give a useful product. The main drawbacks to this approach are large equipment size (since all process steps must handle the full gas flow) and the necessity, in most cases, for carrying out the oxidation or reduction at high temperature. In some processes the product quality is also a problem.

Processes for in situ oxidation are both of the dry and wet (aqueous) type. However, the latter have attained little success. Reduction methods are practically all of the dry type. It should be noted that sorption on carbon could be classed as in situ oxidation because H_2SO_4 is formed directly. However, since adsorption of SO_2 must precede the oxidation it seems more appropriate to include the method with the sorption-desorption types (Chapter 10).

Although in situ methods have received a great deal of research attention, there has been less large-scale development than for the sorption-desorption processes. The dry oxidation types are farthest along in this respect; one is being demonstrated full scale on a power plant, another is in commercial operation on tail gas from a Claus sulfur unit, and others are in use on sulfuric acid plant tail gas.

OXIDATION IN DRY SYSTEMS

The most straightforward method, and the one that has been developed the farthest, is oxidation of the SO_2 by use of catalyst such as used in contact-type sulfuric acid plants. For a power plant this is, in effect, converting the entire boiler system to something like a contact plant of the wet type.

For power plants and smelters the dust must first be removed; an electrostatic precipitator is indicated for this service since most other dust collection methods would not be operable at the high temperature level necessary for catalytic SO_2 oxidation. The degree of dust removal necessary is similar to that in making H_2SO_4 from ore roaster gas, where wet scrubbing methods are normally used to reduce the dust content to a very low level. In tests at the Omuta plant of Mitsui Koatsu in Japan, on an oil-burning boiler, an amount of dust equivalent to that passing through the dust collection equipment (98.5% efficient) over a period of a year did not reduce the catalyst efficiency.

If the process is inserted into a boiler system the usual removal of moisture is not ordinarily carried out because the gases would be difficult to cool and reheat as in standard H_2SO_4 production. The cleaned gas therefore passed directly to the catalyst at the usual temperature (800° to 900°F), through the usual heat recovery equipment, and then into scrubbing or condensing equipment. Because of moisture condensation at this point, mist generation is a problem and high efficiency mist eliminators must be installed.

These considerations apply particularly to application of dry oxidation in a new power plant, where the economics dictate insertion of the oxidation unit into the power plant heat recovery train. For application to existing power plants, sulfuric acid plants, Claus units, and possibly to smelters and refineries, the situation is somewhat different. In each case the gas to be treated is a tail gas for which the producing plant has no further use. For power plants it is usually more economical to add the oxidation equipment at the end of the system rather than to break into the existing heat recovery train; for the others, of course, tail end application is a necessity. Thus, the tail gas must be heated in some way to catalyst temperature whereas in a new power plant (or perhaps in smelters) the equipment can be installed at a point where the gas is already at the required temperature. The various methods of reheating will be discussed later.

Although H_2SO_4 is the usual end product, there are several processes in which SO_2 is oxidized and the SO_3 is then absorbed by some material other than water. Various objectives and techniques are involved in this approach. In most of the processes standard V_2O_5 catalyst apparently has been used. There is some reference in patents to a catalyst that is superior for very low SO_2 concentrations but no reports of use have been noted.

Early bench scale work on dry oxidation was done by several investigators. BCR (Bituminous Coal Research, U.S.), for example, compared platinum and vanadium catalysts, studied condensation of the product acid, and evaluated problems in dust collection. BCURA (British Coal Utilization Research Assn.) studied SO_2 oxidation as a first step in production of H_2SO_4 or $(NH_4)_2SO_4$; the main conclusions were that the presence of water vapor and CO_2 in power plant stack gas has no adverse effect and that commercial V_2O_5 catalyst (promoted with K_2O) is quite effective.

Some work has been done also on oxidation at low temperature, with the objective of treating low temperature waste gases in an add-on process. Poor results appear to have been obtained in most of the work; the reaction is slow and the catalyst tends to become poisoned because of chemisorption of the SO_3. An exception to this is use of NO_2 as the catalyst, which appears to work quite well.

High Temperature

Dry oxidation over a solid catalyst at high temperature, followd by H_2SO_4 production, is a relatively simple method; the major process variations come from the type of gas treated. The principal solid catalyst processes are of the proprietary type. There are also processes in which catalytic conversion of SO_2 to SO_3 is carried out, but the SO_3 is absorbed by some material other than water. The advantage is that a final product can be produced in one step whereas sulfuric acid is always an intermediate.

Bayer Double Contact: The major application of dry oxidation to waste gases is the double contact process developed by Farbenfabriken Bayer in Germany for treatment of sulfuric acid plant tail gas. In effect it involves adding another sulfuric acid plant to the existing one (or a second stage in design of a new one), although integration of equipment makes the system somewhat different from two plants in series.

A flowsheet for a new plant is given in Figure 11.1. The SO_2 stream passes through the usual H_2SO_4 plant equipment, catalyst stages with intercooling followed by absorption of the resulting SO_3 in intermediate strength H_2SO_4. It is then reheated by heat exchange with the absorber inlet gas and passed through a final catalyst stage and then another absorber.

FIGURE 11.1: BAYER DOUBLE-CONTACT SULFURIC ACID SYSTEM

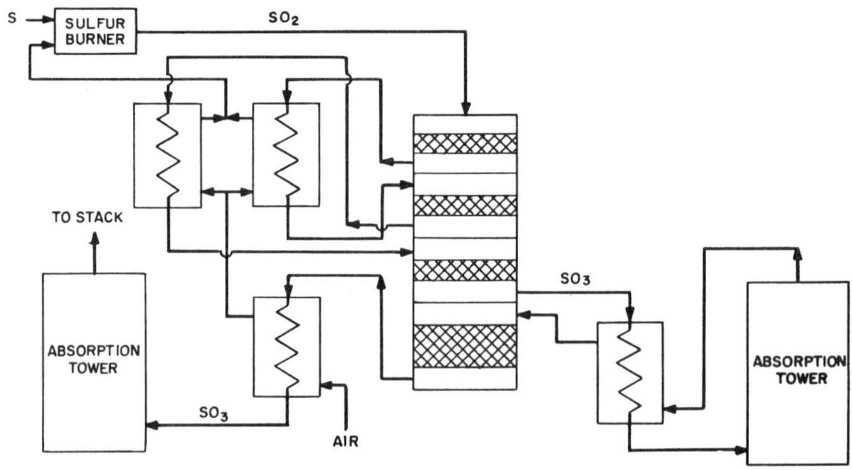

The SO_3 is absorbed in intermediate strength H_2SO_4 in each absorption tower.

In practice the last catalyst bed does not operate at a level of inlet SO_2 concentration similar to that normal in H_2SO_4 plant tail gas, on the order of 0.15 to 0.55%. Instead, the front end is designed to do only about 90% of the conversion, leaving about 10% for the last section (inlet SO_2 concentration of about 1.0 to 1.5%). The equilibrium is so improved by removing the bulk of the SO_3 at an intermediate stage that excellent overall conversion is obtained, up to 99.8% as compared to the usual range of 95.5 to 98.5%.

For a new plant the two sections are designed for these conversions. For an existing one the existing converter can be overloaded to give only 90% conversion and the added section designed for the remainder. Thus, the investment in new equipment is at least partially offset by the higher overall throughput.

Another consideration is that the double contact method allows higher SO_2 content in the starting gas, with consequent increase in throughput or reduction in equipment size. For sulfur burning, the increase is from 10 to 13.5% and for pyrites about 7 up to 10%. The increase is accomplished by using less excess air than normal; the intermediate absorption step gives a higher driving force in the last stage, thus giving good conversion even with the lower initial O_2/SO_2 ratio. It is claimed that this advantage fully offsets the cost of the added equipment.

A further advantage is that good catalyst response can be obtained at a temperature about 50°C lower than normal after the intermediate absorption step (390° vs 440°C). Since the equilibrium improves rapidly with temperature decrease, this effect further reduces emission. The first double contact plant was built in 1964; since then over 30 units have gone into production. The reported SO_2 emission is as low as 100 to 120 ppm under normal load and no higher than 240 ppm at 20% overload.

Monsanto Cat-Ox: The double contact method can be regarded as an add-on process because gas leaving the primary absorber is reheated before passing through a stage of catalytic conversion that reduces SO_2 emission to low and presumably acceptable levels. Generally the process has been applied only in H_2SO_4 plants; however, the same principle is now being applied to one full scale power plant and could also be applied to other SO_2 emitting operations.

The application to power plants has been studied by several organizations. The BCR work was mentioned earlier, and other early investigators also made contributions. The major work began in the early 1960s, when a group composed of Monsanto, Pennsylvania Electric, Research-Cottrell, and Air Preheater began development of a dry catalysis system. A small pilot plant (400 scfm), built to test the technical feasibility of the process, was followed by a 15-Mw prototype unit at the Portland, Penn., station of the Metropolitan Edison Co. This led to a joint EPA-Illinois Power project installed on Illinois Power's 100-Mw Wood River unit 4. Construction started in January 1971 and was completed and placed into service September 4, 1972.

A generalized flowsheet of the method, called the Cat-Ox process, is given in Figure 11.2. The gas is first passed through a hot precipitator to remove ash to a high degree; over 99.5% removal from a gas containing about 4 gr/scf is needed. The hot ash flows freely from the precipitator and can be conveyed by pneumatic means. Practically all the remaining dust is caught in downstream equipment so particulate emission is extremely low. The converter contains catalyst arranged

FIGURE 11.2: MONSANTO CAT-OX SYSTEM FOR SO_2 REMOVAL FROM STACK GAS

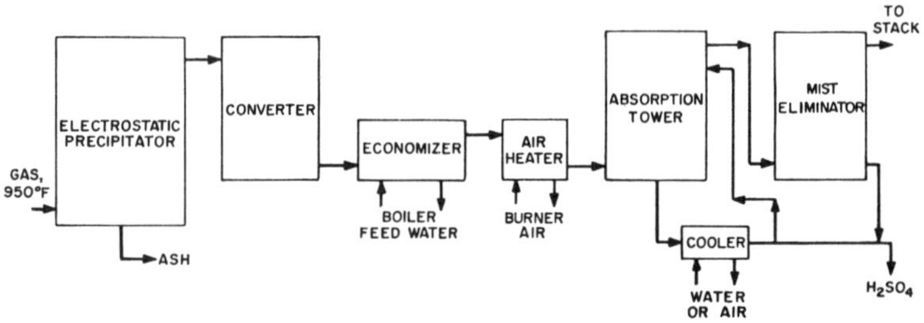

in removable beds so that sections can be removed for cleaning without interrupting operation. With normal operation of the precipitator, it is expected that the pressure drop will build from an initial 1 inch H_2O to 3 inches H_2O in about 2 months as a result of dust deposition in the catalyst. The catalyst would be cleaned at this point, by shaking and screening, to prevent further pressure drop increase and consequent higher operating cost. (Conversion efficiency is not affected). The handling and screening during each cleaning cause a catalyst loss of about 2.5%.

The plant is designed for 80 to 90% conversion of SO_2 to SO_3, depending mainly on swings in the gas temperature. In the prototype unit this range of efficiency was maintained for a year, indicating that catalyst poisoning, plugging, attrition and breakage during cleaning are not significant problems.

The converted gas is cooled in an economizer and air heater to about 450°F, which is above the acid dew point. The air heater is a conventional Ljungstrom type but the economizer (which heats boiler water) can be of light fin-tube construction since the erosive action of dust, which in normal plants requires heavy construction, is not a factor.

The SO_3 is then removed from the gas in a packed tower by scrubbing with a cool stream of sulfuric acid, cooling the gas to about 225°F. The resulting hot acid, about 80% in concentration, is drained from the tower and cooled with boiler water or combustion air to recover the heat. Pressure drop in the packed tower is about 5 inches H_2O.

Condensation of H_2SO_4 from the gas causes formation of a very fine acid mist that is difficult to remove. In the Cat-Ox process a fiber-packed unit (Brink Mist Eliminator) removes the mist down to less than 10 ppm (as 100% H_2SO_4), at the expense of a 13 to 18 inch pressure drop. The vapor pressure of H_2SO_4 at 225°F contributes another 11 ppm, but the total is less than that represented by the SO_3 normally emitted from power plants.

The overall pressure drop in the system is about 42 inches H_2O, or an increase of about 26 inches over normal power plant operation. The flowsheet in Figure 11.2 applies only to a new plant, where the precipitator, economizer and air heater can be designed to fit in with the temperature requirements of the converter and the absorber. The process can also be added on to an existing plant, as at Wood River, in which case the flowsheet is somewhat different because the stack gas must be reheated to converter temperature.

A flowsheet of the Cat-Ox reheat system installed at Wood River is shown in Figure 11.3. As stated earlier, start-up began in September, 1972. Natural gas was used for the in-line reheat burners. The system was operated for 444 hours during this period. When natural gas became unavailable, in October, 1972, it became necessary to operate the in-line reheat burners on #2 oil. The reheat burners were tested with fuel oil and modified until operable during the period November, 1972 to June, 1973. The performance guarantee was satisfactorily completed using #2 oil in July, 1973. Operation in July, 1973 brought the total operating time for the Cat-Ox system to 602 hours. However, it was apparent that the continuous long-term operation of the in-line heaters on #2 oil would cause excessive contamination of the catalyst, thereby causing frequent shutdowns for catalyst cleaning.

FIGURE 11.3: REHEAT CAT-OX SYSTEM

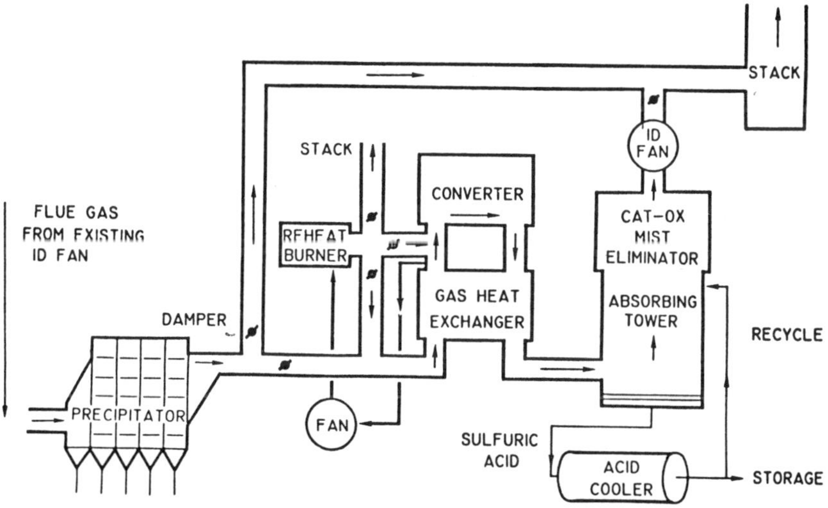

Since this would not be acceptable, an external reheat burner was constructed and installed using #2 oil with the required heat being ducted into the system at the present location of the in-line burners. The installation of the external burner was completed in early 1974, and the unit was placed back in operation. During start-up, a leak in the lead lining of the absorbing tower was discovered. In addition, the water intake screens were plugging with debris from the Mississippi River which required a change in design of the intake structure. It was further found that the impellers in the acid circulation pumps had been thermally shocked and would have to be replaced. These repairs and modifications were made and the system was placed back in operation in August, 1974. However, additional mechanical and structural problems have since developed and the operation of the system has been sporadic. The original capital costs involved in the design and construction of the Cat-Ox system were estimated at $7,300,000 including the cost of providing off-battery services. Final costs are now approaching $8,500,000 resulting in an installed cost of about $84.00 per kilowatt based on the 103 Mw net capacity of the generating unit.

The main advantage of the process is simplicity; no operating labor or raw materials are required and no regeneration step is necessary. On the other hand, the equipment is large and expensive because all of it must accomodate the full flow of gas. Moreover, the product acid contains more water (78% H_2SO_4, 60°Bé) and dissolved impurities than normal commercial acid.

Topsoe-SNPA: Refineries and sour gas purification plants have a special problem in that conversion of H_2S to sulfur in Claus plants is never complete. Since emission of the unconverted H_2S is not advisable, a combustion step to convert the residual H_2S to SO_2 is usually incorporated and the SO_2 emitted.

Haldor Topsoe, Inc., and Société National des Petroles d'Aquitaine (SNPA) have developed a dry catalysis method for treating tail gas from SNPA's Claus sulfur plants at Lacq, France, where SNPA recovers sulfur from sour natural gas. A flowsheet for the method is shown in Figure 11.4. The Claus plant tail gas is heated to about 1000°F in a gas-fired heater; at this temperature combustibles such as H_2S, sulfur, H_2 and CO are converted to the oxidized form. The gas is then cooled to converter temperature in a waste heat boiler and passed into a two-stage SO_2 converter with cooling between stages.

FIGURE 11.4: TOPSOE-SNPA PROCESS FOR TREATING CLAUS PLANT TAIL GAS

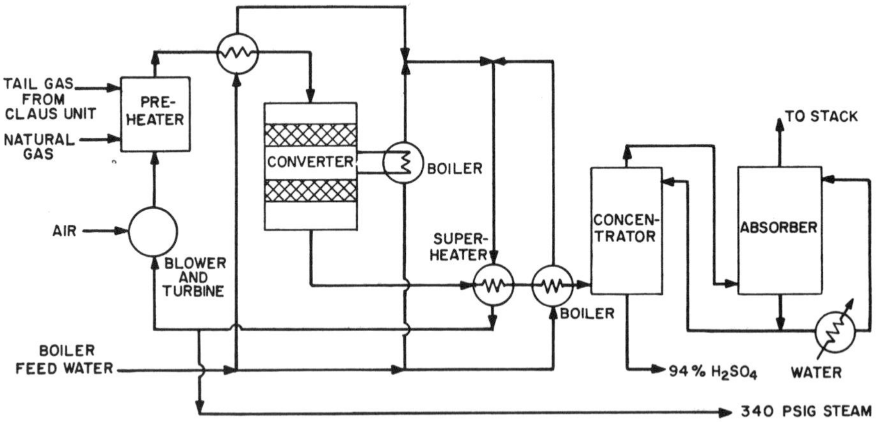

Alternately, the combustion can be carried out by a catalytic incineration method also developed by Topsoe. This allows operation at lower temperature, thus reducing the amount of fuel necessary for heating the gas to combustion temperature and also eliminating the cooler-boiler. The catalyst is a specially developed type (vanadium base). Enough is used so that the SO_2 concentration in the stack gas is less than 400 ppm. Normally, the exothermic heat of conversion plus residual heat from the combustion step makes it necessary to operate the converter with two stages and intercooling between.

Gas from the converter, after cooling, passes to a two-stage absorption-acid concentration system. In the first stage the gas is cooled by contact with acid from the second stage and part of the SO_3 is absorbed. Absorption is completed in the second stage, giving a gas containing about 25 ppm SO_3 (and less than 400 ppm SO_2, as noted). The gas then passes through a mist eliminator and to the stack.

The water required for production of H_2SO_4 is condensed from the gas, as it must be when any gas with a high dew point is processed. However, the two-stage system makes it possible to condense only as much water as is desired in the product acid. The process has been in operation at Lacq since 1965, in a plant with a converter feed gas rate of 9.3 million scf/hr (200 to 500 tons /day of acid). A typical composition of the Claus tail gas is as follows:

Constituent	Mol %
H_2S	0.65
S	0.25
SO_2	0.30
H_2O	33.67
H_2	1.27
CO	0.27
CO_2	5.75
N_2	57.84

Fuel consumption Is about 7 million Btu per ton of 94% acid. About 2 tons of 340 psig steam is produced per ton of acid. For application to power plants the much lower SO_2 concentration (0.05 to 0.3% vs 1.2% equivalent) probably would make some flowsheet changes desirable, as is the case in the Monsanto process, where the lower amount of heat generated in the converter makes the two stages with intercooling unnecessary.

For an add-on process, the direct heating to cover the entire temperature range up to converter temperature, as compared to a combination of direct heating and gas/gas heat exchange, may be advantageous. The boilers used in the Topsoe-SNPA process should be small compared to the gas/gas exchanger. The two-stage absorber system would also be an advantage since it rejects water to the final stack gas and thus allows production of stronger acid. To date, utility interest in the process is very low.

Kiyoura-TIT: The Kiyoura-TIT method, developed under the direction of Dr. Raisaku Kiyoura , formerly of the Tokyo Institute of Technology, and currently Chairman of the Board, Research Institute of Environmental Science, involves catalytic oxidation at high temperature, but instead of absorbing the SO_3 by condensing moisture from the gas, ammonia is added at a point where the temperature is above the H_2SO_4 dew point (about 220° to 260°C) thereby forming $(NH_4)_2SO_4$ in the gas stream. The sulfate is collected and removed from the gas stream in dust collection equipment. The process has been tested (in 1967) in a small pilot plant at a Toyo Koatsu fertilizer plant in Omuta, Kyushy, Japan.

The stack gas came from a plant boiler burning oil containing 3.6% sulfur. After ammonia addition, the gas was cooled in a tubular cooler to about 140°C before passing through an electrostatic precipitator for collection of the $(NH_4)_2SO_4$ (at 100° to 110°C). Plugging of the cooler was a problem, but could be kept under control by blowing with air every eight hours or so. The $(NH_4)_2SO_4$ recovered had high purity, 99.2 to 99.6%. Conversion efficiency was 93% and collection efficiency 97.5%, giving an overall efficiency of about 90.7%.

The process has some advantage if ammonium sulfate is considered to be an acceptable product, because the usual step of reacting sulfuric acid with ammonia and then crystallizing the sulfate is not required. The main drawback, however, is the same as for other high temperature catalytic processes, that is, investment is high because all the process steps involve handling the entire gas stream. As far as is known, no further work has been done on this since the Omuta test program.

Bureau of Mines: The U.S. Bureau of Mines has studied absorption of SO_3 by various materials, including spent oil shale, red mud (by-product of aluminum industry) phosphate rock (ore), and zinc oxide. The first two of these are by-products that under favorable conditions might be obtained at low cost. In each case, regeneration would be accomplished by reduction. Reaction of SO_3 with phosphate rock was discussed earlier. A further step would be required to produce a useful fertilizer. Zinc oxide is reactive with SO_3 (but not with SO_2) at high temperature. One method projected for regeneration of the $ZnSO_4$ is electrolysis.

All of these methods have the drawback that the expensive step of oxidizing SO_2 to SO_3 is required before absorption can take place. Thus the oxidation, absorption and regeneration required makes the process complicated and expensive. However, the gas is not cooled and absorption of SO_3 may be faster and more complete than absorption of SO_2. There does not appear to be much work under way at present on oxidation followed by use of absorbents such as those mentioned.

Callery Chemical: A process for SO_3 concentration, similar in a general way to the many developed for SO_2 concentration, has been investigated by Callery Chemical in a project carried out for EPA. The SO_2 is first oxidized and the SO_3 then absorbed by Na_2SO_4 at 550° to 600°F to form sodium pyrosulfate ($Na_2S_2O_7$). The pyrosulfate is separated and decomposed at 930° to 1020°F to give a gas containing about 6% SO_3, which can be absorbed to give sulfuric acid. In one variation of the method, the catalyst (V_2O_5) and absorbent are combined so that the pyrosulfate can be produced in one step.

One of the problems in any process that evolves SO_3 at high temperature is dissociation of SO_3 to SO_2. In the Callery work, only about 12% dissociation occurred at 1020°F; the resulting SO_2 was recycled to the catalysis step.

Although the pyrosulfate method has some good points, it has the same major drawback as some of those already described—high capital cost for the process sequence of oxidation, sorption, desorption, and final absorption. No additional work is now underway.

Low Temperature

Although oxidation at high temperature has some advantages, there are complications in using the method as an add-on process because of the gas reheating required. Catalysis at low temperature thus would be desirable but there are some important technical limitations. Various low temperature methods have been proposed, including both dry and wet types.

Tyco Chamber Process: The chamber process for H_2SO_4 production has been adapted to SO_2 recovery from power plants by Tyco Laboratories, Inc. The chamber process is older than the contact method and has generally been abandoned, especially in the United States, because of high capital cost and low concentration of the product acid. However, it has some particular advantages for SO_2 recovery.

The basic step in the chamber process is oxidation of SO_2 by NO_2. A simplified version of the reactions involved is shown below. Reaction (3) is the key reaction because it regenerates NO_2 to continue the oxidations; therefore the $NO_2 \rightarrow NO \rightarrow NO_2$ cycle serves as an oxygen-carrying mechanism to accomplish the oxidation. Reactions (4) and (5) are also quite important as they prevent significant loss of nitrogen oxides in the stack gas. Neither NO nor NO_2 are very soluble in H_2SO_4 but N_2O_3 is quite soluble. This makes it essential to maintain an equimolar NO/NO_2 ratio at the point where the product H_2SO_4 scrubs the nitrogen oxides from the exit gas. The nitrosylsulfuric acid ($HNSO_5$) formed is recycled and releases NO_2 (by hydrolysis) to continue the oxidation.

(1) $$SO_2 + NO_2 \longrightarrow SO_3 + NO$$

(2) $$SO_3 + H_2O \longrightarrow H_2SO_4$$

(3) $$2NO + O_2 \longrightarrow 2NO_2$$

(4) $$NO + NO_2 \longrightarrow N_2O_3$$

(5) $$N_2O_3 + 2H_2SO_4 \longrightarrow 2HNSO_5 + H_2O$$

The composition and temperature of power plant stack gas bring in complications in application of the chamber process to SO_2 recovery; the power plant gas has lower values for several operating parameters.

(1) SO_2 content of the gas ($<$0.5% vs 8 to 13%).
(2) Heat generated per volume of gas (from SO_3 absorption).
(3) Inlet gas temperature (about 300°F vs 1600°F).
(4) SO_2/H_2O ratio (1:15 vs 15:1).

The process developed by Tyco (under the sponsorship of EPA) is fitted to these conditions and makes use of the nitrogen oxides already present in power plant stack to accomplish SO_2 oxidation. A simplified flowsheet is given in Figure 11.5. Recycled NO_2 is added to the gas in an amount sufficient to oxidize the SO_2 plus enough to give an equimolar NO_2/NO ratio after oxidation. Since the NO/NO_2 ratio in the incoming gas is high, a large amount of NO_2 must be recycled.

FIGURE 11.5: TYCO APPLICATION OF THE CHAMBER PROCESS TO SO_2 RECOVERY

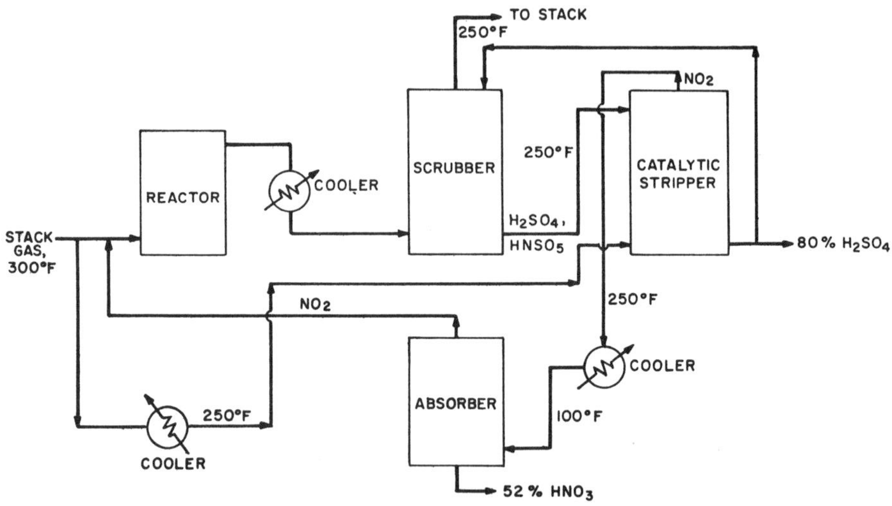

The gas mixture enters the reactor (where oxidation occurs), passes through a cooler, and is then scrubbed with recycled H_2SO_4 at 250°F. The scrubber must also collect mist because heavy H_2SO_4 mist formation occurs in the reactor and cooler as a result of SO_3 reaction with moisture in the gas.

The scrubber is operated at a temperature such that only enough water is condensed from the gas to give the acid strength desired. For the best balance between acid concentration and the problems associated with high scrubbing temperature, the temperature selected is 250°F, which gives an acid concentration of about 80% (at 7.3% moisture in the gas).

The NO/NO_2 equimolar mixture, which converts to N_2O_3, is quite soluble in and is retained well by the H_2SO_4 even at the high temperature. The resulting solution of $HNSO_5$ in H_2SO_4 is stripped with a stream of raw stack gas (about 10% of the total flow), the oxygen from which oxidizes the $HNSO_5$ to NO_2.

$$(6) \quad 4HNSO_5 + O_2 + 2H_2O \longrightarrow 4NO_2 + 4H_2SO_4$$

The NO_2 passes through an absorber where the net amount recovered from the stack gas is converted to nitric acid and the remainder is recycled to oxidize SO_2. The Tyco method has the outstanding advantage of recovering NO_x as well as SO_2. Good recovery of NO_x, however, depends on a high degree of N_2O_3 absorption in the scrubber. The large amount of NO_2 recycled makes good scrubbing essential if a large percentage of the NO_x in the stack gas entering the system is to be recovered.

In an EPA sponsored program, TVA made a detailed cost estimate (in draft form) of the Tyco process but unfortunately, the process was abandoned (EPA stopped its sponsorship) and the cost estimates were never published. Several major process problems were uncovered which rendered its development quite unattractive. First, the oxidation chamber was drastically under-designed, requiring more than five times the original volume for complete reaction. Second, because of the large amounts of NO_2 recycled back to the inlet gas stream, extremely efficient (>95%) removal of NO_x was required. In fact, ammonia had to be converted to NO_2 in the small scale studies because of the lack of efficient removal. These major problems were considered extremely detrimental so development of the program was terminated.

Use of Electrical Discharge: A limited amount of work has been done on use of electrical discharges to oxidize the SO_2 at low temperature. The mechanism is not completely clear; some claim that the SO_2 is ionized to a form more reactive with oxygen and others that oxygen is converted to ozone, which is a strong enough oxidizing agent to convert SO_2 to SO_3.

Work was done on this approach both by the U.S. Bureau of Mines (College Park Metallurgy Research Center, College Park, Md.) and by the Georgia Institute of Technology (Atlanta, Ga.). Good results were reported in some tests, e.g., 96% removal when a pulsed high frequency current was applied to a gas containing 0.98% SO_2 (plus oxygen and moisture). The pulsed current was much more effective then direct current.

There is also some evidence to the effect that the controlling reaction is formation of NO_2 from nitrogen and oxygen, which then oxidizes SO_2 to SO_3. Although this might be an effective way of getting low temperature oxidation, the formation of NO, which would escape with the stack gas, is undesirable. The work so far has been on a small scale only. Further study is required, both to define a process and to determine whether power requirement would be low enough to be acceptable. The Lockheed Missile and Space Co. (Palo Alto, Cal.) is also reported to have worked on an electrochemical reactor for SO_2 removal. No details have been made available.

OXIDATION IN AQUEOUS SYSTEMS

Absorption and oxidation in an aqueous solution is attractive because of its relative simplicity. For example, sulfuric acid can be produced directly, in a single step, by absorption of SO_2 in water containing a catalyst. Such a process can be added to an existing plant, no complicated desorption procedure is involved, and capital cost should be relatively low.

Early work on such a method was done in studies directed to recovery of SO_2 in an acid process for alumina production. The SO_2 was scrubbed from kiln gas (1.0 to 6.5% SO_2) by weak H_2SO_4 containing manganese ion as a catalyst. Oxygen was supplied by aerating the acid with a stream of air containing a small amount of ozone. In this work the absorption and oxidation steps were separated to make operation easier.

The early studies showed that the main problems in applying solution oxidation to SO_2 recovery from dilute gases are (1) high scrubber cost, (2) low acid concentration attainable, and (3) catalyst poisoning from oxidation inhibitors present in power plant stack gas (from burning of coal). In early work at TVA (1956), a small amount of ozone was added to the stack gas and the mixture scrubbed with H_2SO_4 containing manganese ion as catalyst. In tests with simulated power plant gas, the maximum acid strength attainable was about 40%; at higher concentrations, both SO_2 recovery and ozone utilization dropped off. Calculations indicated that at typical power plant conditions the heat in the inlet gas would be adequate to concentrate the acid to commercial concentration levels if the initial concentration were 15% or higher.

However, as noted earlier, the fly ash would complicate a concentration step unless the ash were removed first in a dry collection system. Moreover, when the process was tested on actual power plant gas, both recovery efficiency and ozone utilization were considerably lower than for the simulated gas. Work by other investigators indicates that this is due to catalyst poisoning by gaseous phenolic-type compounds present in gas from burning of coal.

It was concluded that the process is not economically feasible unless the interfering substances can be removed from the gas before the oxidation step. For other types of gas, such as from sulfuric acid plants and possibly smelters and refineries, catalyst poisoning may not be a problem. Work on oxidation in solution by use of manganese and iron has also been carried out in Czechoslovakia and in the USSR. The results apparently were unpromising.

The USSR CO process for H_2SO_4 production contains a step in which the tail gas is scrubbed with H_2SO_4 in the presence of a catalyst. Good conversion and absorption are claimed plus elimination of mist in the scrubber. A standard V_2O_5 catalyst apparently is used. As noted in Chapter 10 activated carbon serves as an excellent catalyst for low temperature oxidation but getting the product acid out of the carbon pores is a problem. In work by BASF in Germany, the SO_2 is absorbed and oxidized in a suspension of finely divided carbon (mean diameter of 30 microns). Production of 20% H_2SO_4 is claimed.

REDUCTION

Reduction of SO_2 in situ, if it were technically feasible, would be an ideal recovery process. No complicated sorption-regeneration system would be required and the ideal material, elemental sulfur, would be the product. Unfortunately, there are several major problems that have kept the method from making much headway. Although work still continues, it is on a relatively small scale and there are not yet any commercial plants, or even any prototype units.

The main drawback is the excess oxygen present in most waste gases that contain SO_2. The oxygen is helpful in oxidation processes but interferes with reduction. Some have proposed operation at lower air feed so as to give a stack gas containing little or no oxygen. Others operate under conditions such that selective reduction of SO_2 is achieved.

Another problem, which is also encountered in oxidation (at low temperature), is the presence of catalyst poisons, particularly in gas from burning of coal. The reducing agents used have been CO and H_2S. The latter has the advantage that it can be used at low temperature, in an add-on process.

Reduction with H_2S

The use of H_2S as the reductant has been tested through the pilot plant stage by Peter Spence and Sons in England. In laboratory tests excellent reaction of H_2S and SO_2 was attained even at stoichiometric H_2S and at temperatures low enough to allow add-on operation. The process visualized involves operation at a temperature low enough (about 100°C) for both formation and condensation of sulfur in the catalyst. After accumulating a loading, the catalyst is removed and heated to remove the sulfur. A fairly heavy loading can be attained without reducing catalyst activity. A hot mixture of CO, H_2O, and H_2 is used to convert product sulfur to H_2S for recycling to the converter.

Although the method was quite promising at the small scale, pilot tests with actual power plant gas have indicated catalyst poisoning by some constituent of the gas. A catalytic process similar to the Peter Spence method was announced by Princeton Chemical Research, Inc. (Princeton, N.J.). Product sulfur is reacted with methane and steam to give H_2S, which is injected into the gas and the mixture passed over a catalyst at 200° to 300°F. The work is reported to have been on a pilot-plant scale. No details are known but further development work has stopped.

Reduction with CO

The Chevron Research Co. (U.S.) has worked with CO as the reducing agent in small scale tests. It is assumed in the proposed process that the combustion unit is operated with little or no excess air and that the gas therefore contains at least the stoichiometric amount of CO to react with the SO_2.

$$(7)\quad 2CO + SO_2 \longrightarrow 2CO_2 + S$$

The reduction is carried out over a catalyst, preferably copper supported on alumina (3 to 10% Cu). At 1000°F the SO_2 removal was 90% or better at residence times as short as 0.35 second (space velocity of 10^4 vol/vol/hr. Two other reactions are important in the process. These are shown below.

$$(8)\quad CO + S \longrightarrow COS$$

$$(9)\quad CO + NO \longrightarrow CO_2 + 0.5N_2$$

Formation of COS, which is undesirable because of its toxicity, can be minimized by holding the CO at stoichiometric or below for reactions (7) and (9) at the expense, of course, of lower SO_2 removal as compared to the use of excess CO. At 981°F and stoichiometric CO, the partial pressures of COS and SO_2 were about 3×10^{-5} and 2×10^{-4} atm, respectively. The reduction of NO_X is a major advantage for the method. Removal approaching 100% was obtained.

The method would be expected to be high in capital cost; in this respect it is similar to in situ oxidation. However, it also has the same advantages as oxidation—simple operation, no raw material or labor requirements, and makes a superior product. The main question is whether or not it is feasible to operate a boiler (or other combustion-type operation) to produce the required amount of CO. Failure to control the combustion closely could result in excess CO formation, with consequent production of COS. Loss of control in the other direction would reduce SO_2 removal. Other problems resulting from operating under reducing conditions are also inherent, particularly with coal as the fuel, such as: increased boiler slagging, danger from flameouts, and potential for rapid and/or accelerated corrosion and deposits. Such operation is highly unlikely to be adopted for conventional boiler operation.

Reduction with CO has been explored also by workers at the University of Massachusetts. As in the Chevron work, reduction is carried out at high temperature over a copper-on-alumina catalyst. Methods proposed for avoiding the COS problem are (1) operate at such a high temperature (over 1500°F) that COS is not formed and (2) split the gas stream, treating a ⅔ portion with enough CO to convert all the SO_2 to COS and then mixing with the other third to obtain reaction of COS with SO_2 with resulting formation of elemental sulfur. No data appear to have been made available on these departures.

Reduction with Carbon

Reduction of SO_2 with carbon, an old process for treating concentrated SO_2 streams, has been applied to dilute stack gases by workers at Pennsylvania State University. A process has been proposed in which the SO_2 is first reduced to H_2S and COS by carbon at 600° to 800°C. Then in a second reactor, at about 150°C the COS and H_2S are oxidized to sulfur in an activated carbon bed. Since the sulfur deposits in the bed, it is necessary to remove the carbon and treat it to recover the sulfur.

It is visualized in this process that combustion conditions would be normal and that excess oxygen would be burned out by the carbon in the first reactor. If all the oxygen were consumed in the first step, it would be necessary, of course, to add oxygen for the second. Only small scale tests, to establish that the two steps are technically feasible, have been carried out.

UTILIZATION OF WASTE HEAT FROM POWER PLANTS 1974

by David Rimberg

Pollution Technology Review No. 14

Energy Technology Review No. 3

Present-day steam-driven turbo-electric power plants in the United States discharge as waste heat an amount of energy roughly equivalent to twice their total electricity-generating capacity. This energy is most difficult to utilize, because it is degraded in temperature. The effluent water is warm, but it is far from being near the boiling point. It constitutes a necessary, but unwanted, by-product of the energy conversion process for generating electricity.

The growing quantities of waste heat discharged, and the increasing ecological anxieties about undesirable growths, energy utilization and thermal discharge problems, have stimulated an examination of methods for productively using energy presently wasted on the environment.

Part I of this book discusses the reasons for present-day ineffective utilization, but it also shows ways and means to promote more efficient energy usage.

Part II assesses the cause, magnitude, and possible effects of heat discharges into water from steam electric power plants and related condenser cooling systems. Also contained in this section is a discussion of the thermodynamics of the electric power generation cycle detailing the reasons for the "inefficiencies" in by-product heat generation.

Ultimately all the accessible waste energy appears as low temperature heat, and the term "waste heat utilization" refers to the performance of useful functions with this heat before it is discharged into the environment. Within the framework of this book, the minimum temperature of the effluent water considered for subsequent use is about 38°C or 100°F. These uses are discussed in **Part III:** Food production in agriculture, hydroponics and aquaculture (fish farming) and the use of heat in wastewater treatment.

In the past the dissipation of waste heat was accomplished by wet and dry cooling towers and cooling ponds, lakes and streams. These methods are now being challenged by some sectors of society, and industry is being forced to consider their environmental impact. In this regard **Part IV** discusses the research needs necessary to equate the complicated interactions of the physical, engineering, biological, and social aspects of this waste heat problem.

This Pollution Technology Review is based on studies conducted by industrial and engineering firms or university research teams under the auspices of various governmental agencies, e.g. The National Water Commission, Oak Ridge National Laboratory Federal Water Pollution Control Administration, Office of Water Resources Research, Environmental Protection Agency, National Science Foundation, and the Department of Commerce.

A partial and condensed table of contents follows here:

ISBN 0-8155-0555-8 **175 pages**

ENVIRONMENTAL SOURCES AND EMISSIONS HANDBOOK 1975

by Marshall Sittig

Environmental Technology Handbook No. 2

Environmental pollution with its far-reaching effects is only now beginning to be understood. In this practical handbook the various modifications a given pollutant can assume, are traced back to their origins, and the intermedia transfers are discussed fully. Intermedia transfers include direct transfer (removal of a pollutant from one medium and its disposal in another) or indirect (pollution created in another medium and usually in another form by a basic change in a process or industry).

Also, an example of insidious conversion is mercury. Metallic mercury is relatively innocuous, but when organic molecules or dead organisms are present in rivers and lakes, mercury reacts with these molecules to form toxic methylmercury compounds which are excreted very slowly by fish and man.

The implications are formidable: Chemical, biochemical, technological, statistical, mineralogical, zoological, pharmacological, medical, legal, legislative and public health involvements are all too obvious.

This volume therefore surveys the origins of both air and water pollution. Significant emphasis is placed on altered pollution which can result when an air pollutant is intentionally or accidentally transferred to an aqueous stream or vice versa.

Sometimes pollutants react with each other or initiate deleterious chain reactions. Yet this same reactivity may be the key to efficient removal: by adding certain chemicals or flocculants or microorganisms, toxic substances may be carried off physically or converted not only to nontoxic substances, but also into useful products.

The why and where is in this book, commensurate with present-day technology, without becoming too sophisticated or losing sight of the all important economic considerations.

Operators or potential operators of processes which produce pollutants will find this volume quite useful. Besides discussing all sorts of pollution sources, it should help to define industry-wide emission practices and magnitudes.

A partial and condensed table of contents follows here. The book contains 200 subject entries and 382 tables with figures and graphs. Descriptions are mostly based on studies conducted by industrial and engineering firms or university research teams under the auspices of various government agencies. As is the case with other handbooks in this series the entries are arranged in an alphabetical and encyclopedic fashion:

Aluminum Production, Primary
Aluminum Industry, Secondary
Aluminum Sulfate Manufacture
Ammonia Manufacture
Ammonium Nitrate Manufacture
Ammonium Sulfate Production
Asbestos Products Industry
Asphalt Roofing Manufacture
Asphaltic Concrete Plants
(Asphalt Batching)
Automobile Body Incineration
Automotive Vehicle Operation
Bauxite Refining
Beet Sugar Manufacture
Boat & Ship Operation
Brass & Bronze Ingot Prod.
(Secondary Copper Industry)
Brick & Related Clay
Products Manufacture
Builders Paper Manufacture
Calcium Carbide Manufacture
Calcium Chloride Manufacture
Carbon Black Manufacture
Castable Refractory Manufacture
Cement Manufacture
Ceramic Clay Manufacture
Charcoal Manufacture
Chlor-Alkali Manufacture
Clay & Fly Ash Sintering
Coal Cleaning
Coal Combustion, Anthracite
Coal Combustion, Bituminous
Coffee Roasting
Coke (Metallurgical) Manufacture
Combustion Sources
Concrete Batching
Conical Burners
Copper (Primary) Smelting
Cotton Ginning
Dairy Industry
Dry Cleaning
Electroplating
Explosives Manufacture
Feedlots
Fermentation Industry
Ferroalloy Production
Fertilizer Industry
Fiberglass Manufacture
Fish Processing
Frit Manufacture
Fruit & Vegetable Industry
Fuel Oil Combustion
Glass Manufacture
Gold & Silver Mining & Production
Grain & Feed Mills & Elevators
Gypsum Manufacture
Hydrochloric Acid Manufacture
Hydrofluoric Acid Manufacture
Hydrogen Peroxide Manufacture
Incineration, Municipal Refuse
Inorganic Chemical Industry
Iron Foundries
Iron & Steel Mills
Lead Smelting, Primary
Lead Smelting, Secondary
Leather Industry
Lime Manufacture
Liquefied Petroleum Gas Combustion
Magnesium Smelting, Secondary
Meat Industry
Mercury Mining & Production
Mineral Wool Manufacturing
Natural Gas Combustion
Nitric Acid Manufacture
Nitrogen Fertilizer Manufacture
Nonferrous Metals Industry
Open Burning
Orchard Heating
Organic Chemicals Manufacture
Paint & Varnish Manufacture
Paperboard Manufacture
Perlite Manufacture
Pesticide Manufacture
Petroleum Marketing & Transportation
Petroleum Refining
Petroleum Storage
Phosphate Chemicals Manufacture
Phosphate Fertilizer Manufacture
Phosphate Rock Processing
Phosphoric Acid Manufacture
Phosphorus Manufacture
Phosphorus Oxychloride Manufacture
Phosphorus Pentasulfide Manufacture
Phosphorus Pentoxide Manufacture
Phosphorus Trichloride Manufacture
Phthalic Anhydride Manufacture
Plastics Industry
Potassium Chloride Production
Potassium Dichromate Manufacture
Potassium Sulfate Manufacture
Poultry Processing
Power Plants
Printing Ink Manufacture
Pulp & Paper Industry
Railway Operation
Refractory Metal (Mo, W) Production
Rubber Industry, Synthetic
Sand & Gravel Processing
Seafood Processing Industry
Sewage Sludge Incineration
Soap & Detergent Manufacture
Sodium Bicarbonate Manufacture
Sodium Carbonate Manufacture
Sodium Chloride Manufacture
Sodium Dichromate Manufacture
Sodium Metal Manufacture
Sodium Silicate Manufacture
Sodium Sulfite Manufacture
Starch Manufacture
Stationary Engine Operation
Steel Foundries
Stone Quarrying & Processing
Sugar Cane Processing
Sulfur Production
Sulfuric Acid Manufacture
Surface Coating Application
Synthetic Fiber Manufacture
Terephthalic Acid Manufacture
Textile Industry
Timber Industry
Tire & Inner Tube Manufacture
Titanium Dioxide Manufacture
Urea Manufacture
Wood Veneer & Plywood Products Industry
Wood Waste Combustion
Zinc Processing, Secondary
Zinc Smelting, Primary

ISBN 0-8155-0568-X

521 pages

POLLUTANT REMOVAL HANDBOOK 1973

by Marshall Sittig

The purpose of this handbook is to provide a one volume practical reference book showing specifically how to remove pollutants, particularly those emanating from industrial processes. This book contains substantial technical information.

This volume is designed to save the concerned reader time and money in the search for pertinent information relating to the control of specific pollutants. Through citations from numerous reports and other sources, hundreds of references to books and periodicals are given.

In this manner this book constitutes a ready reference manual to the entire spectrum of pollutant removal technology. While much of this material is presumably available and in the public domain, the locating thereof is a tedious, time-consuming, and expensive process.

The book is addressed to the industrialist, to local air and water pollution control officers, to legislators who are contemplating new and more stringent control measures, to naturalists and conservationists who are interested in exactly what can be done about the effluents of local factories, to concerned citizens, and also to those eager students who can foresee new and brilliant careers in the fields of antipollution engineering and pollution abatement.

During the past few years, the words "pollution", "environment" and "ecology" have come into more and more frequent usage and the cleanliness of the world we live in has become the concern of all people. Pollution, for example, is no longer just a local problem involving litter in the streets or the condition of a nearby beach. Areas of the oceans, far-reaching rivers and the largest lakes are now classified as polluted or subject to polluting conditions. In addition, very surprisingly, lakes and streams remote from industry and population centers have been found to be contaminated.

This handbook therefore gives pertinent and concise information on such widely divergent topics as the removal of oil slicks in oceans to the containing of odors and particulates from paper mills.

Aside from the practical considerations, including teaching you where to look further and what books and journals to consult for additional information, this book is also helpful in explaining the new lingo of pollution abatement, which is developing new concepts and a new terminology all of its own, for instance: "particulates, microns, polyelectrolytes, flocculation, recycling, activated sludge, gas incineration, catalytic conversion, industrial ecology, etc."

In order to have a safe and healthful environment we must all continue to learn and discover more about the new technology of pollution abatement. Every effort has been made in this manual to give specific instructions and to provide helpful information pointing in the right direction on the arduous and costly antipollution road that industry is now forced to take under ecologic and sociologic pressures. The world over, technological and manpower resources are being directed on an increasing scale toward the control and solution of contamination and pollution problems.

In the United States of America we are fortunate in receiving direct help from the numerous surveys together with active research and development programs that are being supported by the Federal Government to help industry and municipalities control their wastes and harmful emissions.

A partial and condensed table of contents is given here. The book contains a total of 128 subject entries arranged in an alphabetical and encyclopedic fashion. The subject name refers to the polluting substance and the text underneath each entry tells how to combat pollution by said substance:

CARBON MONOXIDE
CARBONYL SULFIDE
CEMENT KILN DUSTS
CHLORIDES
CHLORINATED HYDROCARBONS
CHLORINE
CHROMIUM
CLAY
COKE OVEN EFFLUENTS
COLOR PHOTOGRAPHY EFFLUENTS
COPPER
CRACKING CATALYSTS
CYANIDES
CYCLOHEXANE OXIDATION WASTES
DETERGENTS
DYESTUFFS
FATS
FERTILIZER PLANT EFFLUENTS
FLOUR
FLUORINE COMPOUNDS
FLY ASH
FORMALDEHYDE
FOUNDRY EFFLUENTS
FRUIT PROCESSING INDUSTRY EFFLUENTS
GLYCOLS
GREASE
HYDRAZINE
HYDROCARBONS
HYDROGEN CHLORIDE
HYDROGEN CYANIDE
HYDROGEN FLUORIDE
HYDROGEN SULFIDE
IODINE
IRON
IRON OXIDES
LAUNDRY WASTES
LEAD
LEAD TETRAALKYLS
MAGNESIUM CHLORIDE
MANGANESE
MEAT PACKING FUMES
MERCAPTANS
MERCURY
METAL CARBONYLS
MINE DRAINAGE WATERS
NAPHTHOQUINONES
NICKEL
NITRATES
NITRITES
NITROANILINES
NITROGEN OXIDES
OIL
OIL (INDUSTRIAL WASTE)
OIL (PETROCHEMICAL WASTE)
OIL (PRODUCTION WASTE)
OIL (REFINERY WASTE)
OIL (TRANSPORT SPILLS)
OIL (VEGETABLE)
ORGANIC VAPORS
OXYDEHYDROGENATION PROCESS EFFLUENTS
PAINT AND PAINTING EFFLUENTS
PAPER MILL EFFLUENTS
PARTICULATES
PESTICIDES
PHENOLS
PHOSGENE
PHOSPHATES
PHOSPHORIC ACID
PHOSPHORUS
PICKLING CHEMICALS
PLASTIC WASTES
PLATING CHEMICALS
PLATINUM
PROTEINS
RADIOACTIVE MATERIAL
RARE EARTH
ROLLING MILL DUST & FUMES
ROOFING FACTORY WASTES
RUBBER
SELENIUM
SILVER
SODA ASH
SODIUM MONOXIDE
SOLVENTS
STARCH
STEEL MILL CONVERTER EMISSIONS
STRONTIUM
SULFIDES
SULFUR
SULFUR DIOXIDE
SULFURIC ACID
TANTALUM
TELLURIUM HEXAFLUORIDE
TETRABROMOETHANE
TEXTILE INDUSTRY EFFLUENTS
THIOSULFATES
TIN
TITANIUM
TRIARYLPHOSPHATES
URANIUM
VANADIUM
VEGETABLE PROCESSING INDUSTRY EFFLUENTS
VIRUSES
ZINC

ISBN 0-8155-0489-6

528 pages

ENERGY FROM SOLID WASTE 1974

by Frederick R. Jackson

Pollution Technology Review No. 8

Energy Technology Review No. 1

The solid waste disposal problem is reaching alarming proportions everywhere. The United States alone produces close to 300 million tons of solid waste per year, which is equivalent to about one ton per person.

At the present time the prevalent methods of disposal are dumping and sanitary landfill. Municipal incineration disposes of a small portion only, with attendant high capital and operating costs.

Many methods have been proposed for coping with the problem, such as source separation, source reduction, or material recovery. However, with the energy crisis descending upon us, producing energy from waste is becoming more and more attractive. An estimate currently making the rounds in financial circles is that when the price of crude oil reaches $7.00 a barrel, alternate sources of energy become practicable.

This book is based primarily upon information from studies conducted under the auspices of the EPA. Its foremost topic is burning of solid wastes to create steam directly. The air pollution problem created by burning can be solved quite easily with known technology. Solid wastes are low in sulfur, consequently there is no SO_2 removal problem.

Another technique that may assume more importance in the future is the controlled pyrolysis of wastes, yielding so-called pyrolysis gas or oil. Chapter seven is devoted to this. The final chapter discusses European practice, which is historically far more extensive than that of the U.S. A condensed table of contents follows here:

ISBN 0-8155-0528-0

163 pages

ENERGY IN THE CITY ENVIRONMENT 1973

Edited by Robert N. Rickles

This book is based upon material supplied by Robert Rickles, the Institute for Public Transportation, and the New York Board of Trade.

The important **charts** in the **appendix** present a list of energy conservation measures for the short-term (1972-1975), mid-term (1976-1980) and the long-term (beyond 1980) for the Transportation, Residential/Commercial, Industrial and Electric Utility sectors. The charts also indicate estimated maximum attainable energy savings, possible means for implementing each conservation measure, and pros and cons.

ISBN 0-8155-5019-7

173 pages

POLLUTION DETECTION AND MONITORING HANDBOOK 1974

by Marshall Sittig

Environmental Technology Handbook No. 1

This handbook contains methods for the detection and monitoring of pollutants in industrial effluents, notably air and water.

Such methods are prerequisites for any kind of management of environmental quality. Of equal importance are the data handling systems that must be used to collect and interpret the measurements, regardless of whether manual or automated identifying and monitoring techniques are chosen.

Generally the most important single factor in determining whether manual or automated methods of analysis and level monitoring should be used, is the required frequency interval. In the manual approach costs vary almost directly with the measurement frequency. At monitoring intervals oftener than once a day, or when a continuous watch must be maintained, installation of automatic sensing is more economical, provided reliable automatic sensors are available.

For meaningful control and prevention of pollution the accepted standards and parameters for air and water quality must be known, whether they have been legislated or not. Acceptable levels for each type of impurity from methylmercury to dissolved chlorides or carbon particles must be known by the industrialist or public health official, and this book, with its 1633 references makes a massive attempt to communicate such levels.

As a guideline for the industrial user, data on the toxicity of the various pollutants are included. Such data on the deleterious effects were taken from various government publications. This book is not a treatise of pharmacology or industrial toxicology, but a guide to the accurate evaluation of pollutant levels in a given effluent or ambient substrate.

Whenever applicable, the name of the pollutant is followed by chemical and other information. After this come directions for sampling in air or water or both, as the case may be. Thereafter are given qualitative and quantitative analytical methods suitable for identification and measurement of the degree or level of pollution. This comprehensive information is followed by measurement techniques suitable for repeated or continuous monitoring. Preference is given to methods giving reproducible results in environmental quality control, as recommended by the EPA and other U.S. government agencies. This is followed in each case by many up-to-date references to government publications, patents, and journal articles.

Arrangement is encyclopedic. The book is thus a worthy companion to the author's POLLUTANT REMOVAL HANDBOOK available from the same publisher.

Pollution control is a serious business, and in the design of a monitoring program the specific objectives to be served must be clearly in mind from the very beginning. Otherwise the effort most probably will not result in an efficient and meaningful program. The length of the initial survey, selection of monitoring points, parameter coverage, and sampling frequency all depend on the specific objectives and the timeframe in which the objectives must be achieved. These variables also have a very significant impact on monitoring costs.

Monitoring for the purpose of long-term trend identification, evaluation of standards compliance, and total management of environmental quality must, of course, be a continuing program without end.

It is hoped sincerely that this book, together with its companion volume POLLUTANT REMOVAL HANDBOOK, will do much to establish and sustain such antipollution programs.

A partial and condensed table of contents follows here in which subheadings are indicated below for only the first few entries. The book contains a total of 88 subject entries arranged in an alphabetical and encyclopedic fashion. The subject name refers to the polluting substance, and the text underneath each entry tells how to measure and monitor pollution by said substance:

POLLUTION INSTRUMENTATION
Size of the Market
References

SAMPLING
Air and Gas Sampling
Water Sampling
References

SUBSTANCES AND MATERIALS

ACIDS AND ALKALIS

ALDEHYDES

ALUMINUM

AMMONIA

ANTIMONY

ARSENIC

ASBESTOS

AUTOMOTIVE EXHAUST

BACTERIA

BARIUM

BERYLLIUM

BISMUTH

BORON

BROMIDES

CADMIUM

CALCIUM

CARBOHYDRATES

CARBON

CARBON DIOXIDE

CARBON MONOXIDE

CHLORIDES

CHLORINATED HYDROCARBONS

CHLORINE

CHROMIUM

COBALT

COPPER

CYANIDES

ETHYLENE

FLUORIDES

HALOGENS

HALOGEN COMPOUNDS

HYDROCARBONS

HYDROGEN CHLORIDE

HYDROGEN SULFIDE

IODIDES

IRON

LEAD

LEAD ALKYLS

MAGNESIUM

MANGANESE

MERCAPTANS

MERCURY

MOLYBDENUM

NICKEL

NICKEL CARBONYL

NITRATES

NITRIC ACID

NITRITES

NITROGEN COMPOUNDS

NITROGEN OXIDES

ODOROUS COMPOUNDS
Measurement in Air
Sampling Methods
Qualitative Analyses
Quantitative Methods (Organoleptic)
Quantitative Methods (Instrumental)
Measurement in Water
References

OIL AND GREASE

ORGANICS

OXYGEN DEMAND

OZONE AND OXIDANTS

PARTICULATES

PEROXYACETYL NITRATE (PAN)

PESTICIDES

PHENOLS

PHOSGENE

PHOSPHATES

PHOSPHITES

PHOSPHORIC ACID

PHOSPHORUS

PHOSPHORUS COMPOUNDS (ORG.)

POLLEN

POLYNUCLEAR AROMATICS

POTASSIUM

RADIOACTIVE MATERIALS

SEDIMENTS

SELENIUM

SILVER

SODIUM

STRONTIUM

SULFATES

SULFIDES

SULFITES

SULFUR COMPOUNDS (ORG.)

SULFUR DIOXIDE

SULFUR TRIOXIDE

SULFURIC ACID

SURFACTANTS

TIN

TITANIUM

URANIUM

VANADIUM

ZINC

ZIRCONIUM

FUTURE TRENDS

This book should prove to be very useful as an overall reference volume. The organization is clear, and the text is especially well arranged. The tables and figures are extensive and will be of value to students, engineers, and pollution control authorities.

ISBN 0-8155-0529-9

401 pages